25 to Survive

25 TO SURVIVE

Reducing Residential Injury and LODD

DANIEL D. SHAW & DOUGLAS J. MITCHELL, JR.

Fire Engineering®

> **Disclaimer**
>
> The recommendations, advice, descriptions, and the methods in this book are presented solely for educational purposes. The author and publisher assume no liability whatsoever for any loss or damage that results from the use of any of the material in this book. Use of the material in this book is solely at the risk of the user.

Copyright© 2013 by
PennWell Corporation
1421 South Sheridan Road
Tulsa, Oklahoma 74112-6600 USA

800.752.9764
+1.918.831.9421
sales@pennwell.com
www.FireEngineeringBooks.com
www.pennwellbooks.com
www.pennwell.com

Marketing Manager: Amanda Alvarez
National Account Manager: Cindy J. Huse

Director: Mary McGee
Managing Editor: Marla Patterson
Production Manager: Sheila Brock
Production Editor: Tony Quinn
Book Designer: Susan E. Ormston
Cover Designer: Charles Thomas

Library of Congress Cataloging-in-Publication Data

Shaw, Daniel D. (Fire captain)
  25 to survive : reducing residential injury and LODD / Daniel D. Shaw, Douglas J. Mitchell, Jr.
     pages cm
  Includes index.
  ISBN 978-1-59370-309-7
  1. Fire fighters. 2. Fire extinction. 3. Violent deaths. I. Mitchell, Douglas J. II. Title. III. Title: Twenty-five to survive.
  HD8039.F5S38 2013
  628.9'2--dc23
                                              2013018010

All rights reserved. No part of this book may be reproduced,
stored in a retrieval system, or transcribed in any form or by any means,
electronic or mechanical, including photocopying and recording,
without the prior written permission of the publisher.

Printed in the United States of America

1 2 3 4 5   17 16 15 14 13

To my wife, Kelley, and my kids, Brendan and Nick, for their unwavering support and constant inspiration: I am truly fortunate and appreciate you every day.

For my parents and brothers: you have served as my constant role models, providing guidance that has been invaluable. Especially to my mother, Fran: you are the catalyst that lit this fire in me and brought this book to fruition.

For my many firefighting mentors, and in particular, for Andy Liebno who invested his knowledge and patience in a young volunteer: I am indebted to you. And for Darl "Mickey" McBride, who has been steadfast in his faith in me, providing constant motivation and utmost confidence.

—D.D.S.

For my families:

The precious one at home: my wife Beth, children Madison, Jackson, and Ella, you continue make me so proud of our family. You are the foundation for my efforts.

My parents and brother: especially for my Dad, my hero and inspiration in this noble profession, who spent the "war years" in Brooklyn and the Bronx on rooftops with axes.

The countless brothers at the firehouses in which I have worked: the wisdom and knowledge you imparted molded me to grow as a fireman and fire officer . . . more importantly, as a man, a friend, and brother. Your inspiration and support is unwavering and humbling.

In loving memory of the "Yorkville 9" and the "FDNY 343."

—D.J.M.

Contents

Acknowledgments ix

Introduction xi

Part 1: Combat Readiness
   1. It's All about Me 3
   2. Combat-Ready Fire Apparatus 15
   3. Routine 25
   4. Can I Talk Now? Fireground Communications 31
   5. It's Really the Only Time You Have 39
   6. We Are Aggressive 53

Part 2: Mastering the Environment
   7. Bricks, Sticks, and Straw: Three Little Pigs Construction ... 65
   8. Get Out There 77
   9. On-Scene Reports: Nothing Showing Means Nothing 87
  10. Stairway to Heaven or Hell 97
  11. The Fire Down Below 111
  12. 360-Degree Check 121
  13. Fire Behavior: Taking the Science to the Street 131

Part 3: Engine Company Operations
  14. Preplanned Engine Company Riding Assignments 149
  15. Backup Is Not a Direction of Travel, It Is a Position 159
  16. Where Is My Water? 167
  17. One More Room: Making the Stretch 173
  18. The Last Line of Defense: The Firefighting Nozzle 181

Part 4: Truck Company Operations
  19. A Bull in a China Shop: Truck Work 191
  20. Preplanned Riding Assignments: The Truck Company 217
  21. Look, Listen, Feel: Residential Recognition 239
  22. Ladders, Ladders, Everywhere 251
  23. To the Roof? 273
  24. Probability in Search: VEIS 285
  25. We Save Lives 297

Index 311

Acknowledgments

One cannot look back without also looking forward. Look back at the sacrifices we have made with gratitude, remembrance and honor. Look forward with new strength, new growth, and a never-wavering commitment to the future.

—Lt. Douglas J. Mitchell, Jr.

This book is a culmination of ideas formed from our years of exposure to great individuals in the fire service, far too many to mention individually. So many experienced leaders, mentors, teammates, confidants, brothers, and sisters to glean information from. We were fortunate to have been surrounded by such a great group of firefighters and fire officers throughout our careers.

We must recognize the founders and current staff of Traditions Training, LLC, the cadre of tirelessly dedicated, passionate instructors who have facilitated teaching "beyond the book." This top-notch training company is the platform from which we routinely share our message. There is no doubt that your teachings have saved firefighters' lives and will continue to do so.

We would certainly be remiss if we did not mention our in-house editor Fran Shaw; neither of us is a literary genius, and we prefer the fire floor to the keyboard. Thank you for keeping us on target and on task and re-explaining to us the English language.

A special thanks to Dan Madrzykowski, who provided his profound knowledge of fire behavior and demonstrated his constant desire and dedication to keeping firefighters informed with the right information.

Finally, we must also acknowledge the great working relationship with the many wonderful people at Fire Engineering for believing in us and in our message. Thank you for giving us the privilege to put our material in book form.

Introduction

Dubbing the next 25 chapters a book may be misleading. A more appropriate name might be a firefighter's guide or tactical handbook for fires in residential buildings. We wrote it in memory of all the firefighters who have lost their lives in the line of duty at fires in these specific buildings (fig. I–1).

Fig. I–1. Firefighters monument in Washingtonville, NY, remembers those who made the supreme sacrifice.

The concept behind *25 to Survive* is to offer information to curb the trend of firefighter injuries and deaths while operating at residential building fires. We feel the groundwork to change this trend is presented within the 25 critical lessons and topics in this book. The chapter topics are from information gathered from our firefighting brothers and sisters, and from reviewing fire service resources that provide statistical data from residential fires. We hope the concept and tactical information gives insight to firefighters from rookie to veteran, instructor, company, or chief officer.

"Those who cannot remember the past are condemned to repeat it."

—George Santayana (1863–1952)

When the Spanish-American philosopher wrote this in his book, *The Life of Reason, Volume I*, he certainly was not referring to the fire service. Yet, it is deadly accurate for our profession and has applications for our everyday operations.

As members of the fire service, we are provided with detailed, concise, and sobering documents each year that detail the events of fireground incidents that lead to line-of-duty deaths (LODDs). These reports are printed, emailed, and posted for firefighters and officers in firehouses around the world. Unfortunately, many use them to start a finger-pointing barrage at that particular department or adopt a Monday-morning quarterback position. How many times have we heard, "Oh, that will never happen here."

The question is, what do we do with the information? Once we remove the "skin" of these various reports and examine the "meat" of the information, we begin to see patterns of consistently repeating events. How can we become better at what we do? Can we slow or stop the repetitious cycle of injuries and deaths through knowledge and understanding? Just as many of the causes and types of our fireground injuries are repeated from year to year, so are many of the underlying messages in the book. The repetition is built in by design, both within the chapters and carried from chapter to chapter.

What separates this book from others? While some may merely shine light on the issues, we present fireground-proven tips, tricks, drills, and procedures to overcome them. We address a problem and provide the firefighter the avenue to fix it. In many cases, we explain the *why* behind what we do at the residential fire. Ultimately, our goal is to maintain the high level of professionalism and courage demonstrated each day by firefighters and pair it with a detailed blueprint for success.

Read it cover to cover, or pick a chapter for each day you are at the firehouse and review it with your fellow firefighters. Either way, discuss and perform the drill or wrap-up at the end of each chapter, which we have termed "The Backstep." Above all, keep yourself and your crew combat ready. This attitude starts with the commitment to continued learning. Remember, we are all in this together. We must consistently watch out for each other. Only a firefighter truly knows what it is like to be a firefighter. Only we, who crawl down smoke-filled hallways, know what it is really like when lives are on the line. When things go wrong, only more firefighters like us are coming in after us.

Saving Lives and Property

Our job as firefighters is to save lives and protect property. Since the beginning of organized fire companies, this has been the credo of every firefighter and every fire department. As generations change and new members enter our firehouses, that credo remains intact. It is the cornerstone of the services we provide our citizens. However, in the last 50 years or so there have been major modifications in how we fight fires, both in our tactical decision making and our firefighting equipment.

Change hasn't always come easy for those of us in the fire department. Some have called our response to change *resistance*, *stubbornness*, *laziness*, or *traditionalism*. Our values as firefighters are deeply ingrained in our culture. Sometimes these traditional beliefs have tied our hands as new and innovative ideas passed us by.

The fire service has a philosophy of what we like to call "rigid flexibility." We must hold on to our core values on one hand while realizing that change is inevitable on the other. While we must not be too quick to jump on every new idea and fad, we must see if it would work for *our department*. Just because a tool or policy works for someone else doesn't mean it will work for you. Try it out. If it works, by all means adopt it. Professionals in any trade recognize there is always room for improvement.

The Beginning

Remember Santayana's quote about history? We must never forget our rich fire service past. Much of it has been formed with the blood, sweat, and tears of our predecessors. We need to look to our firefighting past, recognize the changes that have transpired, and understand how it leads us to where we are today.

The early days of firefighting take on an almost romantic aura today. The fire department and their firefighters were larger than life, and whether volunteer or career-staffed, firehouses and firefighters were a focal point of the community.

Elaborate gold leaf decoration was featured on hand-drawn apparatus, fires appeared on Currier and Ives prints, and parades brought civilian fanfare (fig. I–2). It was the day of brave, bearded firemen on shiny hose wagons, and officers with speaking trumpets. Firehouses and firefighters were entwined in the fabric of the community. Neighbors helped neighbors, and friends and families pulled together to protect their livelihoods

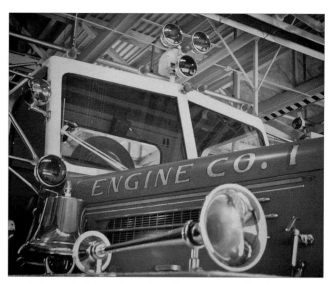

Fig. I–2. Gold leaf decorated antique fire apparatus. (Courtesy of Nate Camfiord.)

from the ravages of fire. Firefighters gallantly did what they could to save lives and property with what they had available. Fires were a true menace and blight to society, difficult to control at best.

As the urban hubs and cities began to swell and become overcrowded, more and more people began moving to the outlying areas. The suburbs sprang up, with single-family home ownership in high demand. Citizens were tired of the cramped, dank urban settings and wanted to purchase their own homes on their own land. In the early 1900s home ownership grew to become the largest recognized feature of success; it was the beginning of the "American dream," and is still is a thriving part of Americana.

More people living in the suburbs meant more residential buildings, which in turn increased the potential for fires in these dwellings. The residential single-family dwelling was then and still is deadly for firefighters. Unfortunately, recent empirical data demonstrates that residential buildings are where the greatest number of fireground injuries and deaths occur for civilians and firefighters alike (table I–1).

Table I–1. Residential building fires 2006–2010. (Courtesy of USFA.)

Year	Fires	Deaths	Injuries	Dollar Loss
2006	392,700	2,490	12,550	7,118,000,000
2007	393,300	2,765	13,525	7,527,000,000
2008	378,200	2,650	13,100	8,124,100,000
2009	356,200	2,480	12,600	7,378,800,000
2010	362,100	2,555	13,275	6,646,900,000

While we cannot forget our past, we also can't live in it. We must change as our response areas change; it has an absolute impact on the delivery of our service. How can we strive to increase our ability to provide the service of saving lives and property?

Moving Forward

Being a firefighter certainly seemed simpler in the past, even with the technology of the day. Units were called for fires, responded, and extinguished them. Today, the role of modern firefighter and fire department has morphed into the proverbial jack-of-all-trades. Sometimes it seems we are the last bastions of hope for civilization itself! Dialing 9-1-1 seems to always end with a fire department response. All departments today, urban, suburban, and rural, are certainly responding and tasked with more than just extinguishing fire.

While we still spend hours in our initial fundamental training (albeit more formal) focusing on such topics as search, hose operations, portable ladders, communications, etc., this does not necessarily complete the skill set you will need to be a modern-day firefighter. Emergencies are skyrocketing throughout the fire service. We respond to events that have us giving input as engineers, vehicle mechanics, social workers, plumbers, electricians, and more. We have also added a multilayer inventory of accessories, "add-ons," if you will, to those fundamental suppression tasks: emergency medical care, hazardous materials awareness, auto extrication, swift water rescue, ICS, and so on (fig. I–3).

Fig. I–3. Jack-of-all-trades, master of none?

Habits, Focus, Mastery

We must strive toward *mastery* in the fire service. While at times this may seem like an unattainable goal, we must push ourselves and focus our training in an attempt to achieve it. And while no specific individual part of our training can be ignored, the grim residential statistics cannot be ignored, either. Firefighting teams must strive for flawless performance, for perfection in execution. Each fire department and each firefighter must have a solid foundation and strive for mastery of the fundamental basics, all the while focusing on the fire events to which they respond with the greatest frequency.

Every fire department and every fire company has a certain level of preset, built-in mastery. For example, based on the frequency of the fire duty in their first due, many companies have reputations for the type and quality of work that they routinely do. Whether urban, suburban, or rural, you know the area in which you serve and the fires that you respond to with the greatest frequency. Whatever the setting, we get good at what we do most. The philosopher Aristotle said, "We are what we repeatedly do. Excellence, then, is not an act, but a habit."

In creating good *habits*, we must also recognize when things aren't working. Specifically, when the tried and true techniques employed on the fireground are no longer producing the desired end result. As mentioned earlier, change is inevitable. As change occurs within our neighborhoods in terms of fire duty, building characteristics, occupancy load, and fire behavior, we must change and adapt our habits and training to meet them. We must consistently examine our fire service statistics and evaluate how they may affect our personnel and our fire department operations.

Do not lose *focus*; do not overlook the statistics found on the charts in this introduction. More firefighters are seriously injured and killed while operating at residential building fires than at any other building type we encounter.

This book identifies 25 prevalent and vital topics necessary for firefighters and company officers to address when called to combat fires in residential buildings, highlighting issues common to fires in those structures. These topics are obviously not an exhaustive or exclusive list, but are compiled from our personal and firefighting teammates' experiences, data repeatedly noted in recent "near miss reports," National Institute for Occupational

Safety and Health (NIOSH) recommendations in after-action reports, and in line-of-duty death (LODD) reports (fig. I–4).

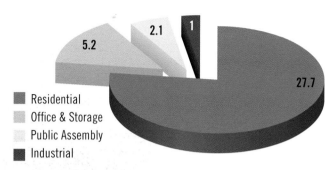

Fig. I–4. Deaths in dwellings 1990–2000. (Courtesy of USFA.)

We address topics ranging from before the response to the need to develop and maintain your department's policies. We discuss the application of various residential operational tactics for engine and ladder companies. This is a direct attempt to increase efficiency, which, in turn, may eliminate many hazards in our operations on the fireground. The mission is twofold. It is for citizens in the communities we serve, but more importantly it is for firefighters like us.

The Residential Building Today

What is encompassed in the term "residential building?" In this book, a residential building is considered a one-family occupied dwelling. In chapter 8, "Get Out There," we dissect various common nomenclatures for a variety of styles of such homes. Legally or illegally, many of these homes have been converted to accommodate the addition of extended family or even an entire other family, effectively making them two-family homes. Other terms such as private dwelling (PD) and single-family dwelling (SFD) are used interchangeably with residential building.

According to the 2005–2010 American Community Survey conducted by the Census Bureau, 61.6% of United States residents live in detached residential buildings. By comparison, according to the same survey, the second most numerous type of home in which our citizens live is the 20+ unit apartment house, which accounts for 8.2% of U.S. dwellings. In 2008 the National Fire Protection Association (NFPA) recorded residential building fires accounting for over $8 billion in fire losses. U.S. Fire Administration (USFA) national statistics show that, on average, 70%–80% of our line-of-duty operational deaths and the greatest numbers of civilian casualties occur in residential buildings.

These grim statistics reinforce the need for all firefighters, from the probationary firefighter to the chief officer, to have a thorough knowledge of these structures and how fire will travel through the residential building. Additionally, all personnel need to know sound strategies and tactics and how to employ them in order to fight these fires successfully. Firefighting is a team effort; every member of your team must be prepared mentally, physically, and operationally in order to best attain the goal of saving lives and property. We must continually analyze, develop, implement, train, and refine our strategies and tactics for fighting fires in the residential structure. We remember our past . . . and will not be condemned to repeat it.

The Future

25 to Survive will serve as a handbook to members of the fire service in each part of that process, ultimately making them the most prepared for battle. The honor and privilege of preparing a handbook for the fire service is humbling, sometimes overwhelming, and not without much credit due to others. None of us were born with vast knowledge of fire service operations. We learned from our formal and informal leaders. We were fortunate to go to fires and run calls with the best and brightest firefighters and fire officers who impacted us greatly. They mentored us. They helped us in our quest to learn our profession and be the best firefighters we could be. Whether through "war stories" that shared valuable lessons or taking the time to continually motivate and challenge us, their impact on us continues today.

To that end, this is a book by *you* and for *you*, developed to continue the trend in the fire service of learning, teaching, and becoming a master of your trade. As you read this book and implement the Backstep drills and discuss the tips and tactics offered at the firehouse kitchen table, remember this quote from Civil War era lawyer Albert Pike:

"What we have done for ourselves alone dies with us; what we have done for others and the world remains and is immortal."

PART 1

Combat Readiness

1 It's All about Me

Instant Effects

Andre Agassi, the famous 1980s and 1990s tennis star, once said in a Canon camera television commercial, "Image is everything." Well, in the fire service, image, while not everything, certainly has its place. Our public image and the public's perception of that image can be quite powerful. Fires and emergency operations have always drawn large civilian crowds. This is not new for us. Today, not only are they watching our operations, they are also recording them (fig. 1-1).

Fig. 1–1. Civilians recording fire operations

Nearly everyone has a mobile phone, most of which have a quality photo camera. As technology moves forward, most phones also have video and audio recording capabilities. Websites such as YouTube and other file viewing sites have become clearinghouses for thousands of fire department operational video and audio recordings. Your next fire may be recorded for the entire world to see, if it hasn't already happened.

Images or moments of fire operations are recorded at the push of a button, most of the time captured without our permission or knowledge. How well did you perform? How prepared were you and your department? While at first just large or noteworthy incidents were recorded, now—no matter how large or small the incident—you *may be* recorded. The public is recording you in your quarters, at the store, at drill, driving down the street, and at operations large and small. Most times the unnamed Internet video and photo poster will never take them down and they could remain viewable by the public forever.

There is a revolution coming to the fire service compliments of the World Wide Web. If it hasn't happened to you—if you haven't been a victim of getting caught on the web—it's just a matter of time. Very often, before you pack the last length of hose back on the rig, or even make it back to the firehouse, your actions and inactions at the incident are broadcast for the entire world to see.

As firefighters and fire officers, we need to use the images published online as training tools (fig. 1-2). They can prove invaluable for reviewing operational tactics and readiness: things to do and not to do in the future.

But guess who else is perfectly able to view these images? Your family; friends and neighbors; your town and local community leaders; and other citizens whom you are sworn to protect. Think about this from a more local perspective. A video snippet of a response you ran last week or last year could impact your fire company's next fund drive or operational budget vote. This is going to be the new norm for our responses. We as the fire service must present the best possible public image for ourselves and our departments or be required to explain the actions recorded.

Sometimes we are our own worst enemies. How many Internet videos recorded from helmet or apparatus dash cameras have you seen placed on the Web by our own members (fig. 1-3)? What sounds like an innocent, gallows-humor remark in the apparatus cab among the crew can cause the department a local or global public relations (PR) disaster when posted online. Those outside the firehouse walls don't often understand our vernacular. Civilians don't understand jocular firehouse phrases such as "That was a great job," or "We ain't saving this place," or "We got a roast in here." Again, recorded and posted, the damage may be done. How many photos or descriptive posts of fireground events on social media sites such as Facebook have resulted in member discipline, demotion, or termination? Has the collective fire service not *gotten the email* on this one?

Fig. 1-2. Firefighter reviewing incidents posted on the Internet

Fig. 1–3. Firefighter helmet camera. (Courtesy of Firehouse Pride, www.firehousepride.com.)

These photos and short video clips often have long-lasting effects. Society places so much emphasis on the *right now*. The social media and Internet evolution has become a double-edged sword. At times online media avenues are used to convey praise for completed operations and a job well done. Unfortunately, for every 100 positive operations, the public seems to only remember the one negative incident or perception. We know these snapshots in time do not fully represent an entire operation. Therefore, we must be prepared to have our operations recorded and broadcast every time we go out the door.

Image is everything. Fire chiefs must put policies in place to make sure their departments project the best image to the citizens they serve. Can you imagine that the new rig you need may be voted down due to a snide remark caught on a rig-mounted dash camera? Public image and positive PR are paramount for support and success in community relations.

As a favorite captain of mine once said about his members, "It's far easier to *keep* them out of trouble than *get* them out of trouble." The chief and company officers should develop and enforce policies designed to best shape their department's image. Rules and regulations must be in place to address posting work-related comments, photos, or videos on social media. If no internal parameters are established to help keep all members safe from this publicity, we are playing with fire and will eventually get burned. While policy can shape the route to a positive public image, the image your department presents to the public starts with you, the individual.

Why Are You Here?

When you decide to become a member of any fire department, you are making a large commitment (fig. 1-4). As the often-quoted New York City Chief of Department Edward F. Croker (1899-1911) stated during his tenure, "When a man becomes a fireman his greatest act of bravery has been accomplished. What he does after that is all in the line of work."

The fire service is family. We all work together to do our job. Becoming a firefighter means joining something larger than just the individual. You are not alone on the line fighting fire, but part of a team. You are "on the job," a part of the fabric that makes up a fire company, and beyond that, a member of a fire department.

When we say "on the job," we are certainly not separating a volunteer firefighter from a paid on-call or full-time firefighter. If you have taken the oath to protect life and property and crawl down at least one smoke-filled hallway, you are a firefighter, plain and simple. The volunteer firefighters have, at times, a much more difficult challenge. As members of the community they serve, many times they know personally the constituents to whom they are responding. There are community ties to each fire response and the aftermath of every fire.

Fig. 1–4. Firefighter ready to take oath

Attitude

As part of the oath you swore, a largely unwritten part lies in your personal attitude. In our community, whether we asked for it or not, the average citizen admires us. As firefighters, we are not ordinary citizens. We are tasked, paycheck or not, to risk our lives defending life and property in the community. The status of firefighters as community leaders and role models is seen in nearly every department. What kids don't want to be firefighters at some point in their lives?

Sometimes internal or external sources can tarnish the luster of that status, but it is our job to keep the public trust. We maintain a polished image by presenting a positive attitude in our interactions with the public. When civilians dial 9-1-1, prompting our response, it is because they can no longer control the situation. While not every call to 9-1-1 may be a true emergency to us, it is to them. Therefore, we must recognize that these incidents may be the worst day of their lives. As responders, we owe them a certain level of respect and understanding. However, that level of respect has to be a two-way street. No one should be allowed to compromise our authority insofar as it impacts our crew's safety or our ability to mitigate the emergency at hand.

In dealing with the public, I always remember what my mother told me:

1. Talk to other people the way you would want them to talk to your mother.

2. Often, it's not what you say, but how you say it.

3. Your attitude will influence the people around you.

In the firehouse, the "you" as you once were in the regular world may not be the same. But firefighting isn't all about you; it's about your role on the team. Your persona is often perceived based upon your attitude. The way you carry and present yourself to others contributes to your success both in the firehouse and with the public. Take a good look inside. Why are you here? What are your motivations? Who inspired you to take such risks?

You must recognize that your attitude is contagious to those around you and affects your crew (fig. 1–5). Now I am not saying that you can never have a bad day, but try to leave your personal baggage at home. No one needs you to be a downer all day long. Use your strengths and all methods at your disposal to be the most valuable asset to that firefighting chain each day. Strengthen your personal weaknesses and recognize the strengths and abilities of those around you. Only *you* have the ability to change *you*. Be a student of the fire service. Strive for perfection and be the best that you can be. You owe that to yourself, your crew, and your family at home.

Fig. 1–5. Your attitude is infectious to those around you.

Combat Ready

There is no doubt that fighting fire is an extremely dangerous and physically demanding profession. Fires do not discriminate between race, gender, paid, volunteer; red, yellow, or green fire trucks; engine, truck, or rescue company members. It will take the lives of firefighters at their first fire and a cagey veteran's all the same.

In this sense, while firefighting isn't all about you, it certainly starts with you. You are the person who dons gear and gets on the rig. You control what you put in your pockets. You are in charge of your mental and physical shape. These play a huge factor in your role as a firefighter on a firefighting team. Remember, a chain is only as strong as its weakest link. We want no weak links in our firefighting chain.

There are many correlations made between the fire service and military commands. Both agencies have rank and file soldiers, company and command level officers, and a command structure for operations (fig. 1-6). Our military must be ready to spring into action at a moment's notice, just as we must. How can we best prepare our crews and ourselves for ultimate success on the fireground? We must make ourselves "combat ready."

We first heard "combat ready" used as a firefighting term coined by two veteran firefighters and fire officers, Lt. Peter Lund, FDNY Rescue 2, and Division Chief of Operations Richard Riley of Clearwater, Florida (fig. 1-7). These two seasoned fire service instructors noted a gap in the fire service training continuum. They noticed a lapse in realistic training beyond the academy. There was a significant lag in getting street proven tactics (specifically, information that may not be noted in traditional training manuals) back to the fire service members.

Fig. 1–6. Reminder above the doors to the apparatus bay

Fig. 1–7. Lieutenant Lund and Chief Riley. (Courtesy of Traditions Training, LLC, www.traditionstraining.com.)

Together they created a training company called Traditions Training, LLC (fig. 1–8). They wanted to fill that perceived training gap, keeping members' skills sharp beyond the academy. They polished old skills and refined and simplified others. One constant, regardless of the class title, was a single concept, the cornerstone of all of their training programs. The concept was called combat ready.

Fig. 1–8. Traditions Training, LLC. (Courtesy of Traditions Training, LLC, www.traditionstraining.com.)

Lt. Lund, R2 FDNY (whose nickname in the firehouse was Lt. Vulcan, "God of Fire") and Chief Riley (also former chief of the Kentland Volunteer Fire Company, MD, and a career firefighter with Fairfax County Fire & Rescue, VA) had certainly been to many fires. What started off as mutual admiration became a lasting friendship. On one of many visits to Lt. Lund's FDNY Rescue 2's quarters, Chief Riley put a few things in perspective as it related to the combat-ready concept.

While he watched Lt. Lund's members going from run to run, he noted similar traits and actions. For those of you who do not know about Rescue Co. 2 in Brooklyn, NY, a little background is in order. In addition to their complement of normally assigned first-due boxes, Rescue 2 was assigned to every working fire in the borough of Brooklyn. But they did not perform fire duty at every fire they were dispatched to. Sometimes when they arrived they went to work; other times they were cancelled en route, or arrived to find the fire extinguished by the first-due companies. The commonalities were that at every run, every time out the door, they were ready to go to work. Their bunker gear was laid out neatly, poised for action (fig. 1–9). When the bell rang out, they effortlessly dressed in all of their personal protective equipment (PPE): hood, radio, ear-flaps down, gloves on.

Do you have such discipline and readiness come fire or not? Lt. Lund's men were ready to work what may or may not have been the fire of their career every single time out the door.

Fig. 1–9. Bunker gear laid out for efficient donning

In Chief Riley and Lt. Lund's teachings, they noted that while members had the skill set to complete the task once ready, it took far too long to get themselves into the action. The cornerstone laid in a combat-ready attitude is that we and our fire companies must be ready to go to work, not for just confirmed working fires, but on every single run out the door. Every response we make, until we determine otherwise, could be the biggest fire of our career.

If we could ask any member killed in the line of duty if he or she expected to die at that fire, I am certain the answer would be "Absolutely not." No one ever goes to a fire expecting to be injured or killed. No one expects the next fire to be the last. Lt. Peter Lund, who retired in 2003 after 30 years of service to the citizens of New York City and the FDNY, never stopped being combat ready (fig. 1–10). In retirement, Lt. Lund continued to volunteer in his hometown as a firefighter. He tragically gave his life for his community after fighting a fire in Woodmere, Long Island, on June 14, 2005.

Fig. 1–10. Lieutenant Peter Lund, FDNY. (Courtesy of Traditions Training, LLC, www.traditionstraining.com.)

Fig. 1–11. Complacency can kill.

Complacency

Complacency kills, and as firefighters we must fight it (fig. 1–11). If it is allowed in your firehouse, it can be contagious and cancer-like in its spread, found among firefighters from career to volunteer. There are many schools of thought about how complacency originates and infiltrates into the culture of our job. The two excuses we hear most for allowing complacent attitudes are "We *never* go to fires" and "We *always* go to fires."

Never go to fires? Well, if you don't go to fires, is that a real problem? A "never go to fires" attitude is an absolute powder keg. If you never go to fires, perhaps it means your fire prevention and education programs are working. That is obviously a good thing. However, if you aren't going to fires, how do your members get their experience?

Fewer fires in the community must be paired with more realistic training opportunities for members. To maintain our firefighting skill level, we must train to be prepared for the next fire. When complacency is allowed in the department, the chances increase that a fireground injury or death will happen to the team. You must commit to maintain the maximum level of readiness for your next job.

If you are fortunate to always go to fires, you must still remain vigilant and repress the temptation to be complacent. While not every run that reports a fire turns out to be one, neither is every alarm bell sounding just a malfunction. We are tasked to respond and be ready to work at every run.

Did you ever go to a fire where everything went exactly as planned or where you came out and thought you did everything right and wouldn't change a thing you said or did? The only consistent thing when it comes to firefighting is that it is different each time. No fire is routine and no two fires are ever exactly alike. While you may practice due diligence in an area where fires are prevalent, fire behavior from building to building, from day to day, warrants constant combat readiness.

Let's talk about you as an individual firefighter for a moment. There is an old adage: "Change begins with the individual" (fig. 1-12). Adopting a combat-ready attitude has to start with you. While department policy and guidance from company officers can encourage such behavior, in reality they merely provide checklists for action. What we are suggesting is that you create good fireground habits for yourself. Merriam-Webster defines habit as *an acquired behavior pattern regularly followed until it becomes involuntary*. Creating good habits and routines creates muscle memory. Dressing for a fire will become habitual if you do it the same way every time. If you have bad habits, correct them now!

Fig. 1–12. Give yourself a good look over.

Habits

Anyone can develop a few bad habits. At a buff show, I saw a Velcro "strappy glove holder" that I thought looked neat, so I bought it and promptly affixed to my turnout coat (fig. 1–13). I thought it was great—no more fishing in pockets for this glove or that. With this strap, they were right on my coat's D-ring for me to just unhook and go. I had the contraption for about a month, and even received a few compliments on it from the other guys working in the firehouse.

One day shortly after sporting my new glove holder, while I had the "can" position in our ladder company, I had a run for an odor of smoke in a high-rise multiple dwelling. As we boarded the elevator and started our ascent, stopping two floors below the reported condition, my officer kept looking at me funny. He looked at me, then my coat, and then back at me. After a few moments, he said to me, "Hey, kid, where's your gloves?" Not understanding what I was hearing as I had watched him eyeball me, without a word I pointed proudly to the D-ring glove holder. I know he didn't miss it. They were in plain sight; my bright gloves were hanging right there. He looked away.

Ten seconds or so later he repeated, "Kid, really, the gloves, where are they?" I thought to myself, *geez, I don't know where he is going with this!* I retorted, "Loo, they are right here, attached on this Velcro band."

Fig. 1–13. Velcro glove holder

He turned away with a small grunt and didn't ask again. When we got off the elevator, the boss led the way up the stairs. I looked to my senior guy with the irons and said, "Guess he doesn't like my strap, huh?"

In a low voice so that the boss couldn't hear, he said, "Kid, he was telling you to put your gloves on your hands, not leave them on your strap, you moron! We are going to a fire. Gloves are for hands, not straps."

The point was taken, and upon returning to quarters I promptly removed the strap from my coat's D-ring. I realized what my boss was telling me. That strap gave me the bad habit of not putting my gloves on my hands for every response. It had become a crutch. I felt that I didn't need to wear them all the time because I knew they were right there. Have you ever had to try to put your gloves on quickly while under pressure, with a fire staring you in the face? It isn't as easy as you think. It is, however, easy to take them off if the alarm turns out to not be a fire.

Something as simple as the way you lay out your gear for action can save seconds in donning and ensure the best possible protection. Whether your turnout gear is stowed in a personal locker or hung on a hook near the rig, place it in the same manner each and every time. Keep items like gloves and small tools on the same side of your coat or pants pockets, and lay it all out for action. Repetition is the key to creating these good habits. Come off the rig ready to go to work. Who are we letting down if we aren't ready to engage the enemy?

Don't Let Me Down

We are part of a team: a company of firefighters (fig. 1–14). We are the civilians' last hope. By dialing 9-1-1 they have effectively given up. Sometimes it feels like we are the last people left on Earth before total anarchy sets in. It is our job to make everything right again. If summoned to respond for an odor of smoke, shouldn't we be ready to find a fire? If we are called for a fire and aren't ready to promptly operate on arrival, who are we letting down?

We have to be the ones who look in the mirror after the fire is out. Will sustaining an injury due to your poor attitude and readiness be acceptable? Be aware of the way your PPE protects you, create good habits when donning it, and know its limitations.

If we are not combat ready, we are also breaching parts of the oath that we took as firefighters. Our stubborn complacency and our inability to adopt a combat-ready attitude might cause us to be unable to reach that child trapped in the back bedroom. How? Well, maybe we leave our hood in the inside pocket of our coat. It is hot today, and, "We never go to fires anyway." Now as we push down the hall, our ears burn and blister and we are stuck in our tracks. We can't take our gloved hands away from protecting our ears to move in, and we can't push past the pain. We are absolutely letting down the civilians we are sworn to protect (fig. 1–15). They may not know it, but can you live with it? You and your team may never forgive yourselves.

Fig. 1–14. Firefighting is a team activity.

For starters, we are letting ourselves down. Our lack of preparedness may cause us to be injured or killed at this fire. Don't you owe it to your family at home to prepare the best that you can? Do you want to explain to your loved ones that the injury you sustained could have been prevented if you had worn all your protective equipment? Do you want to disfigure your hands and never return to fire duty because you wore your mechanic's gloves instead of fire gloves because they fit better? Will it be okay to never be able to hold your young child's hand again?

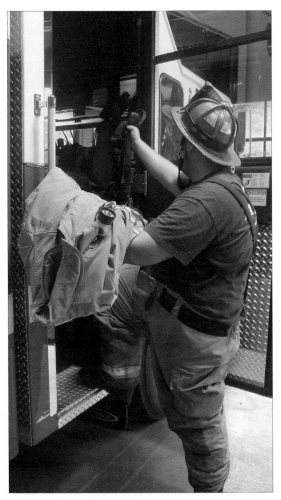

Fig. 1–15. Are you prepared?

Finally, a non-combat-ready firefighter is the weak link in the company's chain. Every firefighter is an integral part of the firefighting team. If you are unable to complete assigned tasks, someone else will feel it somewhere down the line. While it may not be noticed at every fire, it will someday come crashing down on the team. In reviewing multiple NIOSH LODD reports, we've found a vast amount of information supporting this. Many of our firefighting LODDs do not result from one catastrophic event, but instead an assortment of simple failures that, when put in sequence, become tragic.

It is our responsibility to ourselves, our citizens, and other firefighters in our departments to be combat ready every time we roll out the door.

Be a Leader: Push the Curve

In our firefighting careers we have worked in multiple states, visited hundreds of firehouses, and met thousands of great firefighters. There are always a few sayings heard in firehouses. As I heard as a covering officer in the FDNY, from firehouse to firehouse it is often "the same circus . . . different clowns."

This refers to the characteristics of the members in each company. While the names change from house to house, the act is fairly similar. Each firehouse has the *kitchen lawyer*, the *constant naysayer*, the *reverend*, the *topper*, etc. When companies have 50 or 60 members on the roster, invariably there will be different levels of experience and skill mastery. That said, we have noted the standard firefighting bell curve as it relates to firefighting skill sets and firefighters in general (fig. 1–16).

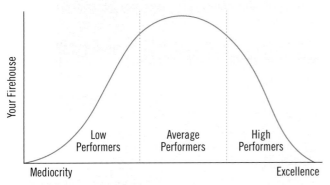

Fig. 1–16. Standard firefighting bell curve. (Illustration by Matthew Tamillow.)

Even the most well-trained and highly educated fire company will have firefighters who excel beyond others. As we look at this paradigm of the firefighting bell curve, think of this curve shape on a sliding scale. This sliding scale can move in two directions: toward excellence on one end and mediocrity on the other. As leaders, either formal or informal, we should push the entire curve in the direction of excellence. While each individual firefighter might not achieve the pinnacle of firefighting excellence, we should always try to push the team curve in that direction (fig. 1–17). The finest fire company may have a few weaker links, but with training and sharing among all members, the whole chain will be stronger in the end.

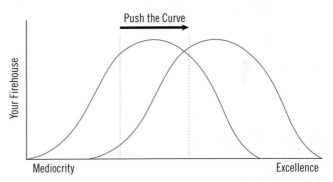

Fig. 1–17. Push the curve toward excellence. (Illustration by Matthew Tamillow.)

Leadership Traits

You do not have to be a formal leader to have influence (fig. 1–18). Leaders come in all shapes and sizes, but good leaders carry similar traits in their approach to getting things done. Some members in your department, whether officers or not, may be born leaders. These are people with an innate ability to draw others toward them. Other members can be taught certain character traits to help them motivate their peers and subordinates. Elected, promoted, or appointed volunteer or career fire officers have an enormous task, not to be taken lightly. Bugles and bars on your collar mean nothing if without the respect of the membership. Leaders often struggle to find the balance between interpersonal relationships and the ability to pass along messages in battle and in the firehouse.

Fig. 1–18. Leaders can be formal or informal, elected or appointed.

First-line supervisors are often placed in the most difficult leadership predicaments. At this level of leadership, line officers are like a layer of cheese in a grilled cheese sandwich. They have chief-level ranks as the upper slice of bread and the firefighters as the bottom slice. Depending on the organizational chart, there may other levels of officers (toppings on the sandwich like tomato, lettuce, or bacon) to mix in with the cheese, such as sergeants, lieutenants, captains, etc. Much like the cheese in that sandwich, line officers feel pressure from both sides. Sometimes it's pushing down from management, sometimes pushing up from the firefighters. All the while, the officer just wants to keep the cheese from bubbling out of the sandwich completely! It is a delicate balance.

We have taken our leadership concepts from various relationships with fellow officers and leaders, both formal and informal. You can certainly learn as much from watching the actions of a poor leader as a good one.

The following is a list of leadership traits (using the acronym LDRSHIP). This acronym was adapted to fire service applications from the U.S. Army Seven Core Values.

Loyalty

Officers must be loyal to the members they represent. It can take years to build trust in others, both on the job and off. We can earn the loyalty of the membership by training well, treating people fairly, and by setting a positive example. Those around you will emulate this positive example. Someone is always watching your actions and interactions. Set the tone for the way you expect your crew to perform, and they will follow you. Loyalty is aided by the number of people you have led in your ascension, not the number of people you have left in your wake.

Duty

Firefighting is not a check-box event. As a leader you must forecast what is going to be needed and prepare your team. There is no minimum standard! We must set the bar at an achievable level and reward completion, but neither accept nor tolerate complacency.

Respect

Treat others as you would want them to treat you. Learn about your members, who they are, and what talents they may have that will accent the company. Seek their input on how they want the company and the firehouse to succeed. Know that there is a difference between "like" and "respect." You don't have to be the most well-liked member of the team, but you need to have the *respect* of the team. General George Patton famously said, "I don't want these men to love me, I want them to fight for me."

Selfless service

John F. Kennedy said, "Ask not what your country can do for you—ask what you can do for your country." Inspire your members to appreciate the commitment they have made to the community, the department, the company, and each other. Let the members know that complaints are to be brought forth only if they are willing to work with you to create a solution.

Honor

Be a professional. Be a PRO-fessional. Take pride in being proficient at your career in the fire service. Honor and remember the sacrifices and rich history of your department. Be there for each other, in celebration and tragedy, for only firefighters truly know what it is that we do every day.

Integrity

Be fair, impartial, and equitable with each other. Favoritism and bias have no place in a good leader and can undermine the critical mission of the profession. Adhere to the ethical standards of the department, and set the tone for those around you.

Personal courage

In a leader, it is not the absence of fear but the ability to put it aside to do what is necessary at the time that defines courage. Leaders have both physical bravery and moral fortitude. They must stand firm on principles and convictions because oftentimes, the courageous decision may not be the most popular.

Fig. 1–19. Company officer drilling the members

We have all had leaders and bosses in our past, whether formal, informal, elected, or promoted, who have influenced us (fig. 1-19). Some were positive, some were negative, but all of them influenced us nonetheless. Remember our mentors, like Lieutenant Lund and Chief Riley. Motivation is what gets us started, good habits keep us prospering, leaders keep our focus keen, and we all should strive to push the curve in the right direction, towards fire service excellence.

Many great books have been written about leadership, and a leadership reading list is included as the drill, "The Backstep," for this chapter. Most, if not all, of these works have fire service leadership applications. Some discuss the human psyche, others delve into management styles, and others facilitation of motivation. All are inspiring and engaging.

Bringing It Home

There are many factors in the fire service that we cannot begin to control, such as budget cycles, types of buildings in your area, time to notification of alarms, etc. We must choose to focus on the things that are within our control. Remember that this chapter started and is now ending about *you*.

What things can you exercise control over in your quest for fire service excellence? Attitude is number one, the driving force behind your appearance and your desire to strive for the best. Have a positive attitude and pick up those who try to drag you down. You control your preparedness, both in the physical and mental aspects of your fire service career. Be a student of the game, expand your knowledge base, and be open to new ideas. Intimately know the tools, equipment, and PPE that you have in your grasp. It is what you are going to be wearing and using at your next fire. Have a good turnout to the apparatus, whether from home or while in the station.

Time is precious in firefighting, and we can't make up time going down the road. We need to make it up in other places. You are part of a firefighting team, a team that has dedicated itself to the citizens it has sworn to protect. Remember who is counting on you to know your job: your team and your loved ones, for they will be awaiting your safe return, fire of your career or not.

The Backstep

An instrumental part of being a leader is a commitment to learn and share knowledge. Knowledge can be gained from the lessons shared by leaders of yesteryear and from today. The role of a leader is to glean what is applicable and share it. A leadership reading list is a personal accumulation of information, but also reflects your commitment to those with whom you serve. Develop your list, share it with your peers and subordinates, and continue to learn!

Leadership Reading List

Deep Survival, Laurence Gonzales

It's Your Ship, D. Michael Abrashoff

Leadership secret of Attila the Hun, Wess Roberts

The Killer Angels, Michael Schaara

Principle-Centered Leadership, Stephen Covey

20,000 Alarms, Richard Hamilton

How Good People Make Tough Choices, Rushworth M. Kidder

On Leadership, John Gardner

Small-Unit Leadership: A Commonsense Approach, Dandridge M. Malone

The Leadership Challenge: How to Keep Getting Extraordinary Things Done in Organizations, James M. Kouzes and Barry Z. Posner

Once an Eagle, Anton Myrer

Five-Star Leadership, Patrick L. Townsend and Joan E. Gebhardt

The West Point Way of Leadership: From Learning Principled Leadership to Practicing It, Larry R. Donnithorne

Drive: The Surprising Truth about What Motivates Us, Daniel H. Pink

Caught Between the Dog and the Fireplug, or How to Survive Public Service, Kenneth Ashworth

Requisite Organization, Elliott Jacques

Leading Change, John P. Kotter

Co-Leaders: The Power of Great Partnerships, David A. Heenan and Warren Bennis

The Center for Creative Leadership Handbook of Leader Development, 2nd ed., Cynthia D. McCauley and Ellen Van Velsor, eds.

The Passion of Command, B.P. McCoy

Be, Know, Do: Leadership the Army Way, Adapted from the Official Army Leadership Manual, Frances Hesselbein and Eric K. Shinseki

Delivering Happines: A Path to Profits, Passion, and Purpose, Tony Hsieh

First In, Last Out: Leadership Lessons from the New York Fire Department, John Salka

2. Combat-Ready Fire Apparatus

Since the advent of the fire service the apparatus which delivers us to the fire scene, along with aiding us in our ability to suppress the fire, has been pivotal to our success or failure. The organized chaos we call the fireground dictates the type and amount of equipment we dispatch to fires. Additionally, the fireground demonstrates the need to specifically place equipment in particular areas of the fireground, e.g., truck on side Alpha, engine and truck on side Charlie, etc. This is all done with the goal of having the right equipment in the right place to aid us in completing our job—putting out fires!

Yet every day we see equipment that is not ready to go to fires and assist the firefighters. Rather we see apparatus that is well equipped to go down Main Street in a parade (fig. 2–1). Like doctors, firefighters cannot complete their jobs without their tools, regardless of how much knowledge they have. Of course, we must straddle that burden between having equipment that is not only presentable to the community that bought the million-dollar ladder truck for us, but also is ready to go the biggest fire of our career at a moment's notice (fig. 2–2).

Fig. 2–2. Engine and truck company operating at a house fire. (Courtesy of Kentland Volunteer Fire Department, www.kentland33.com.)

Think about your own career and all the fires you have fought. How many times has a functioning, well-equipped engine, truck, or rescue squad made your ability to complete your job much easier? Having the apparatus ready prior to ever leaving the apparatus bay floor eliminates this thought process from your brain when you are faced with fire showing, people trapped, and parents in the front yard screaming for you to save their baby.

In the last chapter we introduced the theory of combat readiness. Like any other mantra, it is nothing but words unless we embrace the essence of the theory. The combat-ready approach is really a way of running your professional life as a firefighter. It means always being prepared, always being a professional whether you are compensated or not, never accepting complacency, and always doing the job for which you took an oath to the highest level. That highest level is always being redefined because you are constantly searching and learning more about your job, making you the best (fig. 2–3)!

Fig. 2–1. Engine company adorned with Christmas lights

Fig. 2–3. Military personnel preparing for an engagement. (Courtesy of Sean O'Neill.)

The combat-ready theory is not singularly applied to just you and your firefighters but also to the apparatus we operate. As stated earlier, the apparatus is pivotal to our success on the fireground so we must prepare it for the firefight like we prepare ourselves (fig. 2–4).

Fig. 2–4. Combat-ready engine company with tools and equipment, clean and ready for work. (Courtesy of Kentland Volunteer Fire Department, www.kentland33.com.)

Where Do We Start?

During our classes, we ask the students the all-important question of who decides what goes on your apparatus? As you can imagine, the answers range from the fire chief to the chief engineer to the guy who is considered the most competent. All of the answers are a reflection of our paramilitary structure and respect for the chain of command, yet they miss the mark. While each of those aforementioned titled rank positions may make the final decision, the hope is that this person has adopted a combat-ready approach.

Specifically, the combat-ready approach means that the fire chief, chief engineer, or whoever is your "go-to guy" knows that no *one* person decides that all-too-important answer. The answer is three fold:

1. The area you serve
2. The citizens you serve
3. The skills and abilities of your members

These three factors are what decide what equipment you will need, where you place it, and nothing else. Let's discuss each one.

The area you serve

Picture for a moment that you work in an area with towering skyscrapers, yet the company which protects the area has nothing taller than a 35-foot portable ladder. Preposterous, right? Absolutely! We should never lack a piece of equipment that could assist us in completing our job.

The same theory applies to the residential structure. If the area you serve is residential, is your apparatus still set up for what the area looked like in the 1960s and 1970s? Has it "always been that way?" If so, the trend of society may have passed you by and left you handcuffed into not being able to complete your job efficiently and effectively. According to the National Association of Home Builders, the average square footage of a home increased 93% between 1970 and 2009, ballooning to an average size of 2,700 square feet. This means that a "McMansion" was not even in our vernacular when your apparatus was initially built (fig. 2–5).

Fig. 2–5. A McMansion

Thirty to forty years ago the front of every house in your area may have been merely 40 feet from the apparatus, and you could reach the farthest room with 150 feet of hoseline. With the introduction of the McMansion, the residential home could be 80 feet deep and it may be necessary to cross 200 feet of yard to reach the front door (fig. 2-6). The McMansion is just one hurdle we face. Add to that the hoarder who has created an obstacle course for our operations, or the home that has been remodeled into a maze-like layout.

We must ensure that our handlines can overcome the greatest setback and our ladders can reach the tallest window where a victim is begging for assistance.

The citizens you serve

Each one of us who has taken the oath to be a firefighter has committed ourselves to saving lives, regardless of the gender, age, sex, or religion of the citizens. We demonstrate every year, through our somber statistics, that we give our lives for perfect strangers reportedly trapped by fire.

To ensure our apparatus is combat ready we must know our customers and know that our apparatus is ready to help us maintain our commitment to our oath. For instance, if we have always served a bedroom community, we have a demographic in our mind that if a home catches fire at 0300 hours, we will encounter a family with two adults, two kids, a dog, and everyone will be in the front yard when we arrive at the tree they designated as their meeting point in their family emergency plan.

The reality is that the home owned by the Smiths might now have become an assisted living facility for six non-ambulatory adults. It could be run by a nursing staff who did not attend your fire prevention talk at the local elementary school (fig. 2-7). Or since the local college enrollment has grown, the beautiful Victorian home owned by the nice elderly gentleman who stopped by the firehouse each day is now a frat house with twenty drunken young adults (fig. 2-8). Each scenario clearly demonstrates that we must know our citizens and know that our equipment is ready to help us maintain our combat-ready approach.

Fig. 2–6. A McMansion with a long setback of a yard

Fig. 2–7. Single-story ranch home turned into assisted living facility with front handicap ramp

Fig. 2–8. Large Victorian turned into assisted living facility. The small front sign indicates the occupancy change.

Fig. 2–9. Firefighters discussing strategy and tactics at kitchen table

The skills and abilities of your members

The area you serve may now have some residential buildings set back 200 feet, and you need an additional 200 feet to get to the seat of the fire. The quick answer to this problem is that you will now switch the hoseline off the back of your engine company with a 400-foot line and all is good. Or maybe you have hatched in your mind the game plan that when you run this area you will stretch the longest crosslay and extend the line with your high-rise pack.

All great ideas, *if* your members have been trained in and refined these skills. The foundation of our success has always been and always will be in the training we perform every day. To be a success, we must ensure that all who will be charging down that dark and smoky hallway have been well trained. There can be no huddle in the front yard when you decide to roll out your new hose-stretching idea. This conversation must take place at the kitchen table over a cup of coffee and then be reinforced by the actual performance of the skill afterward (fig. 2-9). That is what being combat ready is all about.

The Next Step

We have made the transition to becoming combat ready and now we are on our quest for learning more and becoming an everyday professional. So the question is, where do we start?

The first and easiest step may be to utilize some things that are in every firehouse: the kitchen table and the local newspaper (fig. 2-10). Nestled within each paper delivered to every community is a real estate section that can provide some vital information about our area.

Fig. 2–10. Firehouse table with local newspapers

The real estate section is chock full of information on new developments being built. For instance, the area that used to be "Farmer John's field" (fig. 2–11) is soon to be a sprawling campus of garden apartments with 300-foot setbacks. The days of every house being built on an acre of land with a white picket fence are gone in most of the country. More and more homes are going up on less and less land, which equates at a minimum to unusual designs, long setbacks, and water supply constraints.

Fig. 2–11. Farmer John's field is to be turned into a large plot of new garden apartments.

Reflective of the economic conditions in 2013, the real estate section may also give you a glimpse of what homes or parts of the area you serve are now vacant. Most newspapers will list the foreclosures, short sales, and auctions which should be a tip to you that what was an occupied family home is now vacant or occupied by squatters.

Reading the newspaper is not the only way to becoming a tried and true professional. We must now take it to the streets. The next element to building this knowledge base is to get up from the table and out into your area. On the return trip home from the next run, when you go out to pick up a meal, take time to evaluate the area you serve and stop at that development you just read about. Get out of the rig and walk around the structure. Talk about the construction and build a plan for how you will overcome the obstacles you observe (fig. 2–12). While we never advocate trespassing, arriving in a big fire truck wearing your uniform and explaining that you are merely keeping your community safe will get you in anywhere.

Fig. 2–12. Personnel performing a lap around a residential structure under construction

All of these techniques fall under the mantle of pre-planning. This can be the physical preplanning of walking the structure or neighborhood to discuss tactical items, or it can be the preparation of a document that can be distributed and read by all your members. While we cannot get every member of the organization into the engine or truck (unless you have a clown car), we can still get the relevant material out to each person. Develop and mandate a standard form for pre-plans in your area. A quick, informative document will not only be an introduction to your newest member of the area they will serve and a testament to your company's commitment, but it will also serve as a life-saving blueprint to the incident commander when he or she is faced with a Mayday and needs to find the quickest access point to save one of our own.

The Combat-Ready Apparatus

Much of what we have talked about so far has been centered on the personnel who will be riding the apparatus. This is because these are the people who will be deciding what equipment we will purchase and where we will place on the apparatus. We must have personnel with this frame of mind making decisions that are consistent with being combat ready.

Akin to crazed football fans who wear team jerseys every Sunday, firefighters do the same with other departments. For instance, think about how many firefighters you may see at the next conference in Small Town, Nowhere that will be wearing FDNY Rescue 2 shirts. Did all of the members of Rescue 2 fly to this conference? Probably not, but we have firefighters who are fans of what Rescue 2 does and the image they have built.

We want to apply the same mentality when we design our combat-ready apparatus. While carrying an FDNY high-rise nozzle simply because the FDNY does so will not apply to an area with one-story ranchers, carrying a married set of irons does (fig. 2–13). You should adopt ideas that work for your department and not just because the cool guys use it.

Fig. 2–13. Married pairs of tools such as the "irons"

A true professional is always on the hunt for new information, regardless of where it originates. He or she will gather the specifics, bring it to his or her company, test it to see if applies, and then institute it if it works. Ultimately, we should strive every day to eradicate the complacent mantra "We have always done it that way." If you can't explain why you have a tool or hoseload on your apparatus, it is a good indication that you are not training enough or that it's time for that item to go.

The blueprint for success starts with a basic foundation for your equipment and then builds up. For instance, an engine should be equipped like an engine, not a tower ladder or a rescue squad. There may be some instances where you have to combine tasks due to staffing or long response times, but ultimately no fire is going to be extinguished if an engine does not put water on it. So begin your combat-ready apparatus blueprint by defining the scope of what you want your apparatus to do. Engines bring water, so what do we need to accomplish this task? Trucks perform forcible entry, search, and ventilation (fig. 2–14). What do we need to accomplish this task?

Fig. 2–14. Engine and truck company personnel operating at a house fire with stretched hoselines and placed ladders. (Courtesy of Nate Camfiord.)

Once we have formed this blueprint, we can begin to assemble our combat-ready apparatus. We have done a comprehensive review of the area we serve, identified the shortcomings of our response and/or staffing needs, adopted some ideas that work, and listened to the input of our end-users, the firefighters. Don't forget them!

Throughout this book are tips and tricks that will save you seconds, even minutes, and make you better at your job, all with the goal of designing this elite piece of apparatus. The "marrying" of tools will ensure that the firefighter can go to *one* compartment to grab tools to complete the designated job.

For instance, pairing the 8-pound axe and Halligan with a simple Velcro strap, affectionately called "the irons," will place the two effective tools for forcible entry together (fig. 2–13). While we would all like to bash down doors every day, the reality is that it may be some time between fires and even greater time between opportunities to implement conventional forcible entry.

To help yourself out, consider notching your Halligan bar forks with a depth marker (fig. 2–15). This simple notch on the side of the fork will serve as the reminder of the average depth of a doorjamb. While your fellow firefighter is tapping the fork of the Halligan into the door to attempt entry, you can watch this marker. When the marker is at the door you know that you have successfully forced the fork into the jamb and past the door frame and are ready to gain entry.

Fig. 2–15. Halligan bar notched on the fork end for assisting in judging the depth of a door jamb

The tools we use and prepare are useless unless the firefighters holding them are competent and professional. We can design and implement the greatest piece of apparatus in the fire service, but without a combat-ready operator they are nothing but items on display. How we achieve this final piece of the puzzle and achieve success is the key.

Our tools and our people should be ready to work at a moment's notice. For tools, this readiness is centered on maintenance, care, and operation, but for people it is centered on training. The analogy that many would rather be the guy in the casket then the guy giving the eulogy is true for most firefighters with respect to instructing. They may never want to be an instructor. Their fear of standing in front of a room of people may be greater than having to vent, enter, isolate, and search (VEIS) a window belching black smoke. Regardless, we know that every firefighter is and must be an instructor, albeit in different forums. Some will take to the stage, others are best at one-on-one interaction with a probie on the bay floor (fig. 2–16). Platform aside, the point is that we all leave a legacy because the fire service never retires but we will, and how we conduct ourselves will define our legacy.

Fig. 2–16. Regardless of the venue, all firefighters must be instructors sharing knowledge and experience. (Courtesy of Traditions Training, LLC, www.traditionstraining.com.)

You have taken the strides to perform the background work of learning the area you serve and designed and implemented a combat-ready apparatus. Now you must ensure the longevity of the product. How do we get our firefighters to carry this legacy forward? The answer may not be as hard as it seems. Every member who rides on your apparatus should know the capabilities and limitations of the equipment they operate. This requirement is the first step to maintaining the combat-ready code.

Maintaining the Legacy

All the work you have done has been focused on the present, but it has long-reaching effects well after you have run your last call. Pride is not new to us. Think about what we do every day. We proudly wear our company patches on our uniforms or have a hat that boasts your engine number. This is not false pride, it is built out of your dedication to the job and desire to be part of something great. This is not limited to the fire service and is seen every day in society. We want to tap into that theory to ensure the legacy you create is not short-lived.

Consider for a moment if I offered you the two watches below to purchase (fig. 2–17). They are both priced at $10. Which would you buy? Why?

on the side of your combat-ready apparatus. Think about what was discussed earlier in this chapter regarding the large number of shirts you will see at the next conference that advertise for the FDNY, Boston FD, etc. What does the image of that patch stir in the wearer's mind? Integrity, tradition, and good firefighters? Yet they may have never even met a firefighter from that organization in their lives.

The question you have to ask yourself is, what does your patch represent? It may define your legacy and direct your purpose each day that you place your gear on the apparatus. What have you done to build positive brand recognition (fig. 2–18)?

It all starts with you and continues with the development of your combat-ready apparatus and firefighters.

Fig. 2–17. A comparison of two watches. One is a no-name watch and the other a Timex.

I venture to guess that a large percentage of you will choose the Timex watch. No great surprise, but the reason *why* you did is what we want to capture for the fire service. You have chosen the Timex because you know that for $10 you are getting a quality watch. You are getting a watch that will help you to complete an Ironman triathlon, scale tall mountains, or challenge Usain Bolt in a footrace. Timex has worked very hard to build their brand image and attach it to their product. The other watch? Well, it's just a watch, and it will tell you the time.

All of this is called "brand recognition." What does this logo stir in the minds of the potential consumer? If they have done a good job of developing brand recognition, then you will purchase their product based upon an assumption of a quality product through proven results.

The same should apply to when someone sees the patch attached to the side of your job shirt and painted

Fig. 2–18. Patch of Company 8 in the apparatus bay

The Backstep

Combat-Ready Equipment Book

Develop and implement a combat-ready apparatus equipment book. Design a universal template for your equipment book that provides a detailed synopsis of every piece of equipment carried on your apparatus. Assign each firefighter a specific tool and the template sheet to go with it (fig. 2–19). Not only will you develop a comprehensive reference for your apparatus and a legacy based on being combat ready, but you also will be developing the teaching skills of your peers. Like figure 2–19, a simple one-page document may be designed for each piece of equipment that outlines:

- Name
- Specifications
- Capabilities and limitations
- Tactical application
- Maintenance

Once you have assigned each firefighter a tool, they will complete this document in the format provided. Additionally, have the firefighters present the research they have done to the entire company as a drill. Once the document and drill are done, place the completed information sheet into your combat-ready equipment information binder. The informative book will now be in the hands of every member who comes in contact with your combat-ready apparatus.

Think back to when you started in your company. Wouldn't it have been great if when you walked through the door the captain met with you, discussed expectations, and handed you a book that you could study every day? This book would become the foundation of your knowledge for how to become an asset to your fellow firefighters, and maybe even save your life!

Tower Ladder 08
Combat-Ready Equipment Information & Drill Book

Equipment name:
Hydra-Ram forcible entry tool

Specifications:
Weight: 12 lb
Length: 13″
Constructed with aircraft seals to prevent leaking of fluid
Stainless Steel jaws (rated to 220,000 lb tensile strength)

Capabilities and limitations:
10,000 lb of force will be exerted when 138 lb of force is applied to pumping, each stroke will spread jaws 3/4″ up to maximum of 4″
HR–1 = 4″ spread   HR–2 = 6″ spread

Tactical application:
One-person operation

Can be used in any forcible entry application where a metal frame with a metal door is present.

Can be placed in any position without affecting operations (upside down to force a bottom lock)

Takes 7 pumps to achieve full extension (depending on age of unit)

This is important for detecting any failures in the system in a smoke-filled environment.

Maintenance:
Limited; hydraulic fluid reservoir is on front plate of Hydra-Ram

Test by placing unit under load while extending the arm and ensuring it can support the load for a length of time. An example to conduct a weekly test is to:
- Place the Hydra-Ram under the firehouse dumpster
- Pump for a few times to slightly lift dumpster
- Leave the unit in place to ensure it holds the load and does not fail

Fig. 2–19. An example of the tower ladder equipment book page for the Hydra-Ram tool

3 Routine

A typical day in firehouse unfolds with all of the members sitting around the table sharing a cup of joe and stories until the house tones drop (fig. 3–1).

Fig. 3–1. Firefighters sitting at the kitchen table drinking coffee and talking

> "Engines 22, 75, 42, 33, Tower 33, Truck 8, Rescue Company 4, and Battalion 4 respond for the reported house on fire at 217 East 34th Street. Report of a house on fire across the street, unknown if everyone is out."

Everyone rushes to the apparatus, quickly dons their turnout gear, and races down the road. You are dispatched as the second arriving engine and your boss confirms the assignment once he has reviewed the supplemental information provided. You can see the plume of smoke on the horizon, confirming that this is going to be a fire. You are making your final preparations on your gear and running through your mental checklist of tasks. You know the first engine should be marking on-scene in a moment, verifying the fire location and what operational tactics they are going to employ.

> "Engine 22 on the scene of a routine-looking two-story, single-family dwelling with a routine amount of fire and routine smoke showing from the second floor. We will be doing some routine work, copy?"

Sounds ridiculous right? Of course it is, but how many times have our actions or the actions of other firefighters mirrored the above statement? There is nothing routine about what we do in the performance of our jobs. No two fires, or emergency calls for that matter, are the same. Even if we don't run many calls and don't have much fireground experience from which to draw, we can use the Internet as our educational tool to demonstrate that there is nothing routine about our trade. A quick search of YouTube will offer you plenty of examples of "routine" fires that suddenly went terribly wrong.

They go wrong because we are complacent in the performance of our duties *and* they go wrong because the fire did not behave in the routine manner that we expected. To overcome this mentality, the first step to eliminating complacency is to revert back to our foundation: training.

Understanding Why You Do What You Do

Is our training focused on merely "checking the box" that we trained today and kept the chief off our backs, or is it focused on the needs of our personnel by identifying their weaknesses to build stronger firefighters? Is it focused on the performance of our duties in the area we serve? While there is an attraction to the image of being a firefighter in a city setting, scaling fire escapes and stretching lines up tenements, focus on what you have if you don't serve this type of community. The fact that almost every community has residential structures not only demonstrates why this is the most common fire we run, but also the one where we are the most complacent in our operations (fig. 3–2). Focus your training on what and how your company will operate at the residential structure, then explain *why* you operate that way. We want firefighters to understand why fire is not routine and demands our highest level of attention and dedication every time.

Fig. 3–2. A residential neighborhood picture that could be taken anywhere in the United States.

Create Your Legacy

The next step is to get your operations down on paper! The legacy of a firefighter's dedication and commitment are best served not by folklore but by what you share. If you take the time to train, develop the firefighters with you, and eliminate the routine behavior, you will develop operational manuals that carry on through many generations after your career is finished. The development of an operational manual is not a strict game plan of what you do on every single fire. Rather, it is a blueprint to success, a guide to the immensely complicated battle site we refer to as the fireground. The operational manual provides strategies and tactics to accomplish the ultimate goal of saving lives and extinguishing fire. It is up to the firefighters and fire officers of the future to update it to match the environment they face. The environment we face today is a perfect example of how we must stay vigilant. We know that fires burn hotter; we know that fires grow exponentially by the second, we know that fires emit chemicals that can render a firefighter dead in seconds without an SCBA. All of that information does not change our end goal of saving lives and extinguishing fire. Rather, it ensures we consider nothing routine and we employ the correct tactics.

Developing operational manuals is merely a cog in the development wheel that makes us combat ready, eliminating complacency and providing a base for a strong foundation. If we view our entire day, shift, or tour when we are at the firehouse, a review of the operational manual may be the nighttime ritual to build upon the base of training and experience we learned that day. The question is how do we build our experience "mental slide tray" so that when we run that 3:00 a.m. house fire with people trapped, we are not trying to recite the bullet points of our operational manual (fig. 3–3). The solution is to get out into your community and pre-plan long before the fire occurs.

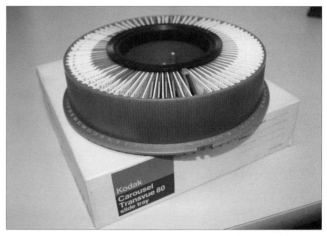

Fig. 3–3. Slide tray

Nearly every fire department in the country, with a big shiny fire truck and a measure of confidence in their step, can walk into the commercial building to conduct a tactical pre-plan (fig. 3–4). A simple meet-and-greet with the store owner would suffice, stating that this is not an inspection but an opportunity to ensure that, if a fire ever occurs, the fire department can quickly extinguish it and save the business. Check off the commercial building! We have built our game plan, all the firefighters know why and what we will do when the building is on fire, and all of the hazards have been identified. Now we just wait for the alarm to come in. Unfortunately, the commercial fire is the low-frequency, high-risk event that does not statistically occur too often. Since the residential structure is the most common fire for us, let's pre-plan all of our response area homes. Sounds like an unrealistic and daunting task, but like many other challenges in our occupation, if approached correctly, it is manageable.

Conversely, we know that the same technique we used to pre-plan the commercial building is not going to work on the residential structure. As much as our public likes us, they are not likely to open their doors to a group of firefighters who show up and want to supposedly "pre-plan" their home. Add in the economic crisis currently in the country and many people are even more hesitant to let anyone in their home. They may have converted rooms into illegal apartments or other non-code approved renovations that they do not want us to see (fig. 3-5). Therefore, we must find an alternate avenue to pre-plan the residential structure.

Fig. 3–4. Firefighters preparing to conduct a commercial building inspection

Fig. 3–5. Due to the economy and desire to earn money, many residential structures are illegally separated into apartments or single-room occupancies (SRO).

Seize Every Opportunity

The national average for call types in the United States consistently hovers around 70%–80% EMS related calls, with the remaining percentage attributed to false alarms, nuisance alarms, public service, and an occasional fire thrown in. That translates into the fact that we run many calls for service that allow us to enter a private citizen's residential structure. Think back to your own experience of the last medical call you ran or smoke detector in the home that has been chirping for the last two days. You probably got into the apparatus and thought, "Man, another BS call, wish we could run a fire," and proceeded to the call with a preformed image of what the person would look like and what they would say.

Let's take the medical aspects for this situation. You arrive with your crew of four on the engine company to assist the ambulance from a neighboring community. You grab the cot for the ambulance crew already inside assessing the patient. Your boss tells you and the probie to go in and help the ambulance crew while he waits out in the yard. You walk into the front room and find the ambulance crew talking to an elderly female who is short of breath. They advise they just need you to bring the cot inside, and you can go in service. You and the probie go outside, grab the cot, bring it inside and place it next to patient. You are feeling generous today so you even help the grandmotherly woman onto the cot, and shortly after you are out front telling the boss that you are okay to go in service. Everyone jumps back into the apparatus and heads back to the firehouse to stand vigilant for the next run. Sound like a typical medical assist? Of course it does, but in reality it was a missed opportunity to learn critical information that could save your life and the lives of the citizens you serve.

Six hours later you awaken from sleep in your bunk to the dispatcher toning out a reported basement fire with persons trapped. As everyone is rubbing the sleep out of their eyes and quickly donning their turnout gear, you hear the captain state that this is the house next door to the one where you ran the medical assist earlier that night. As you arrive on the scene, thick black smoke is pushing from every side of the home, obscuring the structure from view in an envelope of acrid smoke (fig. 3-6). As you remark on the fact that the house looks just like the one next door that you ran the call on earlier, the father runs up to your crew frantically stating his son is still in the bedroom. You assure him that you will get him out and don your facepiece to prepare for entry. Your probie pops the front door and is ready to conduct the primary search. Now, which way to the bedroom? Is it straight back? To the left? To the right? We don't want to waste critical time aimlessly fumbling around the home while every second this young child's survivability is plummeting.

Fig. 3–6. Basement fire with smoke enveloping the home. (Courtesy of Kentland Volunteer Fire Department, www.kentland33.com.)

The key information that could guide us to make the grab of this child and demonstrate our commitment to risk our lives to save others should have been gathered a mere six hours ago. We must seize every opportunity to learn and, more importantly, recognize the situations that present us with information from which to learn. What may have appeared as a mundane medical assist call was in reality the pre-fire inspection of that home. The opportunity was presented to see the floor plan of the home in perfect visibility, stamping into your subconscious the locations and quickest path to each room.

Nothing We Do Is Routine

The definition of *routine* according to *Merriam-Webster's* is, "a habitual or mechanical performance of an established procedure." Nowhere in this definition is the word "complacency" found, nor does it declare the level of seriousness that would lead us to believe we have paired those words with routine. We have declared that a routine fire is not serious and that putting out one of those fires is all muscle memory. We know, just from our close calls and line-of-duty death reports, that no fire is routine. Therefore, we must define what is routine for the fire service, and that all begins with how we train our fellow firefighters and ourselves. We must turn our nuisance calls, medical assist calls, and fire alarms into routine training opportunities with the focus not being on the redundancy of the call type but on the continual opportunity to learn more about the building before it is ever on fire. You, as a firefighter, company officer, chief officer, or mentor, must start that trend and define it for your fire department. That can begin with this chapter's drill in "The Backstep."

The Backstep

Remove the Blinders

Seize the opportunity at your next nuisance alarm, false alarm, or medical call in a residential structure to not only mitigate the situation but also remove your blinders and observe the following:

- Topography affecting tactical decisions (one story in the front, two in the rear) (fig. 3–7)
- Location of all entry and exit doors (including an exterior walkout for the basement)
- Locks and security devices that may slow our entry
- Location of stairs to upper floor or basement
- Location of bedrooms where persons may be trapped
- Location of kitchen, where most fires occur in the residential structure
- Compartmented rooms or large open areas
- Location of main hallways that will serve as the path of fleeing occupants and our path to the fire location
- Presence of attached garages, exterior entrances to the basement level, etc.

Fig. 3–7. Due to topography, this house has one story in front and two stories in the rear.

Once you have mitigated the emergency, observed the details listed above, and have gotten back in your apparatus, wait until you are out of sight of the residential structure and test your fellow firefighters. Ask the following basic questions:

- *What is the address of the home we just went to?* You want everyone listening to the radio all the time when going to a emergency. Information shared before you get there may save your life or affect your tactics.
- *What kind of locks are on the front door?* We want to build a catalog of different locks and security devices in our minds, along with the strategies to overcome them.
- *Where is the kitchen?* This is the leading location of fires in the residential structure, so statistically when we go to a fire in a residential structure, we are most likely headed there.
- *Where are the basement and the stairs?* While not the most common fire we run in the residential structure, it is the most dangerous. Time is of the essence. We cannot waste time crawling aimlessly around upstairs trying to find the residents. We must get to the door, control it, and, if applicable, make the advance down to extinguish the fire.
- *Where are the bedrooms?* The bedrooms and hallways leading from them are where we typically find trapped occupants at fires in the early hours of the morning. We have taken an oath to save lives so we must do all we can to eliminate time wasted and quickly direct our search to the correct area.

If you find you want to ask additional questions, go for it! The goal is to breed a behavior that removes the blinders and always focuses on our environment. Many departments have names for this. Whether it is termed "situational awareness" or "operational awareness" does not matter, what is important is to understand *why* we want to do this. We do it to make ourselves the best firefighters possible, preserve the commitment to our oath, and make our fireground safer.

4. Can I Talk Now? Fireground Communications

"That's one small step for [a] man, one giant leap for mankind."

–Neil Armstrong, July 20, 1969

Astronaut Neil Armstrong delivered those 12 simple words upon dismounting the *Apollo 11* module and walking onto the surface of the moon, words that have been etched in every history book and were delivered from 238, 857 miles from Earth. While a monumental day in the history of the United States and NASA, one may ask what does this have to do with the fire service?

What made Armstrong's statement so powerful was the fact that it was received. Imagine if all Houston heard was a garbled message that sounded like he was possessed by an alien, how the course of history would have been changed. It is amazing that in 1969 we could transmit a message with clarity from 238,857 miles away. Yet on the modern fireground, we cannot transmit a message with any clarity from the basement of a 1,900-square-foot residential structure to a chief in the front yard 60 feet away. While this chapter does not dissect the technology of radio communications from yesteryear to the modern day fireground, it does focus on the essence of our communication—the message!

The Message

Technology has afforded us the ability to make famous transmissions like Armstrong's along with advising a truck company performing vent, enter, isolate, and search (VEIS) of a quickly advancing fire. Conversely, it has opened a Pandora's box of issues that limit our ability to communicate, such as limited channel use, building interference, and high-dollar bargaining for bandwidth. All these are issues that affect the end users—the firefighters fighting fire—ability to deliver a message. Recognizing that communication technology can be a lifesaver on the fireground as well as a detriment is the first step in developing a strong communication policy (fig. 4–1). Much like when an issue arises on the fireground with a hoseline or pump, we must have a strong infrastructure to address and overcome the communication problem. The fire will not wait or give you a time-out to work out the issue, so we must quickly address and overcome it.

Fig. 4–1. Communication between the IC and company officers is paramount. (Courtesy of Roger Steger.)

Assuming you have the opportunity to have a radio communication system that is modern enough to handle your needs, the next step is to formulate a strong and clear communication policy. This is not the most exciting portion of our job, but think about the importance of having clear and concise radio traffic.

> "Engine 220 to Command, w-------hhhhhhhaave a vic----ttttttiimmm and neeeee---d help!"
>
> "Command to Engine 220, repeat your message, could not copy."
>
> "En----- 22-, wbbbbbbbbbblllllaaaaaaahhhhhh, help."
>
> "Command to Engine 220, do you need help? Are you calling a Mayday?"

How many times have you been on an incident where the above exchange has occurred? Frustration mounts as the chief is trying to determine what and who needs help, the rapid intervention team (RIT) is readying for a deployment, and the family is eagerly waiting in the front yard for their missing child to come out in the arms of a heroic firefighter. Just one simple message delivered and the resources could be deployed, the situation mitigated, and the child removed safely. Ultimately, what we want to achieve in a strong communication policy is to remove all of the stressful elements we can control in our fireground so that all that is left is a clear and concise message.

Does your department have a standardized policy for how to perform an on-scene report, confirmation of a working fire, or when to call a Mayday (fig. 4–2)? If not, when one of the above occurs you can expect to receive a multitude of variations, leaving you to focus on the sender and not the message they want delivered. A good example of a strong communication policy is for the working fire in a residential structure. In a perfect fire department world, the following would occur:

- The first-arriving engine delivers a clear and concise on-scene report outlining type and size of structure, conditions present, and initial actions. Shortly thereafter, there is confirmation of fire knocked down.

- The second-arriving engine confirms water supply is achieved and a backup line deployed.

- The third-due engine or second truck checks and communicates conditions on side Charlie of the structure.

- The fourth-due engine confirms RIT assignment and location of team.

- The first-due truck confirms completion of the primary search.

- The second-due truck confirms deployment and placement of ground ladders.

In seven clear and concise transmissions we have run a working fire in a residential structure from dispatch to fire knockdown with primary search completed. This limited yet useful and relevant information has freed the channel for any additional or emergency traffic that may occur. The pertinent information is delivered from company officers who know their tasks and communicate only when tasks are completed, in progress, or when additional resources are needed to complete them. What is noticeably absent in this fictitious yet perfect fire is the fourth-due engine calling command in the midst of the firefight to confirm the officer can see smoke in the sky on approach, or the driver of the second engine calling the driver of the first engine repeatedly to know if they are ready for water when they are separated by 300 feet and there is no visual obstruction. Such useless radio traffic serves more as a comfort mechanism for the sender. While the hope to eradicate much of these nonessential radio transmissions is far fetched, the development of a strong communication policy is the first step in educating and mandating compliance.

Fig. 4–2. The IC must be able to have a message understood by all operating units.

Asset or Liability to the Fireground?

The foundation of building a strong communication policy during the most stressful of incidents, whether it is a working fire, Mayday, or trapped victim, is to mandate compliance on the most mundane of incidents.

In the modern fire service, many pieces of apparatus are equipped with some variation of a computer-aided dispatch system (CAD) that denotes en route, on-scene, in service, etc. This simple tool should eliminate many useless transmissions that are instead handled by a simple keystroke. In fire departments without this tool, the transmission of this information is important and needed, but for departments with the computer system, consider why you are talking.

Regardless of the affiliation or pay status, how many times have you listened to a radio transmission that sounded more like an appeal for approval of a decision versus a command statement of action or resources needed?

"Truck 8 to Communications, when Engine 27 arrives have them go to side Charlie and check it."

Why would you want a person sitting in an office chair 30 miles away with no visual idea of what you are dealing with to relay an order for you that is not clear in its purpose (fig. 4–3)? Is the above order to check that there is a backside of the building or if it is on fire? What would the dispatcher respond with if Engine 27 asks what they want them to check? All are valid questions that steal more radio traffic because we did not deliver a clear directive and we involved a third party (dispatch). Whenever a firefighter, company officer, or chief officer decides to reach down and key up the microphone, consider the classic Mark Twain quote, "It is better to keep your mouth closed and let people think you are a fool than to open it and remove all doubt."

Developing a Communication Policy

The development and implementation of a communication policy is the foundation of ensuring clear and concise radio traffic on your fireground. Once the policy is in place, the end-users, firefighters, must exercise radio discipline on every incident. Consider whether the message you want to deliver brings any positive action to the fireground or are you just stating the obvious? This theory does not apply only to the new members of the fire service or the fledging company officers, it applies to all personnel. Chief officers can inundate a company that is working diligently to stretch a hoseline down a smoke filled, heat infested hallway to knock down a fire by calling them 20 seconds after they entered to get a situation report.

While the burden of running a fireground is great and the responsibility of accounting for all members overwhelming, it must be balanced with common sense. The difference between being burdensome and being an asset can be paper thin, but the results will be exceedingly better when we are an asset. When the incident commander calls with a request, he demands clear and concise traffic so he can move to the next unit assigned a task. He wants to confirm the fire is being mitigated, personnel accounted for, and the channel free for any unexpected emergencies. So when we key the microphone up to answer his request, do we know what and how we want to say it so that the communication is clear and concise (fig. 4–4)?

Fig. 4–3. Dispatchers are remote from the fireground. They should be relied upon for assistance, not tactical decisions.

Fig. 4–4. Fire service radio operations should be brief and to the point rather than like a teenager carrying a conversation on a mobile phone.

Many times we are confronted with having to deliver a message while overcoming punishing fireground conditions, completing a task such as stretching a hoseline or throwing a ladder, and thinking about what we want to say. The demand to complete the task many times cannot be eliminated, so a helpful mnemonic can ensure the message is consistent and concise.

CAN Reports

CAN reports allow the sender to quickly condense a message into three parts that can be delivered concisely and consistently back to the incident commander. CAN stands for:

- **Conditions**—What do you have on the fire floor?

  "Engine 8 to Command, we have zero visibility and moderate heat on the 5th floor."

- **Actions**—What are you doing?

  "We have a handline stretched and a knock on the fire."

- **Needs**—Do you need more resources? Do other tasks need to be completed to aid in completion of your task?

  "I will need the truck to open up the windows on the Bravo quadrant and we can handle with the engine and truck in the fire apartment."

With the aid of a simple mnemonic device and continual practice, any fire officer or firefighter can deliver clear, concise, and consistent messages via the radio. Our job is already inherently dangerous. Why would we want to throw another element into the mix to make it more difficult? Whether it is CAN or some other useful device you develop, adopt a clear communication strategy.

Aside from the daily transmissions on the fireground, we should have a process in place for the most stressful of situations on the fireground. Specifically, we should have a well understood and communicated plan for aiding a firefighter in trouble. It does not matter if it is called a Mayday, priority traffic, or emergency traffic, but we *must* have a plan. As was true with the CAN report, there is no need to reinvent the wheel since many departments have developed polices, helpful mnemonics, and tips for when and how to call these distress messages. Ultimately, without a policy in place, the fire service cannot expect compliance because there is nothing tangible with which to comply. This dangerous proposition leaves the most important message you may ever deliver via radio to chance. Not a good alternative when we are talking about our firefighter's lives and our own (fig. 4–5).

Fig. 4–5. Firefighters operating at a house fire, demonstrating the inherent danger of our job.

Other Reports

One example that can be a helpful guide is the system that the FDNY utilizes to differentiate between urgent and Mayday traffic. "Deputy Chief's Wife" is the memory aid for remembering urgent traffic. While comical (and you surely won't forget it), it is effective and detailed:

Urgent

Discontinue interior attack
Command of channel
Water loss
Injury (non-life threat)
Feared collapse
Extending fire

Specific to the FDNY, they prefer to separate situations encountered that will be different between a Mayday and urgent, but you may not. In the interest of developing your system we should review and determine if it works for you. The Mayday mnemonic is the simple phrase, "I Owe You My Miserably Pathetic Life." Charming, but yet again, very effective and you won't forget it.

Mayday

 Imminent collapse
 Occurred collapse
 Unconscious member
 Missing member
 Mask emergencies
 Personal safety system deployed
 Lost/trapped member

Regardless of whether you adopt another fire department's system or develop your own, the key element is to have *something*. A foundation to an effective fireground is to develop a successful blueprint, and one of the building blocks is communication.

The Backstep

Lego Communication Drill

National statistics tell us that fire duty is down across the country, so the opportunity to practice and hone our communication skills on working fires is constantly decreasing. Regardless, the need to work on communications is paramount, as our injuries and line-of-duty deaths do not seem to decrease in comparison.

This leaves the firefighter with the quandary of how to improve, practice, and refine communication skills without running fires. Fortunately, with a little assistance from a couple of Lego toys and some engaging acting on the part of fellow firefighters, it is not that difficult.

The Lego communication drill is a practice in incident command and tactical operations and only requires the following materials (fig. 4–6).

- You want sets of between five and eight pieces. You will need two portable radios, one for the firefighter and one for the incident commander.
- A blindfold or self-contained breathing apparatus (SCBA) with blacked-out facepiece
- An engaging audience to recreate fireground noises and serve as a nuisance to the incident commander and firefighter

Fig. 4–6. A Lego toy designed for four- to six-year olds works well for the preparing the firefighter to work in zero visibility conditions with simulated fireground sounds.

Once all of the tools are assembled, follow these simple instructions:

1. Provide the incident commander with a location remote from the firefighter and an instruction sheet for building the Lego toy.
2. Provide the firefighter an adequate working space, preferably a table to perform the task.
3. Perform a radio check to ensure the incident commander and firefighter are operating on the same channel.
4. Once communication is confirmed, place the blindfold or blacked out face piece on the firefighter. Explain to the firefighter that the task is to assemble the toy in zero visibility and with the assistance of the incident commander.
5. Place the disassembled toy in front of the firefighter. Instruct the firefighter not to touch or begin to assemble the piece until the start command is given.
6. Advise the incident commander to instruct the firefighter on how to correctly assemble the piece over the radio using the instruction sheet.
7. Inform the incident commander the incident has started, and initiate a 20-minute timer. This time period is used to simulate the approximate time a firefighter will deplete an SCBA.

While the firefighter is assembling, simulate realistic fireground stressors, such as:

- Loud sounds made by forcible entry tools
- Playing recorded audio of incidents over a speaker system
- Spraying water from a small water bottle on the fireground to simulate an advancing engine company.
- Jostling by other firefighters as is evidenced on the fireground while additional units advance on the fire.

- Whatever your imagination can conjure up to increase the stressors (fig. 4–7)

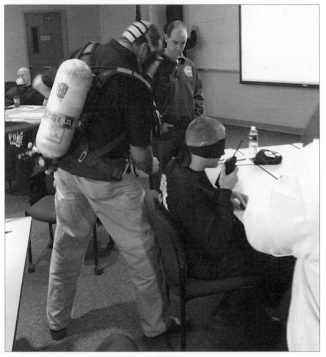

Fig. 4–7. A realistic fireground is full of external stressors like fellow firefighters operating.(Courtesy of Traditions Training, LLC, www.traditionstraining.com.)

Fig. 4–8. A simple Lego toy can prepare the newest to the most seasoned firefighter for conducting proper radio discipline. (Courtesy of Traditions Training, LLC, www.traditionstraining.com.)

The list of possible fireground realities is endless, but remember that you might be building the next Lego. This drill is guaranteed to produce some frustration, anger, but more importantly some laughs. Ultimately, there is a core message to be pulled from the drill. Communication on the fireground can and will be difficult and, in some situations, non-existent. This does not diminish the need to communicate on the fireground. Rather it stresses the importance that when communication is needed it must be clear, concise, and consistent (fig. 4–8).

Many of the lessons learned in this drill parlay into our daily fireground. As the incident commander did we give clear, concise, accurate, and descriptive orders? For instance, did we tell the firefighters what we were building first so they could visualize the item? We have to tell them not only what you want them to do but also who and what they are responsible for while completing the task. This lesson is learned when we tell another officer to take the Charlie Division and that is the end of the order. What do you want them to do on Charlie Division? Who is working with them? How will they know when the mission is complete in the Charlie Division?

When you do have the opportunity to talk, which on many firegrounds is very limited, don't treat it like a phone conversation. Say what needs to be said and clear the channel. The introduction of the CAN report mentally allows you take a message and break it into three clear and concise parts to deliver.

Lastly, having mastery of a skill makes it much easier to communicate what you want done as an incident commander. Would it be easier to tell you how to build the Lego if I built if beforehand? Absolutely! If we transition that to the fireground, is it easier to communicate my incident plan to the units operating if I have served in each of those capacities? Absolutely! In the modern fire service, it seems much easier to fly through ranks, maybe even fly *over* a couple of the ranks. Regardless of your ascension, you must learn each facet of the jobs performed on the fireground to be able to lead firefighters. Remember, the most important asset we deploy on the fireground is people! We must be experts and professionals.

5 It's Really the Only Time You Have

Who knows how long a 60-minute SCBA will last at your next fire? Who knows how long a 45-minute SCBA will let members operate on a top floor job? Who knows how long a 30-minute SCBA will last in a basement? The only true answer is YOU!

You must know yourself. While we are all firefighters, that's about where the similarities end. When it comes to the SCBA used on your face and the one on the face of the firefighter next to you, we are not the same. While all SCBA cylinders are created equal, each firefighter is not. We know that all SCBA compressed air cylinders of the same size contain the same finite amount of air inside. The variable is us. Each individual firefighter will consume that compressed air at different rates, based on a variety of factors (fig. 5–1).

Much of how quickly you deplete your air supply is predicated on your level of comfort in wearing the SCBA ensemble, your physical fitness level, and your mental status. The only true time indicator of how you will breathe down your cylinder is to record your "time on air" in varying firefighting situations. In doing so with varying levels of stress imposed, you will have a better understanding of just how long you can operate in your SCBA. Please refer to the Backstep drill for this chapter and find out how you use your SCBA and how long your air will last on the fireground.

Comfort Level

You must be as comfortable in your firefighting SCBA ensemble as you are in your own skin. You must be intimately familiar with your department's brand of SCBA. You must study how it works, inside and out. You must know the noises it makes, the sounds it makes when you breathe, the warning and alert sounds of the pass device, and the meanings and locations of the lights in the heads up display (HUD). While we discuss checking the SCBA later in the chapter, what we are talking about here is something completely different.

The familiarization we are discussing here must be accomplished on an individual basis. We cannot tell you how to feel more comfortable in your SCBA. It is, however, the SCBA ensemble that is the basis for every firefighting operation that we attempt. It is what allows you to perform all the other fireground tasks that you have arduously trained yourself for. If you are not intimately familiar and ultimately comfortable with yourself when using your SCBA, you cannot be a fully combat-ready firefighter.

Fig. 5–1. Firefighter using an SCBA. (Courtesy of Traditions Training, LLC, www.traditionstraining.com.)

SCBA and Straps

Brand specifics aside, each and every SCBA is equipped with both adjustable shoulder straps and a waist belt (fig. 5–2). Shoulders and hips are the points of contact to the body. The two attachment components were set into the SCBA design for necessity, not as options. Like the outdoor backpack, the SCBA pack frame is designed to be worn with all straps connected. If they weren't needed, they wouldn't be there. In fact, as in hiking, the majority of the weight (approximately 60% of the load) from the SCBA pack frame is designed to be carried on the firefighter's hips. But why do we constantly see pictures and video with firefighters working around the fireground with the waist straps not connected (fig. 5–3)?

Fig. 5–3. All SCBA straps should be connected to the user. (Courtesy of Nate Camfiord.)

If you are not wearing your waist belt, your upper body is now carrying a significantly increased load. Again, think about how nearly 60% of the SCBA weight should be on the hips. If there is no waist strap secured, 100% of the weight is carried on the shoulders. Failure to secure the waist belt and tightly cinch the shoulder straps will cause increased fatigue to the arms and shoulders. This will cause you to work harder to accomplish tasks. It will cause your muscles to weaken and cause you to breathe harder, deeper, and faster, expending that vital air in your SCBA cylinder. In addition, those two tails that are supposed to be connected to your waist will now act as two grappling hooks, swinging and dragging along at your sides when in the fire building. Shoulder straps should be snug, but not over-tightened to allow your arms to move freely as needed.

Have you ever been to a fire where everything went right? Think about this from a personal perspective. Have you ever come back from a fire and not thought that you wouldn't change something you did or said at the operation? One way that we can continue to improve ourselves as firefighters is to not only participate in company level after-action reviews and "backstep" or "kitchen table talk," but also think about yourself and what you did at the fire. What did you do or not do, what did you say or not say, what would you do differently at the next fire? Personal and company level after-fire reflection is paramount to addressing and eliminating future problems on the fire-ground. I can honestly

Fig. 5–2. A series of attachment points connect the SCBA to the firefighter.

tell you that there is not one fire I have been to where everything I did was exactly the way I wanted. In order to prevent ourselves from repeating those miscues, we must learn from them.

Since we are on the topic of the SCBA, I want to share with you a lesson I learned pertaining to SCBA waist straps. I wasn't always a believer in the importance of securing my SCBA waist straps at fires, and I hope that you take heed regarding why SCBA straps need to be secured. All you have to do is look around you to see that many of us still aren't. I see it all the time, both at fires and in photos and videos of fires as they appear on various media outlets. As I mentioned in the last paragraph, we must put some self-reflection into ourselves as firefighters and look at what we do as individuals. As with nearly every fire I fight, I learned valuable lessons after this particular job.

I had the forcible entry position in my ladder company at this particular fire. We were originally sent on the run ticket as the firefighter assist search team (FAST), a.k.a., rapid intervention team (RIT) (fig. 5–4). Upon pulling into the block, we heard the battalion chief asking for the location of second-due truck but they weren't there. The fire was roaring out two windows when we pulled up. I let our covering officer know that while we were assigned as the FAST truck, there was no second-due ladder company on the scene. The officer conveyed that message to the chief, and he immediately assigned us to the floor above and made the third truck FAST.

Fig. 5–4. RIT/FAST stickers help identify where equipment is stored on the apparatus.

Masked up on the half landing of the fire floor, it was lights out in the hallway. It was as if someone had switched our masks to the blacked out facepieces we used for training purposes. There was no light. It was so dark that after I turned my light on, I had to look to see if the light was broken, because I didn't see the beam. It was on, but you could only see it when you looked directly into the flashlight lens itself. The beam was non-existent beyond 6 inches.

We had to do everything on the floor above by feel. There were two apartment doors on this floor. The one to the left was unlocked. It would be our refuge apartment if things went bad as it was adjacent to the door of the apartment directly above the fire.

"Get working on this door," the boss shouted. Finding the knob, I felt the door and door frame that we needed to force open. Shock! Shock! I pounded on the door with the fork of the Halligan tool. We all wanted out of that hallway, as it was getting warmer and warmer. With my face to the door and my hand in the jamb, I felt it, a small gap. I grabbed Halligan, threw the forks in the space, turned and shouted "HIT!"

The boss said, "Eh, try to throw the rabbit in there kid, we can't see nothing!" As ordered, I pulled out the rabbit tool and put it in the space . . . pump, pump . . . bang! We got it, it opened, and we were in.

As nasty as the hallway was, the apartment had just a light smoke condition inside. Such a drastic difference from that hallway we were in. It certainly reinforced how a closed door can provide some refuge from smoke and fire. I was just so happy to be out of that hallway. Emotionally still excited from being in the zero-visibility hallway, I went instantly to the first window I saw, and smashed it out clean, sash and all. Just as I was doing so, I thought to myself, "What is that orange glow outside?"

Just then the boss yelled, "What are you doing?" and called for the can man over to our position. The can firefighter made up temporally for my mistake by using the water can to keep the fire from coming into the apartment. The thermal pane windows had done a great job by keeping the fire out, but in my haste, my thoughtless ventilation move negated that in two quick swings. The boss called for a hoseline to be brought to our position on the floor above as we all anticipated that there might be some fire extension, either from me or the smoke now pushing from the baseboard moldings.

Now I know you are wondering what does this have to do with SCBA waist straps? Okay, here we go. I told you earlier that I wasn't religiously fastening them at every fire, right?

We finish the primary search in the apartment and it's negative. The members operating below knock down the bulk of the fire. Now I head into the kitchen, one room removed from where I smashed out the window earlier, to just open the top of the window and remove some smoke. I reach up and grab the curtain rod above the window. The rod is holding a thin, yet long polyester lace-type curtain. As I hold the rod and curtain in one hand, I hear behind me, "Tick tick tick tick," in rapid succession. Odd I thought. What is that noise?

As I turn, curtain rod in hand, I catch a glimpse of the stove in the corner of my eye. Whoosh! The curtain erupts in flame in my hand, leaving me holding a rod of fire! I start to shout, but then was quickly doused with water from a likewise shouting can man. The curtain, now melted and stuck to most of my PPE, is out! I'm shouting, the boss is shouting, and so is the can man! Everyone is yelling at everyone!

What happened? Well, remember those unbuckled, untethered waist straps? As I reached for the kitchen curtain rod, one of the straps somehow wrapped itself around the knob of one of the burners of the stove. That "tick tick tick" sound? In case you don't have an electronic ignition gas stove, this was the igniter for the burner coming into operation. While we laughed about it after the job was over, I could have injured myself or started a secondary fire in the apartment.

It was a lesson learned, and now shared. We must learn from our mistakes. As George Santayana stated in *The Life of Reason, Volume I*, "Those who cannot remember the past are condemned to repeat it."

I have also witnessed unbuckled waist straps become wedged under doors. I have seen them get stuck in various fire debris (bed springs, chairs, etc.) (fig. 5-5). The waistband strap also keeps the cylinder properly positioned on your back. If you do not fasten the strap, you are allowing the SCBA's movement to be unrestrained both laterally and vertically. While searching, if on all fours, the cylinder has a tendency to constantly ride up your back toward your head (figs. 5-6 and 5-7).

Fig. 5–5. Unsecured SCBA waist straps can act like grappling hooks.

Fig. 5–6. SCBA cylinder sliding up the firefighter's back

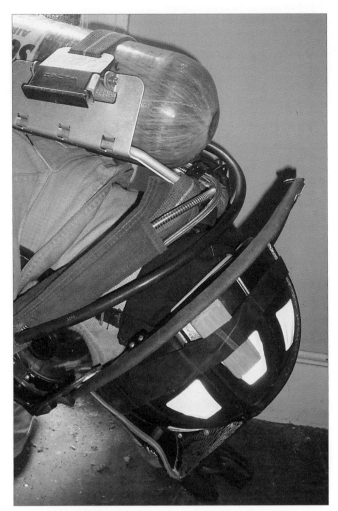

Fig. 5–7. A firefighter's helmet can be pushed forward, causing problems with the facepiece.

items as you can. If you have a positional riding assignment and SCBA for the day, tour, or just for the hours that you will be at the firehouse, it saves you seconds on the next fire. Take out and put on your PPE, place the SCBA straps on your back and set the waist straps to fit your waist, buckle, and replace in the seat. All we need to do now as we are going down the road is grab the buckle and click, done. Shoulder straps can be gently tightened as we get down the road in our seats.

Shaving Seconds

We know that seconds count as firefighters. If we can shave off a second here and there, we may add up just enough seconds that can save a life or stop a small fire from growing. Saving time by presetting straps does not end with the ones found on the SCBA harness. You should pre-set the straps on your facepiece as well (fig. 5–8). Many fire departments are in the process of issuing members personal facepieces. If you have your own personal facepiece, or if you have a riding assignment with an SCBA for any length of time, you should have all the straps preset for *your* face.

Fig. 5–8. Firefighter presetting SCBA facepiece straps

The danger here is that the top of the SCBA cylinder can slide forward and dislodge your helmet, causing it in some cases to fall off, whether your chinstrap is properly affixed or not. In other cases it may push the helmet brim down into your SCBA facepiece and interrupt the hood/facepiece/helmet interface, breaking the delicate seal, which brings the potential for thermal burns. In the worst-case scenario the helmet may push forward to the point that it can dislodge the facepiece from the member's face, increasing the potential for respiratory compromise and loss of precious air escaping from the SCBA.

The best policy is to wear the straps as they are designed. There are no advantages that we know of not too. Their placement is not done so by accident. If manufacturers could leave them off, they would. In the effort to streamline the process of putting the straps on to fit you, take a few moments to pre-set as many of these

Some members preset all straps with the exception of one (usually the lower strap on their dominant hand side). In doing so, you will allow for quicker times in masking up. There will be no time needed for fooling around trying to find all those little ends (some SCBA facepieces have six) with your firefighting gloves on. When leaving the firehouse, remember to reset the SCBA harness and facepieces back to their regular positions.

In our years of conducting firefighter training, without fail, the longest delay in beginning firefighting operations results from the time taken to don the SCBA facepiece and create the facepiece/hood/hemet protective ensemble (fig. 5–9). As mentioned throughout this book, seconds count. Going "on air" should not set you and your crew back in the time management department. There is no excuse for anyone taking more than 30 seconds to go from SCBA on back to fully encapsulated, facepiece on, hood applied, and helmet with chinstrap in place breathing supplied air. If you cannot do this in that amount of time, this must become your training regimen until you can!

Fig. 5–9. Firefighter donning the SCBA facepiece

You may consider yourself to be an excellent tactical firefighter, a great searching firefighter, and the best at the nozzle, but when timed it takes you a minute and a half to get your facepiece/hood/helmet ensemble in place? Really? Is this acceptable? We would argue absolutely not. The drill is simple and free to do. Get a spare SCBA and practice donning and doffing your SCBA and creating the hood/facepiece/helmet interface until you can do it without thinking about it.

There are a few techniques that are taught to maximize this application, find the one that works best for you and master it. Take the spare mask home if needed, get comfortable with it. Your comfort and your acclimation in your SCBA is your life. It is also what allows you to extinguish fires and save the lives of other citizens. It allows you to be a team member with your fellow firefighters in that immediately dangerous to life and health (IDLH) event. No one can tell you when you will be comfortable with your SCBA. Only you know when that will be. It is your life we are talking about here, many around you are counting on you to perform. SCBA air is free to refill and the SCBA tool is used at every fire you fight. Practice with it until you know you are ready.

Physical Demands

Firefighting is a physically demanding profession, some may say on pace with demands placed on world-class athletes. We must maintain a high level of fitness to perform at our peak on the job. Many career departments mandate varying levels of physical fitness training (PT) while at the firehouse. We have also seen a growing trend in many departments, both career and volunteer, that provide health screening and annual fire department medicals. Having a healthy firefighting crew is good for all. Good for the needs of the department, good for the citizens, and good for the individual firefighter.

The onus for fitness is not nor should it be placed on the fire department. It falls directly on you. While it is beneficial for both the department and the firefighter to keep yourself in shape, it has to start with the individual. Department sponsored and encouraged fitness programs are great to help those persons who need that little extra boost or motivation to get started, but remaining physically fit needs to be one of your personal goals. Being fit will give you the best start at being the best that you can be for you and your team. Are you the firefighter that you would want to respond to save a member of your family trapped by fire?

Being out of shape (and yes we know that "round" is indeed one shape) puts added stressors on the heart and lungs. As your body works harder, it needs to send more oxygen to the blood cells to carry out the work. This will increase your breathing rate. Increasing your breathing rate reduces the length of time the air in that finite SCBA cylinder that we carry on our backs will last, so stay fit!

Arguably, no other piece of firefighting equipment is as important as our SCBA. Even the greatest of firefighters cannot function in an IDLH environment without it. The days of "taking a feed" and "smoke eating" are long since behind us. The smoke conditions created from the fires we fight today are laden with documented carcinogens. The smoke itself has also changed, now largely based on the increases in petroleum-based plastic products in the home.

We must be certain to adapt mask usage policies for our departments. Heavy penalties should be handed out for failure to comply. We owe it to ourselves and to our families to ensure that we are as comfortable in this vital piece of equipment as we are living in our own skin. Without a functioning and operating SCBA, the game is over, and the smoke and fire will win. We now become a liability to our team and ourselves.

Mental Demands

Your mental state of being has nearly as much impact on your air utilization as your physical conditioning in your body's demand for oxygen. It is a safe assumption, based upon historical firefighting texts, that most firefighters were taught the skip breathing technique. This technique was utilized when you encountered a low-air situation and you were either lost or a great distance from clean air. The basic belief behind the technique was that you could prolong the supply of air in your SCBA by altering your breathing technique and holding your breath for specific number of seconds. Fortunately, this technique has been identified as antiquated at best and dangerous at worst. There is no way you can trick your mind and body about the involuntary process of breathing and its constant need for balance of oxygen.

By holding your breath you have starved your body of the oxygen it needs to operate and upset the balance of oxygen and carbon dioxide. The moment your brain registers this alarm, you will increase your respirations and pulse to re-achieve the balance, negating any efforts to save air. While we are fortunate that this technique was eliminated in most probationary training, it is unfortunate that no technique was implemented to replace this process. The good news is that in the last few years, Kevin J. Reilly and Frank Ricci have done exhaustive research and offered a new proven technique dubbed the Reilly Emergency Breathing Technique (R-EBT).

In 2008, Reilly and Ricci penned an article for *Fire Engineering* magazine titled, "Rethinking Emergency Air Management: The Reilly Emergency Breathing Technique." The basis of the article centers on the new technique they discovered through comprehensive testing and training. Their work demonstrates that normal breathing techniques take approximately 4–6 seconds per breath for inhalation and exhalation. With the goal of extending this period of breathing and preserving the quantity of air in the SCBA, they devised a technique in which when you exhale, you hum.

This simple hmmmm will extend the single breath from 4–6 seconds to 12 seconds with the majority of that time preserved during the exhale. While 6 seconds does not seem like much, when it is applied to 1,000 to 1,200 psi of air that is left in the SCBA when the low-air alarm activates it can equate to an additional 5 minutes of air, according to Reilly and Ricci.

Assume the five minutes of extra air was in the best possible conditions and under less stress than an actual low-air emergency encountered at a residential fire, we would reasonably decrease the time you could actually preserve. While a decrease may be expected it would still offer an increased measure of air in a toxic environment, possibly saving our lives. When you pair this valuable information with the data presented in the report completed by Captain William R. Mora in 2003 titled *U.S. Firefighter Disorientation Study, 1979–2001*, in which he found during a series of fatal LODD fires that firefighters were disoriented anywhere from 10–40 feet from the point of entry, you can appreciate how valuable it is. Most firefighters could, under a measure of duress, be able to quickly crawl 10 to 40 feet in various directions and hopefully be able to exit. This is just one scenario where a few extra minutes could save your life.

Whether it is due to a Mayday situation, low-air emergency, or disorientation, the fact remains we operate in a toxic and deadly environment in which we need every square inch of air in our SCBA. Having a technique that you can employ that has been honed through training could save your life in our dangerous and heroic occupation.

The SCBA Check

SCBA checks are a must (fig. 5–10). In the FDNY and in Fairfax County, our respective departments, SCBAs are checked on a regular basis. In the FDNY they are checked twice daily, at the start of each day and night tour. In Fairfax County it is done daily with the incoming shift at the start of their day. These checks are done in addition to ones that occur after every use (drill or job) and at FDNY's weekend multiunit drills where companies take the apparatus to a predetermined drill site and check tools and equipment in greater detail.

Fig. 5–10. Firefighter checking out an SCBA

Often it is the job of the junior firefighter working that tour to check the tools for cleanliness and placement on the apparatus. This same person may also complete the SCBA checks for every position when they come in for that tour. While this may be true, it cannot be assumed. It is no one's responsibility but yours to check your own SCBA. You must individually check one of the most important pieces of equipment that will let you work in the IDLH. While the junior firefighter may check it initially, it is your responsibility to check it again. If you have a positional riding assignment, you must check that position's SCBA.

SCBA checklists, whether daily, weekly, or monthly, must be kept current. No failure of one piece of equipment can lead to quicker and more dire emergencies for firefighters than a failed SCBA. Not only may it prevent you from getting in the game if noted initially, but gone undetected might result in the quickest and deadliest end game for you. SCBA emergencies are real and do happen. As Ben Franklin, one of the original pioneers for the fire service, said, "An ounce of prevention is worth a pound of cure." Figure 5–11 shows an example checklist.

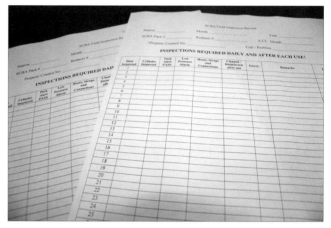

Fig. 5–11. SCBA checklist

While this checklist is highly detailed, there are three main checks that should be performed to ensure basic functionality of your SCBA. They are:

1. Check cylinder and mask assembly and facepiece for damage

   - Frayed, cut, or discolored straps. Broken or deformed buckles/clips. Check all low- and high-pressure hoses, cylinder hydrostatic test date, chipped or cracked cylinders threads, "O" ring present, cracked, burred, out-of-round cylinder necks.

2. Check air status

   - Check air level in cylinder gauge. Charge system with air and check/verify remote gauge for accuracy. Listen for acute air leaks in system in hoses and purge valve and facepiece interface.

3. Functional operation test

   - Breathe air through facepiece. Check HUD. Check and activate purge and demand valves. Activate manually and passively the audible pass device. Shut down system and allow low air alarm to activate listening for any air leaks.

Wear It or Use It?

Smoke is fuel (fig. 5–12). Modern residential building fires are producing toxic byproducts on a level never seen before. The residential fires we fight today bear little resemblance to our father's fires, nor our grandfather's fires. They are indeed different. The dwellings the great majority of us live in and respond to fires in most frequently have changed. They have changed on a few different fronts. The composition of structural components and their associated internal furnishings are composed of many different materials than yesteryear. Gone are single pane window glass and the natural fibers, cotton fabric, and wooden furniture frames padded with horsehair stuffing. We are now faced with energy-efficient thermopane windows, and a plethora of hydrocarbon-based plastic and fabrics in all our furnishings.

Fig. 5–12. Residential fire with smoke showing

We know that the SCBA is our only protection against the products of combustion. It protects our respiratory system from harm. There is a big difference, however, between wearing your SCBA and using it. We would never go into a fire without wearing one on our backs, but sometimes, perhaps for a small incipient fire or in the overhaul stages, we don't wear it. How many times have you been at a job where once one firefighter takes his or her mask off, shortly thereafter, everyone else takes theirs off, too?

If you are operating in a smoke condition, have the facepiece on your face! If the fire is knocked down and you are opening up and washing down, keep using your SCBA to protect yourself. Often it isn't just one acute exposure, but the culmination of several years of smaller chronic exposures to these toxic fire byproducts. These exposures can increase your odds for certain cancers and other negative health effects. Take a moment to review the study that UL conducted on particulate matter in their smoke study at www.ul.com for further evidence to support the use of SCBA in overhaul stages of the residential fire. If the fire is out, your cylinder is going to need to be refilled if you have half of the rated air supply left or if you have a quarter of it remaining. Use the air in the cylinder to protect yourself.

"Free" Air

While there is the notion that using the air in your SCBA is "free," it should not be wasted unnecessarily. Yard breathing is not encouraged (fig. 5–13). When fighting residential fires and operating in smoky conditions or IDLH environments, use you SCBA to its fullest capacity. If you are not in one of those areas, conserve your air supply. This may include operations on floors below the fire and at initial outside defensive operations.

Fig. 5–13. Utilize you SCBA when in an IDLH, not on the lawn. (Courtesy of Nate Camfiord.)

How many times at fires and during your SCBA check have you seen the cylinder valve opened, the integrated personal alert safety system (PASS) fire up, only to be followed immediately by the sound of pressurized escaping air? As trivial as it may seem at the time of release, that free flowing air cannot be put right back into the cylinder.

The sound of that lost air may well be the sound of your last breath escaping at the next fire. It's that important! After each functional test and SCBA check, you must be certain that the purge valve is properly seated closed and that the demand valve is fully depressed for the next time that cylinder is turned on. Five seconds of free flowing air could be 15 more breaths a lost firefighter may just need to get to a window. It could be the difference between life and death. While air is free when drilling, it will only be refilled before or after the fire, not during the attack phase.

SCBA Entanglement

This next statement may sound simple in premise, but it is rarely seen. When conducting drills, doing it with your SCBA on your back is the only way to do it and reap the greatest benefit.

How many times have we practiced a fireground skill without the SCBA on our backs? Whether it was advancing lines or moving portable ladders, bunker gear, helmets and gloves were the accepted level of protection. Are we going into the fire dressed like that? Obviously, we are not. The SCBA's increased weight and bulkiness must become a standard profile that we as firefighters learn to be comfortable with. Anything that hangs off our PPE, including SCBA straps, hoses, bells, whistles, gauges, and anything else are all possible entanglement items. Practice the way you play!

We will be wearing it into every residential fire we go to, why then should we not train with the SCBA worn on our backs as well (fig. 5–14)? Movement while wearing and using an SCBA cannot be imitated nor should it be. If you are worried about damaging "in service" breathing apparatus on heavy duty drag type drills, get a few older spares and or condemned frames and cylinders and begin to create a cache of training SCBAs for use while training. Movement with SCBA must be honed just as the other skills we are working on.

Fig. 5–14. Have firefighters wear and use their SCBA when drilling.

Have the carpenter in the firehouse make a stud wall of scrap wood materials, or perhaps ask your local hardware store for a lumber donation. You can add wire, electrical outlets, sheetrock, and so on to increase realism to a wall breach station.

Have your members practice the SCBA carrying modifications that can allow us to get in and out of places. Practice the reduced profile, low-profile maneuver, swim move, and various disentanglement techniques at the firehouse and on the training ground. Again, as we have stated with all levels of training, know your audience. If you have recruits, start your training in a clear environment without even the facepiece on. Explain the "why" behind each step that we are teaching them to do. As you begin see increased understanding and mastery, increase the intensity and realism. If you are conducting exercises with a seasoned crew, the drill can start with a low-intensity review. When beginning the exercise, start the intensity level of the activity at the level compatible with the competency of the skill set you are looking to have them master.

SCBA Air Emergencies

While it is certainly easy to get ourselves tangled up with all the hoses and gadgets that compose our SCBA, most times we can quickly free ourselves and carry on the task. An air emergency, a failure involving the delivery of the system that supplies you the fresh air you need in the IDLH, is an enormous problem. As such, we label this a true Mayday event. While not all that common, SCBA air emergencies do happen. They can happen for a multitude of reasons, but due to the toxic nature of smoke, they can be immediately deadly if not properly identified and mitigated. Of these malfunctions, most arise in areas within the SCBA which are beyond our checklist control and often neither discernible nor discoverable during our daily or weekly checks. Be certain to keep your SCBA up to date with its preventative maintenance schedule as prescribed by your manufacturer.

We have talked many times about written policy for your department. Every department should have written policies to address functional failures while operating that address your specific brand of SCBA. This will serve as a template for firefighters to study and train to get past them and get out alive.

Air emergencies can immediately thrust firefighters into perilous situations. Operating in the IDLH, firefighters need to be aware of and recognize the specific failures (primary, secondary pressure reducers, regulator, and SCBA face-piece failures) and how to respond if any should occur. They should be well versed and know the actions required to successfully mitigate these various SCBA failures. We will not discuss the specific operations, or failure prone points of any particular brand of SCBA, and this is done on purpose. You need to investigate the documented failures and recovery options for your specific brand of SCBA. You need to focus on the brand and model that you utilize. One thing that is the same regardless of the brand you use, an SCBA air emergency is one that should be listed as a parameter for the issuance of a Mayday radio transmission. Practicing again with increasing levels of realism and pressure to simulate real world fire conditions is the best policy.

There are several excellent resources that can be reviewed that drive several of this chapter's SCBA points home. Current Alexandria Virginia Fire Chief Adam K. Thiel writes one such document that specifically looks at SCBA air emergencies under the auspices of the USFA. (Thiel, A. K., *Special Report: Prevention of Self-Contained Breathing Apparatus Failures*, USFA-TR-088, 2001, U.S. Fire Administration, Emmitsburg, MD.)

This report documents several issues related to the SCBA and their operation and failures. The following section is an extract of the key issue summary from page two of the report. Please review the full document for further information about the issues and specific references to LODDs and significant injury reports that spotlight the SCBA as a contributing factor.

Summary of Key Issues

Failure to use

One of the most common failures of the SCBA system (i.e., SCBA+firefighter=system) is the failure to use it. Even with the current emphasis on firefighter health and safety, and the expanding knowledge of the hazards posed by the products of combustion, some firefighters still fail to use SCBA during interior operations in smoke-filled environments, especially during salvage and overhaul.

Hardware reliability

SCBA that are tested and certified according to the requirements of the *NFPA 1981: Standard on Open-Circuit Self-Contained Breathing Apparatus (SCBA) for Emergency Services* are extremely durable and rugged. Properly used and maintained by well-trained personnel, according to the manufacturer's recommendations, they should provide years of trouble-free service with little potential for hardware failure.

Catastrophic failures

Catastrophic failures of SCBA resulting in death or injury to firefighters are very rare considering the number of routine uses by firefighters each day. Even if such a failure should occur, the fail-safe design of the SCBA may allow it to function long enough for a firefighter to escape the hazard area.

"Low-order" failure

Some failures of the SCBA system do not directly result in firefighter death or injury, but may reduce efficiency and hamper fireground operations. This type of failure is relatively common and most often attributable to operator error, physical abuse or neglect, or inadequate preventive maintenance procedures. Examples include difficult or slow donning of SCBA due to a lack of familiarity or infrequent practice, free-flowing regulators, blown O-rings during cylinder changes, and improperly connected hoses or regulators.

Operator training

Many low-order failures can be prevented through proper operator training. Before entering hostile environments, firefighters must be appropriately trained in all aspects of SCBA inspection and operation. Continued drilling and practice under realistic conditions must be emphasized until complete familiarity is achieved and maintained. Complete knowledge in the use and limitations of SCBA must become second nature to all firefighters to prevent failures.

Preventive maintenance

Regularly maintained and tested by competent, properly trained and certified technicians using the appropriate tools, replacement parts, and testing equipment, following procedures recommended by the respective manufacturers can greatly reduce the risk of SCBA failure. Fire departments should establish preventive maintenance programs for SCBA to ensure firefighter safety and compliance with applicable regulations. *NFPA 1404: Standard for a Fire Service Respiratory Protection Training* can be used to provide guidance for fire department preventive maintenance and training programs.

Upgrades

NFPA 1981 contains realistic, updated procedures for testing and certification of SCBA. Several changes made in these editions were prompted by failure incidents mentioned in this report. Fire departments should upgrade their existing SCBA to meet the current edition of *NFPA 1981* to minimize the risk of repeating past tragedies. SCBA that cannot be upgraded can be replaced with newer models.

Pushing the edge of the envelope

Despite the fact that modern, NFPA-compliant SCBA are extremely durable, the materials used in their construction have physical limitations. Firefighters and maintenance personnel must understand that SCBA are not indestructible, and that the potential exists to expose SCBA to factors in the environment that may contribute to or produce failure.

The Facepiece

Recently, there has been talk about the last point in this report, "Pushing the Edge of the Envelope," as it refers to the SCBA face-piece. While we mentioned the delicate balance of the interface between the helmet, hood, and the facepiece in a few of the other chapters, the focus of this section is specifically on the facepiece. A recent study has taken a look at a trend that seems to be recurring. While improvements in bunker gear have allowed us to move deeper into areas of the residential fire than ever before, have we discovered a weak link?

The NIST/USFA facepiece study

Take a moment also to review the report titled *Fire Exposures of Fire Fighter Self-Contained Breathing Apparatus Facepiece Lenses*, conducted by the National Institute of Standards and Technology (NIST) and the USFA (November, 2011), available at www.hsdl.org. This study noted deformity and even failure of SCBA facepieces under various intensity fire conditions using different brands of facepieces. Of note, the authors are still in the process of really taking the data and giving firefighters the next key. They need to take the time, temperature, and exposure limit and provide that data to the manufactures, standard writing agencies such as the NFPA, and end user firefighters and fire officers. We found that this statement from the document sums it up for us:

> "Data on the limits of the equipment would be valuable information for the fire service to help prevent further injuries and fatalities related to SCBA equipment failure."

Conclusion

The world's greatest firefighters can only work in an IDLH to save lives and property for as long as their SCBA lets them. It is the brain of our firefighting protective ensemble. To say firefighters know their SCBA is to say they are as comfortable wearing and operating in it as they are living in their own skin. They know this device inside and out, its pluses and minuses, its failure points and how to overcome them. The user is mentally and physically fit, which affords the maximum utilization of the finite amount of air in each cylinder. You must know you. You cannot fool yourself or cheat while wearing your SCBA and each breath you take is precious. You must know how you individually perform in your brand of SCBA. Check, test, perform the required maintenance and updates to your SCBA. Make sure that it, and you, are the best they can be when the bell goes off.

The Backstep

SCBA Air Recognition Drill

Our SCBA is the one of the most vital tools that allows us the opportunity to perform our job. Without it we would be unable to cross the threshold into most residential building fires due to the toxicity of the smoke. Understanding this fact underscores the importance of every firefighter knowing how much time they have when donning and operating a SCBA. This drill provides each firefighter with:

- Volume of air in the cylinder at the start (it should *always* be full!)
- Time from start until low air alarm sounds
- Time from low air until completely out of air
- Actions to take when air is completely exhausted from SCBA and firefighter is still in an IDLH

The steps to this drill are to build a course like the one depicted below with the normal physical and mental obstacles we face (fig. 5–15). The entire course is performed in zero visibility, full PPE, and SCBA. Once the firefighters' low air alarm activates, remove them from the course and place them in a separate area with the instructions not to crawl and to concentrate on R-EBT actions to conserve their air. Once their air is exhausted have them demonstrate the proper actions to take when they face this issue in an IDLH environment.

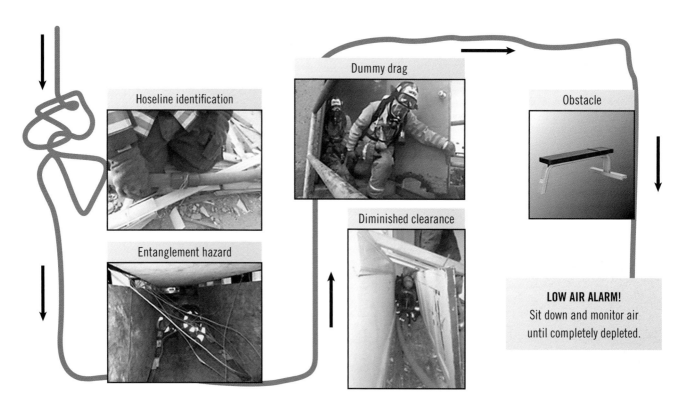

Fig. 5–15. Air recognition drill

6 We Are Aggressive

The term *aggressive* has been gaining notoriety in fire service circles. It has been attached to individual firefighters, specific fire companies, and various fire departments. Perhaps it's a term that's been applied to you. Sometimes it's even self-proclaimed. What does *aggressive* mean to you? What images does this term conjure up for you with respect to the fire service? What is it that makes an aggressive firefighter or fire company? Just where did this notion of aggressiveness come from, and is it something new? Why has this term evolved into countless discussions of who *is* and who *is not* aggressive at fires in the fire service?

We can tell you one thing for certain, when it comes to the term *aggressive*, fires are exactly that (fig. 6–1). Fires are all consuming, nondiscriminating, equal opportunity killers of firefighters and civilians alike. The fact that a single candle's dancing flame, if not properly monitored, could reduce a residential home to ruins and take a life in the process speaks to fire's aggressive demeanor. Fire's uncanny ability to search for sources of oxygen, reaching out to take more fuel to feed its need to sustain its existence is, without a doubt, aggressive.

Fig. 6–1. Fires are aggressive by nature. (Courtesy of Kentland Volunteer Fire Department, www.kentland33.com.)

Firefighters, by the nature of our work, put themselves in the midst of these aggressive factors instigated by fire. Fire is an aggressive combatant. Fire's hazards are present at every fire we go into, regardless of how big or small. Since we have realized that fires are aggressive opponents, must we not match an opponent with equal and opposite aggression?

We must train to fight an aggressive opponent. We must be prepared to fight it aggressively. There is no such thing, no such tactic as passive firefighting. We are *fighting* fires. It is called that for a reason.

But does the word *aggressive* have the wrong connotation in the fire service? *Merriam-Webster's* defines it as "Marked by combative readiness (an aggressive fighter)." The word *aggressive*, as it pertains to the fire service, must not be considered synonymous with reckless or careless.

Aggressive firefighting has its roots in training. Being known as an aggressive firefighter or an aggressive fire company has nothing to do with rushing in carelessly with blinders, cowboy antics, or operating with reckless abandon. We say that aggressive firefighters and fire departments are formed with personal and company level training that perpetuates a state of being that is marked by combative readiness.

Look to Training First

What are we really teaching our new firefighters? Are we setting them on the right path from the start when it comes to understanding the word aggressive and its implications for the fire service? Let's take a moment to speak to this point before we move on. We need to set the tone with the word *aggressive* here, explaining what its true meaning is.

In most departments there has always existed an unofficial "you have to do it this way while you are in the academy, but out in the field" mentality to fire department basic training (fig. 6–2). Why is that? We know that we need to set strong foundations and have new firefighters learn the basic tenets of firefighting operations, tools, and equipment. But why do we confuse the issue with a perceived double standard? What is the real lesson we are teaching them? Fires in school aren't real? We use the skills and techniques we are teaching them in the academy when we are out in the real world of street firefighting. Are we sending them out to the firehouse destined to fail?

Fig. 6–2. Fire academy training ground. (Courtesy of Traditions Training, LLC, www.traditionstraining.com.)

We understand that many academies pride themselves on having order and discipline within their walls. No firefighter or fire officer will argue with the fact that the fire department needs disciplined firefighters. There just seems to be a certain disconnect between the way we teach (talk at) and talk (instructor to student) to young firefighters in the fire academy, as opposed to the way we do it out on the street and in the firehouse (peer to peer). While a strong discipline-based academy is understandable, is it required at all times? This is especially true when fire-matic lessons are taught, where questions may not be asked for fear of pushups.

We are sending firefighters from the academy to the street in a different time than our predecessor's fire department. With fire volume down dramatically in most areas of the country, young firefighters are have less and less "on the street" fire time. We are also finding more time between when they are released from the academy and their first fire, and subsequent fires, and so on and so on. In the past, we relied on the fact that new recruits would have plenty of not just on-the-job but at-the-job learning. This in no longer the reality of cases most times. It is more a fallacy than we would like to admit.

We recently had a conversation with a command-level chief officer from a career fire department who, after a recent fire, experienced something startling. While talking to the lieutenant of the engine company after a good first-due job at a residential dwelling for his crew, the officer said that his firefighters did "good, especially since it was really their *first* fire."

The chief was perplexed. He knew the all the members riding that tour, and none of them were straight from a recent recruit school. How could this be? On further investigation, the chief noted that the firefighters had nearly three years on the job. It had been three long years before they had gone to a first-due fire!

This was not a 30-calls-per-year fire company; this engine company ran more than 2,000 calls on average during the last five years. It was a suburban fire department with a full career firefighting staff. Take a good look at the people riding with you. When was the last fire in which your crew got to stretch and operate a hoseline or conduct a primary search? When was the last time your company went to work? You might be surprised when you sit down and think about it; then again, you might not. Who are we really fooling anyhow?

Are we fooling ourselves? More time between fires and greater times between the end of the fire academy and the first real job must make us all take serious notice. In fire academies today, do we need to refocus our training? Are we finding a greater gap between the academy way and what we are expecting on the street with new firefighters? If so, and we believe it to be true, we need to provide more time in the academy to train them as if they were on the street.

To start with, what are the majority of young firefighters taught in their Firefighter I program or in their initiation to the firefighting academy (fig. 6–3)? It all revolves around time and hustle, right? We have to get in there. Time is the enemy. Go, go, go! The instructors created virtual chaos and pandemonium at every chance! Instructors and team members alike shouted at the crew, "Move, move, get in there . . . hurry!" Is this where we have started to set our firefighters up for failure?

Fig. 6–3. Instructor and students at a fire academy. (Courtesy of Traditions Training, LLC, www.traditionstraining.com.)

Is this where the aggressive and reckless rush-in attitude is wrongly introduced? Foolishly and carelessly rushing into fires unprepared absolutely causes injuries and deaths on the fireground. Do we teach them to put their heads down and barge in headfirst? Move till you hit something and then change direction? Do we scream at them to move in without telling them why?

We must eliminate the tendencies to verbally push students without explanation. We talk about creating habits. We need to instill good habits, not careless rushing habits. Members need to move with purpose to accomplish tasks, not run to and from fires and emergencies like headless chickens.

Yes, we know time is of the essence in fighting fires, but we must make sure that we train our crews to rush in once we are prepared. If not, we are doing nothing more than setting them up to rush into certain peril. What instructor wants to be part of a study on young firefighters where a direct correlation for fireground injuries stem from the lessons learned on the training ground? We must not make the academy environment one in which we erroneously push our young firefighters to create bad habits.

Harness the Energy

No recruit firefighter wants to let his instructors down. If the instructor says, "Jump!" They will reply, "How high, sir?" Remember, though, if the training staff did not teach them fireground self-control in the fire academy, who is responsible to control them when they get out on the street? This rushed approach is not without fault. This is not the aggressive behavior that we want on the fireground. Where does the fault lie, in the student or the teacher? Are we setting them up to fail?

Most young firefighters are obviously eager to get to work. They cannot wait to go to fires. Most can recite along with Lt. Ray McCormack from the FDNY when he stated in his 2010 FDIC keynote speech, "My name is Ray, and I like to go to fires." All firefighters want to go to fires, but it is necessary to harness that energy. It is our responsibility to mold that young firefighter so that the energy isn't put into places to relieve aggression, but to be prepared to face an aggressive opponent.

We need to mentor and to mold that positive energy into the creation of good fireground habits. We can't blame our recruits for being eager. It's good when firefighters are ready to learn. Accept that eagerness as it is, and share with your experiences and knowledge. Be a mentor and role model to them. Make them the best firefighters they can be.

How to Be Aggressive

Aggressive firefighters and fire departments are formed through personal and company level training and a marked state of combat readiness. I would certainly say we should add the word "fire" to the dictionary definition of aggressive. Aggressive (fire) fighters must absolutely have a marked level of combat-readiness.

How do firefighters and fire companies get the reputation of being aggressive? Is this a title that you want attached to you and your department? We would have to say absolutely, unequivocally, yes. We have to understand that we are striving for excellence in our profession. We must pursue increased knowledge in our craft and our training. We are firm believers that no firefighter is invincible. Firefighters are no better than their training and equipment prepare them to be.

As is eloquently carved in stone aside the New York City Fireman's Monument at Riverside Drive and West 100th Street, firefighters fight "Fire in (a) war that never ends." This was written nearly 100 years ago, and it is as true today as it was then. We have seen the ravages of fire. We must be aggressive in our approaches to extinguish them. Fire will only cease if we extinguish it or if we allow it to consume everything in its path. We must never forget that this is the calling we have chosen. It is our sworn duty and we must make it happen because lives and property depend on it.

Aggressively placed hoselines and aggressive search techniques are done through skills derived in and from your training (fig. 6–4). The engine company knows that well placed, timely positioning of hoselines allows for quick knockdown and extinguishment of fire. No greater number of lives are saved than those from the quick and effective coverage provided by a well-placed hoseline. From historic ladder company duties, such as quickly locating the seat of the residential fire due to knowing your local buildings, or aggressively focusing searches employing techniques such as vent, enter, isolate, and search (VEIS), can and do save lives.

Fig. 6–4. Members conducting hands-on drills. (Courtesy of Traditions Training, LLC, www.traditionstraining.com.)

As mentioned earlier, fires in residential buildings are the number one killer of civilians and firefighters. Imagine how many lives are spared by aggressively operating engine, truck, and rescue companies and their members.

That said, aggressive interior fire operations are not for the untrained or under-prepared. This is where we find that no one has explained why we must be aggressive in our response to fire. The true aggressor in our fight, the uncontrolled fire, will punish all units and members who think that they are aggressive solely by rehashing the nature of the word or relying on the legacy or the reputation of those who came before them.

We routinely operate in a rapidly deteriorating, aggressively consuming, physically challenging arena. Punishment in our profession comes with severe ramifications. The damage caused by fire is deeply charring to the structure, but also to the mind, body, and soul.

We must strive to prepare all firefighters and fire companies to increase effectiveness and efficiency on the fireground of the residential building. We must train ourselves mentally and physically to meet the challenge of our aggressor. We must be like the Boy Scouts and always be prepared.

Assistant Chief of Operations Richard (Ricky) Riley of the Clearwater, Florida, Fire & Rescue coined three words that sum up our ability to be known as aggressive, "Preparation, practice, and anticipation."

Preparation, Practice, and Anticipation

Preparation

We know that preparation includes a multitude of avenues for mastery. I can remember my father always talking to me about five "Ps" with regard to preparation. He would say that poor planning produced poor performance (hence the five Ps). Preparation leads to the formulation of a plan. We all know the benefits of preparation. It increases our chances for successful outcomes, both for our home life and at our home away from home, the firehouse. Prepare for the worst, but hope for the best. Set yourself up for those successful fireground operations by recognizing that you must create good habits for yourself. These good habits in pre-fire preparation are both physical and mental in nature.

Your mental preparation, your "game face," will affect how you carry out your jobs on the fireground. You must be ready for the fire of your career, until proven otherwise. You must become a student of the game. You must expand your firefighting knowledge base (fig. 6–5). Never in the fire service history has there been such an opportunity to review so much information.

Fig. 6–5. Firefighters must stay well read.

Nearly on a daily basis, through the wonders of the World Wide Web, we can review fires almost instantaneously. We can read other departments' after-action reports from previous fires. We can listen to recorded fireground audio clips, watch online video, and evaluate training tips. We can pick up new and interesting training ideas from around the world. I know that not every generation of firefighters has the same level of expertise in navigating the Internet, but I guarantee there is someone in the station who can point you and your crew in the right direction. There is so much information at our disposal thanks to the Internet, please take and digest what you read with a careful eye.

Not every author, comment thread, or narration of a video clip is necessarily reflective of doing the right thing, nor should it lead you to believe that a firefighting revolution is upon us. Just as not everything you hear at the kitchen table is the truth, neither is everything you see on the Internet. Review all the actions, positive and negative. Remember that there are just as many lessons to be learned from watching bad decisions as there are for good ones. Take the information for what it is, then decipher what you can use as a learning tool for you and your members. If warranted, use what you see and hear to create a positive change in your department.

Don't fall prey to the tactics of name calling and public bashing that folks like the fictitious "Internet Safety Police" (ISP) employ. The ISP is just one type of anonymous Internet commenter who lambasts the strategy and tactics employed by other departments in derogatory terms. They often point out what they deem as violations, most of the time without really knowing all the facts. Try to keep an open mind and not get caught up in the miniscule, trivial, or soap opera blog banter.

Firefighting is an inherently dangerous profession. Injuries and deaths do happen at fires. Those are the facts. While the premise of this book and many others is to prepare you in order to reduce your risk of injury and death, they will continue to happen. Our attempt is to reduce the occurrences. Proper preparation is the beginning of it all. I am sure that every reader can share a story about how he or she was either injured or knows of an injury that happened to a firefighter as a result of actions at a fire.

Residential building fires hold the dubious title of killing and injuring the greatest numbers of firefighters and civilians. I can guarantee that no firefighter has ever gone to such a private dwelling fire thinking this would be the fire that would result in a serious injury or death. We must prepare ourselves so that every fire that we go to is the fire of our career until proven otherwise. We must mentally prepare ourselves to do battle. We know this dangerous aggressor. We must prepare ourselves to meet it. Preparation allows us to adapt in the ever-changing battle. Good habits, created with a combat-ready, fire-of-your-career mentality, will set you on the right path in fireground preparedness.

Preparing tangible items such as our bodies, our protective ensembles, and our apparatus in that same mind set will keep us operating more effectively at fires. Firefighting is a physically demanding operation. We must prepare our bodies to meet the challenge of the aggressive fire opponent. Physical body preparation starts with a healthy diet and regular exercise. Not only are heart attacks and heart disease the leading causes of death in society at large, but they are also what kill the most firefighters.

Regular exercise and yearly physicals are a must. Again, preparing our bodies to go work at a fire is not like preparing ourselves to file papers at a desk. We must make ourselves physically fit for duty. Being out of shape not only puts you at risk during the operation, but also can make you the anchor on the team, causing everyone else to have to work harder since you are too out of shape to do your job.

As we talk more about preparing tangible items, our personal protective equipment is next on the list. A daily inspection and verification of tools in your pockets is a must. If you are going to be at the firehouse for some time, or are assigned a riding position for the time you are there, this is extremely important. Prepare and set up your turnout gear each time you are at the firehouse, and do it the same way every time (fig. 6–6). Be sure that when you are done, you put it back the same way as well.

Dress out fully in your PPE for every run. We owe it to our families back home and our firefighting team members to be ready for action. It is certainly always easier to take it down a level if we get there and there is no fire. You can always remove your gloves and hood and open your coat once you have determined conditions don't warrant them. Have you ever had to put your hood or gloves on under pressure? Have you had to do it under extreme pressure, for instance on an RIT operation or with a person trapped? We know that it will inherently take more time to accomplish even such seemingly simple tasks when placed under pressure, if you remember to do the tasks at all.

We believe that both good and bad habits, over time, become routine. This is why bad firefighting and/or bad PPE dressing habits must be corrected *now*. This is why we stress the importance of creating good habits, especially early in your career. Repetitive routine, which is really what we do when we respond to fires, becomes muscle memory. Muscle memory is the concept that repeated habits become routine, natural, and recur as such a regular phenomenon that the firefighter completes the tasks nearly subconsciously. Increasing your level of comfort through repetition will undoubtedly cause the task to eventually be committed to muscle memory. Under stress, we tend to fall back to our level of training and what we routinely do, making muscle memory that much more important to have good habits in place and ready to use.

While we just spent a few chapters on getting yourself prepared for firefighting, our apparatus and the tools on them play just as vital a role. As mentioned in chapter 2, "Combat-Ready Fire Apparatus," knowing your rig and the locations of the tools (and knowing the tool's limitations) add to the preparation in the aggressive operations of your fire department.

Paired tools such as the irons (8-pound axe and Halligan) and similar forcible entry tools should be located on the rig in close proximity to each other to speed the operation. Positional riding assignments, which also aid in our approach of preparedness, are addressed in future chapters for both engine and truck company members. Firefighters who know what tools to bring based on their assignment for fire operations will be able to get to the fire and extinguish it faster.

Fig. 6–6. Firefighter ready for action

Practice

Practice makes perfect only if it is perfect practice. All teams practice, including my five-year-old child's soccer team. Why do I take my kids to practice twice a week? It's because the players and the coach want each person to master their skills and perform better on game day. In short, they want to be a better team. Firefighters are part of a similar team, only we are playing with fire. Find a firefighter who doesn't want to practice and I'll show you a liability. How can legitimate firefighters balk at practice? Have they forgotten that they are performing duties in an IDLH environment?

Why is there a tendency to look down on training? Should we not keep training on the front burner, keeping all crews ready to put the plan into action when the bell sounds? Who are we letting down if we allow complacency to enter our firehouses? There are many people on that list.

First, you are letting yourself down. The grim reality is that it is you who will have to look yourself in the mirror when your uncompleted or poorly fashioned fireground skills leads to the injury or death of a civilian or co-worker. Is that something you want on your conscience for the rest of your life?

How about your family and friends, your parents, children, and spouse? How will their lives be permanently changed because of your actions or inactions, your bitter stubbornness against bettering yourself? Will they be able to handle the new injured you, the new normal that you created for them?

What about those citizens who summoned you to their home that was on fire? Are they not counting on you to save their loved ones and protect their property? You took an oath didn't you? They have supported you either in payment of some form of taxes and/or in community fund drives. They now are asking you to pay them back. Are you up to the task? Have you practiced your skills to the best of your abilities?

Finally, look around the firehouse. What does failing to practice your skills mean to the rest of your team? Your goal as a member of that team is to have no weak links. Practice your skills to the point of mastery. The reflection of those skills will come back to you, your company, and your department tenfold.

We must practice in the most realistic atmosphere we can create. Our proverbial office is oftentimes shrouded in zero visibility. Therefore we must practice as we work, with limited visibility. We also need to be performing skills in the level of PPE that we would normally be utilizing at the incident, which means having hoods on, firefighting gloves on, chin straps fastened, etc. Our "practice as we work" approach helps to build good habits, therefore leading to excellence. It will also aid in creating positive muscle memory for your crew. In fast-moving situations like firefighting, having positive muscle memory recall of actions performed will shorten the time needed to make effective decisions on the fireground.

Overall, the number of fires to which we are responding is down. If fires are down, how are our members receiving the on-the-job training that they received in generations past? It is no longer up to the training officer to try to forecast and accommodate all the training for the entire department. Each member must take the time to practice pulling lines and using hand tools. We have to keep moving forward to master our craft. We must encourage and support all levels of training for our members and our departments.

Anticipation

How does anticipation play a role in forming the aggressive firefighter? Is not anticipation defined as waiting for something to happen? How can we anticipate problems at residential building fires? I know sometimes it may seem as though the chief has a crystal ball and just *knows* what is about to happen. How can all firefighters increase their awareness to fireground anticipation? Fires are unpredictable foes. They burn with the intensity and creative license of the scientist and artist.

Fireground anticipation is the ability of firefighters and fire officers to forecast what is going to come next. It has loose correlations with maintaining situational awareness (SA), that is, knowing what is going on around you. But where SA is more about keeping tabs on the now, fireground anticipation looks to the future. It encompasses the preparation and practice needed to fight fires. It molds those two entities together. It is a fusion of your readiness and your rehearsals. It is ultimately the outcome of what you are expecting to occur.

Your fireground anticipation can be further tweaked and refined. This happens all the time and you don't even realize it is occurring. As we gain firefighting knowledge and further prepare and practice using that knowledge, we create in our minds an anticipated reaction. For example, if I am on the top floor of a private dwelling and poke a hole in the ceiling and find fire in the attic, I should call for a hoseline. If I were to open the whole ceiling before that line is in place, I would aid fire spread. Actions and reactions—that is what anticipation is all about.

Experience is an enormous advantage to help you anticipate on the fireground. We can use our previous fire experiences as a reference and recall outcomes. Again, refer back to the typical building construction in your response area. Based on your mastery of your buildings, your brain has seen this fire and instantly recalls similar fires in similar structures. You have prepared for and practiced the firefighting techniques over and over. Now those past experiences will help guide you. You can anticipate outcomes based on past successes or failures.

We can also tweak our level of fireground anticipation by being a student of the job. Keep yourself and your crew up to date on changes that affect our operations. Several new studies in fire behavior, as mentioned in chapter 13, allow us to learn about changes in the phenomenon of fire in dwellings today.

As we see changes in both housing components and the makeup of the homes' contents, we must be well versed in what that means to us as firefighters. Facts discovered in these studies may change our anticipated results. While we must look back at history to understand how we got to where we are today, we must continually look to the future of fire science to make sure that the tried and true tactics of today do not need updating.

- Anticipate fire and smoke movements in your houses because you know your buildings.
- Anticipate personnel and company actions at the fire. Have a plan of attack for which everyone has prepared and practiced.
- Anticipate the reaction of the fire due to an action that you are performing. Be ready for the changes.

Understand the Word

Aggressive is not a dirty word in the fire service (fig. 6–7). Aggressive firefighters are not careless or foolhardy. They have their roots firmly entrenched in the knowledge of their own abilities. When they move, they move with purpose and direction. They are not fearless, indestructible, or infallible. They make mistakes and learn from them. They strive for perfection and always take the time to stay physically and mentally sharp, combat ready. Aggressive firefighters comprehend the fact that civilian and fellow firefighters' lives are depending on them at every fire they go to. As such, they always want to do the right thing. Aggressive firefighters are prepared. They are well practiced and can anticipate the next move.

Fig. 6–7. Ladder company member ready for vent, enter, isolate, and search. (Courtesy of Traditions Training, LLC, www.traditionstraining.com.)

Aggressive fire companies absolutely save lives. Aggressive fire companies expect fire. They have plans made and members trained. They have guidelines and policies to dictate operations. Aggressive companies want to be the best they can be. They tolerate no weak links in their chain. They work as a team, over and over, in and out of fire after fire . . . together. They save lives through the determination in their planning and rehearsing. Being aggressive in fire operations is not for the weak or the meek. It is one of the most important ingredients in delivering the sole mission of the fire service: saving lives and property.

The Backstep

Prepare, Practice, Anticipate

Aggressive fire companies have roots firmly entrenched in task-based training. They strive for flawless performance. They are no strangers to "sets and reps". Regardless of how many runs or working fires they respond to, they are well prepared. They have a pre-determined, yet flexible plan in place. They are not huddling up in the front yard discussing who is doing what, where, and when. They have prepared a plan of action and are ready to meet the tasks at hand using the most efficient methods possible.

They are ready, due in part to the preparations that they have made before the alarm. But to be truly flawless they must practice the plan, over and over, until it becomes habitual. These sets and reps are not viewed as annoyances, but as the way for them to master their craft. These training scenarios are normally conducted with the most realistic fireground conditions possible. We need to practice our response in an environment similar to that in which we will be operating. As such, the plan should be practiced and completed with full PPE, firefighting gloves, limited visibility, noises, etc. We must have a well-practiced plan. This repetitive training also allows time to review the plan while not under duress, making the necessary changes to make our operations more efficient.

Over time, under watchful personal and company level inspection, the repetitive cycle of events on the fireground and in well thought out training drills leads to good habits. Well-prepared and practiced teams learn to predict the moves of their fellow firefighters. They are aware of fire travel paths in their buildings. They anticipate the expected outcomes for their actions in and on the fireground.

Where to begin? How can we attain the lofty goal of firefighting mastery? Well, it starts with a plan. If you do not have an established policy for initial operations for a residential building fire, start there. These are the fires we respond to in the greatest frequency and find ourselves most likely to be killed and injured within. We must have a written standard for members to use as a guide. This should become the framework for several weeks of company level training. You are creating the plan, the ensuing drills will become the practice, and members who can anticipate what is expected of them, based on their assignment knowledge, will be the result. The combination of these three will lead to increased individual and company efficiency, flawless performance, and a safer fireground. No one can argue those benefits.

Write out the typical residential response for your area. Assign each arriving company initial operational tasks. Base the tasks for your operations on the typical crew sizes that you face in your area. Remember, you are doing this for you! If needed, refer to later chapters for typical engine and truck company tasks required at every residential building fire. Make the plan or discuss your existing plan and examine your current assignments. Are they the best fit to your demographics? Take each arriving company and discuss the challenges that you may face. Below is basic template for pre-arrival assignments to get you started:

- *1st Engine*—On-scene report, take initial actions to establish water supply, 360-degree evaluation, effect obvious rescues, attack line (proper length and size) to fire floor

- *2nd Engine*—Ensure 1st engine water supply completed, assist 1st line to seat of fire, if not necessary stretch 2nd line (proper length and size) initially to protect the 1st line, may later cover floors above

- *3rd Engine*—Take steps to establish 2nd water supply, attack line to floor above.

- *4th Engine*—Ensure 2nd water supply completed, backup line for 3rd engine

- *1st Ladder*—Search fire floor, horizontal vent (VEIS if applicable), portable ladder sides A & B

- *2nd Ladder*—Search floor above, assist in horizontal vent, portable ladder sides C & D

- *Heavy Rescue*—Establish RIT

PART 2

Mastering the Environment

7 Bricks, Sticks, and Straw: Three Little Pigs Construction

Every Sunday we don our favorite football jerseys, watch endless hours of pregame analysis, and then gather around to watch four hours of men generating blood, sweat, and tears for victory. Ah, the return of football, but what does this have to do with the fire service or building construction?

A great deal, actually. The professional football player is really no different from us in his preparation. Each week he will study the enemy's actions (fire behavior training for us), work diligently rehearsing plays to prepare for the fight (daily training to engage our enemy fire), and when Sunday arrives, he brings all of the training and practice together into his well-tuned mind to begin his battle for victory. This occurs in the front yard of the house on fire for us. We see our enemy, observe our enemy's actions, assemble our gear for the firefight, and then engage it for victory (fig. 7–1).

Fig. 7–1. Firefighters masking up in the front yard of a house on fire. (Courtesy of Nate Camfiord.)

Imagine if during the course of this entire preparation and development of the blueprint for victory that the football player never took the time to look at the football field. He never took a moment to notice how long or wide the field is, where the first down markers are, or what all of the lines mean on the field. This is a ludicrous thought, because the football field is the office in which he operates, his battleground. It has become part of his subconscious. He has burned the field into his brain so that when the fight is on, he does not think, it is already in his mind and his body reacts.

The office in which we operate is the single-family dwelling and building construction is our battleground (fig. 7–2). According to *NFPA 2010: Fire Estimates*, we know statistically that almost 80% of the fires to which we respond are in a single-family dwelling. Yet do we dedicate arduous hours of training and preparation to learning all we can about our battleground?

Fig. 7–2. The residential house is our office, as it is the most common structure in which we fight fire. (Courtesy of Kentland Volunteer Fire Department, www.kentland33.com.)

When we begin to learn building construction, we are pulled in many directions. Some will direct you to the most comprehensive handbook for us penned by the late Francis L. Brannigan, *Building Construction for the Fire Service*. Others may direct you to the guy in the firehouse who is a carpenter and can relate his work experiences to you. Some may even tell you it is unimportant.

Regardless of the path you take, you *must* learn building construction. The foundation of learning building construction is to understand that ultimately we want to learn how the building will act when it is on fire. Fire is not an expected "guest" for the single-family dwelling. The home is built for families to occupy, not to continually be subjected to fire. We want to know the building well enough to determine how the fire will travel and whether the smoke issuing from two windows will aid or hinder the path of the fire.

> "Engines 08, 10, 28, 26, Trucks 8 and 10, Rescue Squad 28, Battalion 04, respond to the fire reported in the bedroom at 10 Main Street. Occupants report mattress is on fire in the bedroom. All occupants are out of the house."

You have probably been dispatched many times on calls like this one. You hear the report of fire in a bedroom, you know this area is all two-story homes, and you are forming a mental image of the fire (fig. 7–3).

Fig. 7–3. Firefighters preparing to enter a residential structure fire. (Courtesy of Kentland Volunteer Fire Department, www.kentland33.com.)

You pull into the block, dismount the apparatus ready to deploy your hoseline, and look at the structure. In your mind, you are probably already running through your checklist:

- Is this matching what I thought it would look like when I was dispatched?
- Where is fire now? Still in the bedroom?
- Where is the bedroom?
- Where are the occupants?
- Where is it going?
- What are the immediate exposures?
- Which way will I enter the structure and advance towards the fire?

All of these questions must be answered before you even walk across the threshold of the home. What has influenced this decision-making process? If you are not thinking like this, how can you learn to think this way?

The one clear advantage we have when it comes to a single-family dwelling fire is that we usually live in the same types of structures in which we fight fire. We know where the rooms are and can subconsciously commit the floor plan to memory. This is the foundation of learning building construction, not just the reciting of terms like "lally column" or "lintel." Truly understanding building construction begins with breaking the house down into two tangible bits of information: its floor plan and its construction.

The Guts

If I told you my kids were in their bedroom and I live in a traditional Cape Cod style home, where would you immediately direct your search? You know that the bedrooms of a Cape Cod are on the second floor where the dormers are located (fig. 7–4). You know that the stairs are most likely in the kitchen or as soon as you enter the front door. All of this should be known *before* you leave the apparatus bay.

Fig. 7–4. Dormers on a Cape Cod house

The first part of understanding building construction is to understand the "guts" of the building. More appropriately, we need to understand the building construction home styles and their respective floor plans. In the fire service we typically encounter five types of home styles.

Cape Cod

Coined by Timothy Dwight in 1823, the Cape Cod is a smaller, single-family dwelling with low ceilings to conserve heat from the multiple fireplaces within the structure. The typical interior layout is a staircase centrally located in the front of the home upon entering. The first floor consists of a master bedroom and kitchen/dining area, with the second floor having two smaller bedrooms. The most notable exterior feature is the steeply pitched roof to help in the harsh New England winters. The pitch prevents snow from accumulating on the roof and lessens the possibility of damage. The addition of the dormers was not until later in the Cape Cod's life when additional light, ventilation, and/or cosmetic appeal were needed (fig. 7–5). This added feature created the void space, referred to as "knee walls," for hidden fire to travel unimpeded (fig. 7–6).

Fig. 7–5. Cape Cod

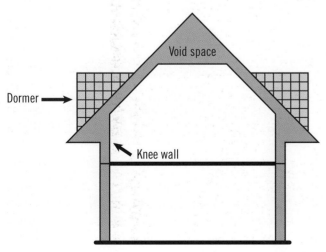

Fig. 7–6. Knee walls in a dormer. (Illustration by Matthew Tamillow.)

Colonial

This is the home style that every child draws when tasked with drawing a house, the classic square home with a fireplace (fig. 7-7). The colonial is the typical American home with windows stacked in a straight line. The home is centered around the front door with the main stairway directly in line with the front door. The rooms and hallways within the home all branch off of the stairway. The first floor typically consists of the living room and dining room in the front portion and the kitchen in the rear of the home. The second floor usually houses all of the bedrooms, with a shared bathroom at the top of the stairway. The ceilings will typically be higher then those of a Cape Cod.

Fig. 7–7. Colonial-style home

Split level

While the specific jargon for these type of homes tend to be geographic, we use this term to explain the marriage of the colonial and rancher (fig. 7-8). In its time, this style was considered the best of both worlds, accentuating the most desirable aspects of the two styles. The entrance of this single-family dwelling leads directly into living and dining room areas with little compartmentalization, consistent with the rancher design. The kitchen is usually in line with the front door and has stairs that lead down to the basement level and may also have access to the garage. Off of the living room floor is a small section of stairs that lead up into the common hallway where all of the bedrooms are located. Access to the attic space may be in this common hallway via a scuttle.

Fig. 7–8. Split level home

Split foyer

The split foyer is yet another home style that may change names depending upon the area of the country. The telltale description of this home is the presence of double-hung, full-sized windows on the basement level visible from the street level (fig. 7-9). The front entrance is up a set of stairs. Upon entering the foyer, which is between the floors, you are greeted with another set of stairs, up and down. Downstairs is a family room or "in-law suite," with windows set high on the walls. Upstairs is the kitchen, living room, dining room, and bedrooms.

Fig. 7–9. Split foyer style home

Rancher/rambler

The rancher/rambler is a home style that gained popularity during the post-World War II housing boom. The simplistic rectangular one-story style construction offers an open floor plan with little compartmentation and, typically, exposed beams (fig. 7-10). Upon entering the home, you are greeted with the kitchen, dining room, and great room, often only separated by half-walls that allow free travel of smoke and fire. The bedrooms are located off the common hallway.

Fig. 7–10. Rancher/rambler style home

McMansion/estate home

Noticeably absent from these classifications is the McMansion, or estate home (fig. 7-11). As discussed in chapter 2, the single-family home has grown 90% in square footage over a 29-year period. Pair this with the overzealous banking practices that allowed people to purchase larger and larger homes, and it is a no-brainer that we ended up with these structures. Some may even argue that these structures qualify more as commercial structures than single-family dwellings. Given their fire load and sheer square footage, commercial firefighting tactics may be appropriate.

Fig. 7–11. Due to society's demand for more square footage, the McMansion was born.

The McMansion is characterized as a home with more than 5,000 square feet total, stretching up to 20,000 square feet in the largest structures. These homes will have multiple hip roofs, a large number of windows, and grand entrances with atriums. There are often chandeliers present, which have an unfortunate habit of crashing down on advancing hoselines during fires. The rooms, and there will be numerous rooms, usually have higher ceilings.

Why Does This Matter?

Understanding all of these characteristics will enable you to make more sound tactical decisions on the fireground and provide a safer fireground. How? We have already discussed that the single-family dwelling is the office in which we operate, so having knowledge of our work area is paramount. Understanding floor plans and interior characteristics should provide us with several key bits of information for tactical operations:

- Depending upon the reported location of the fire (e.g., fire in the kitchen), we will have already devised a path and probable seat to the fire for the engine company.

- Have an understanding of the path of smoke and fire within the structure based upon the lack of, or presence of, compartmentation. For instance, if we are dispatched to a reported fire in the bedroom of a colonial style home, we expect that the fire will enter the common hallway and will threaten to travel into the common attic space. We prepare our tactical assignments based upon this suspected fire progression and confirm them when we arrive and verify the fire.

- Focus search and rescue operations and prepare for aggressive tactical operations. If we are combat ready and consider ourselves professionals, we plan to employ VEIS long before we arrive. We listen to the reported fire location and incorporate information like time of day and neighborhood to maximize the usage of this technique. When we arrive at the two-story, split-level and see a large volume of fire issuing out the front door, and the parents are confirming that their kids are still in their bedrooms, your pre-planned information comes flooding back into your mind and allows you to execute to perfection. Fire has cut off the main stairwell, survivability is decreasing by the second, and viable citizens are trapped in a upper floor bedroom = VEIS (fig. 7–12)!

Fig. 7–12. Having prior knowledge of the floor plan of a structure before the fire can assist in making a sound and rapid tactical decision such as VEIS. (Courtesy of Roger Steger.)

The Frame

We have discussed the inside floor plan of the building, the proverbial guts of the structure. Now we must incorporate the other portion of knowledge we must have in building construction for the single-family dwelling. We must understand the "frame," or the manner in which the home was constructed. Understanding the technique employed when the structure was built will assist us in understanding fire travel and the way we expect the building to act under fire conditions. Say the words "balloon frame construction" to an experienced firefighter, and you're almost guaranteed to get the reply, "It's in the basement and check the attic."

Why?

History confirms that fires in balloon frame style homes, which have no inherent fire stops from the basement to the attic, often start low (basement) and show high (attic and gable end vents). To tackle the frame styles, let's break it down into a few styles we see most often.

Balloon frame

This style of home was very popular in the early 1900s and continued into the post–World War II era (1950s). It was a well-received construction practice since it required long pieces of lumber, which were plentiful, to construct multistory homes. Pair this with the lack of need for fine carpentry work since it require little sawing and mortise and tenon work, and it was an efficient building style (fig. 7–13). The detriment of this style is that it creates an unimpeded fire lane from the lowest floor to the highest floor due to the lack of fire stops. One can look from the attic space down a stud channel and see clear into the basement. If you can see that freely, then fire can travel just as easily in that same channel.

Ordinary

Often referred to as a lumberyard that is surrounded by four block walls (stone, brick, or masonry), the ordinary construction home poses a significant fire hazard for us (fig. 7–14). While the unimpeded fire travel of balloon frame is negated in the ordinary home by the presence of solid block walls, fire travel is still an issue. The entire interior of the home, including floors, roof, and structural members, is wood. Since the convenience of running utilities through open stud channels is gone due to the solid block, builders must find other ways to route pipes, wires, etc. These void spaces and pipe chases are otherwise known as fire pathways. Fire travels through any avenue it can find with little resistance, all the while attacking the structural members on which you are operating in the home.

Fig. 7–14. Ordinary construction style home

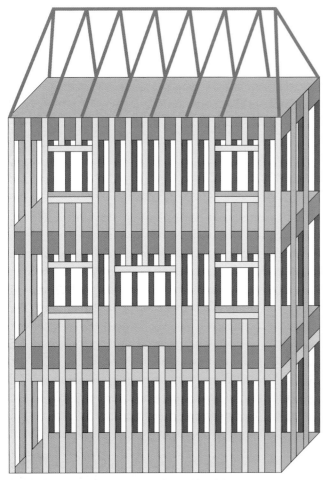

Fig. 7–13. Balloon frame construction residential structure. (Illustration by Matthew Tamillow.)

Platform

As demand for homes increased along with overall cost, a demand was placed upon the building industry to do more but make it cost less. This led to the introduction of the platform frame style of construction that built homes in a "box on box" style (fig. 7–15). Essentially, the builder builds the frame of home in a square or rectangular design, often serving as the first floor. Off this wood base, typically made up of dimensional lumber (2×6, 2×8, etc.), the floors are laid on top to form a walking platform of plywood sheets. The walls are constructed on top of the platform, again of 2×4s, and then finished off with a top sill. The top sill is a solid piece of 2×4 upon which the next platform is constructed. This pattern of floor, walls, floor repeats until the home is built to the desired height. The platform is far superior to the balloon frame in fire stopping since each level has inherent fire stops. Ultimately, these can be negated post-build when utilities are routed and create chases and voids for fire travel.

of engineered lumber, the actual stability under fire conditions is decreased more. Ultimately, the industry is trending toward building more homes with this combination of sawdust, glue, some compression to bind it, and lightweight metal fasteners.

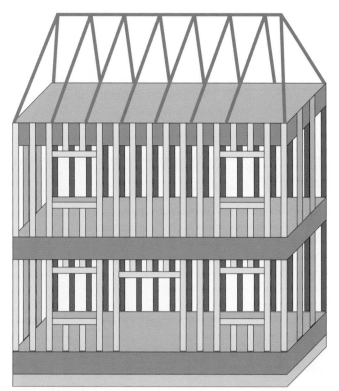

Fig. 7–15. Platform frame construction style home. (Illustration by Matthew Tamillow.)

Fig. 7–16. Lightweight wood frame style home

Fig. 7–17. Lightweight components are supporting a structure that will fail under fire conditions.

Lightweight construction

The platform frame, while allowing for increased production at reduced cost, also introduced a long-time enemy to the fire service: lightweight components. Lightweight framing followed the trend of the platform by using even cheaper materials, smaller dimensional lumber, and now even engineered lumber (Ply-I's, Glulam, etc.) which are structurally sound, when they are not on fire!

Lightweight construction is not a new enemy to us (fig. 7-16). It has been in practice for well over 40 years, yet we are amazed when it fails so quickly. While the walls may be of the same construction style as the platform, the floors we crawl on that are belching black smoke and roof beneath which we operate are dangerous. The fasteners that secure the lightweight components are weak and prone to failure under fire conditions (fig. 7-17). Additionally, with the advent

Old vs. New

The preceding four styles of frame construction are not the entire list of construction styles that we may face during the course of our firefighting careers, but they are the most prominent. In the fire service we strive to follow the KISS rule (Keep It Simple and Straightforward) because we deal in a high stress environment that demands rapid (and correct) decision making. Building construction does not lend itself to this principle with the multitude of terms, applications, and calculations for strength, but we must know and understand the key principles since they can have a deadly impact on how we fight fire.

When we examine the construction styles, structural components, and other factors, we can see that one prominent factor does exist in regard to our firefighting tactical decisions: the age of the structure. Specifically, is the building old or new construction?

Across the country we see the experience level of fire service members decreasing, which equates to less time in fires and experience with how a building reacts under fire (fig. 7–18). Think for a moment of how much easier it would be for the new fire officer who arrives at the scene and has to make a split decision regarding whether we go or don't go, if they could make that initial assessment asking, "Is it old construction or new construction?"

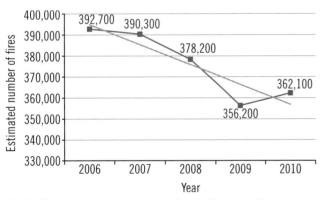

Fig. 7–18. Graph showing a downward trend of residential structure fires from 2006 to 2010. (Courtesy of United States Fire Administration.)

This by no means releases the fire officer from learning more about his response area, building construction, and the types of fires we face today (fig. 7–19). Conversely, we know that when a department adopts the protocol of giving on-scene reports, the announcement of information is not for the person giving it. Rather it is for all the other incoming units, allowing them to formulate their operational plan and paint a picture of the incident. Additionally, it serves as a marker of what the building looked like when you arrived. If we could adopt a system of merely determining the approximate age of the structure, deeming it old or new prior to entry, we can begin to build a timeline for how long operations can continue. We can then deploy our units to positions within the structure with particular assignments in a proactive nature. This is solely based upon our knowledge of building construction and how we believe the building will act under fire.

Fig. 7–19. Firefighters must always be students of the profession and commit to learning; it could save their lives.

The key factor to remember concerning building construction is that there is no resting on our part. We must strive for more knowledge, not only of terms, but also learning what is built in our area and staying on top of the newest trends.

The Usual Suspects and Some New Enemies

Vinyl siding has been our adversary for some time, often forcing the inexperienced fire officer to make decisions that are not based upon actual fire conditions. Vinyl siding is an exterior siding composed of polyvinyl chloride, which in our terms equates to a solid layer of gasoline slapped on the exterior of the home (fig. 7–20). The NIST study titled "Residential Structure Separation Fire Experiment" demonstrated that when a fire that occurs somewhere within the home autoexposes out a window, it can be disastrous. From unscientific studies we have conducted coupled with the experience of running fires where the exterior of the home has caught fire, we can tell you that vinyl siding burns quickly and impressively. They can present the illusion that the entire rear of the home is engulfed in sustained fire. The reality, depending the material underneath, is that the fire may burn this impressively until the vinyl siding is completely consumed, often in only a couple of minutes. If the next layer of construction material underneath the siding is some sort of engineered wood (chipboard, plywood sheating, etc.), then the fire may have enough fuel to become sustained. We must be able to recognize the hazard regardless of this, identify whether the siding is involved in fire, and be able to mitigate it quickly. Often the tactic of deploying a handline to quickly sweep the burning siding may be enough to eliminate the hazard and you can advance into the seat of the fire within the home.

Fig. 7–20. When vinyl siding is involved in fire it can appear to incoming firefighters as a much larger fire. (Courtesy of Nate Camfiord.)

Engineered wood is a generic term that can be applied to any part of the burning home you are entering. Engineered wood is the scrapple of the building construction industry. Essentially, the process takes all of the wood scraps and blends them together with strong adhesive to press into a usable form, in most cases a long beam. While the engineered wood is a wonder of ingenuity and is impressive in its ability to carry a load over great distances, it is not so impressive when it is on fire. These beams can fail quickly, as demonstrated by the UL study, "Structural Stability of Engineered Lumber in Fire Conditions."

Glulam is another member of the family of engineering feats with multiple pieces of wood pressed and held together with a strong glue. Glulam can be constructed in any shape and/or design that a client would like. Even more alarming is the repair process, which in the case of a structural beam consists of boring a hole, inserting a steel rod for stability, and then filing the hole with epoxy. As with every concept offered in building construction, we have to ask how would this act when it is on fire? Glue, epoxy, pressed wood, and a steel rod is not a recipe for success for us.

Regardless of the impressive nature of engineering solutions that the construction industry invents, we must be vigilant students of the profession. We must learn about the characteristics of the material and build an expectation of what that material will do when on fire. The industry never stops, offering items like structural insulated panels (SIP) that eliminate the need to build a house in a bit-by-bit fashion. The exterior sheathing comes glued to a thick core of foam and then is finished off with another layer of interior sheathing. Houses can be constructed with limited manpower in a short time frame, much like a child builds a wood block home. The only differences are that we don't live in our children's block homes and we must fight fires in these "foam core" houses.

Along with the introduction of SIP homes, the building construction industry welcomed the birth of structural insulating sheathing (SIS). While in the same family as the SIP homes, this technology has lessened the actual weight and size of the exterior wall. The SIS is merely one half to one inch of foam attached to a thin slab of resilient material that serves as the entire exterior wall (fig. 7–21). This material replaced the sheets of oriented strand board finished in a house wrap. That entire process is replaced with one a single layer of 4 ft ×9 ft to 4 ft × 10 ft sheets, providing sufficient R-value and wind stability.

Fig. 7–21. The SIS is easily identified by the blue wrapping of the entire home, unless the hole has been covered in an exterior siding.

With this sheathing, nails or staples inserted a minimum of one inch apart secure it to the exterior wall studs. To cover the voids and add to the stability, construction tape is used, providing a fast and efficient means for home builders to reach acceptable insulation and stability benchmarks.

Unfortunately, these benchmarks do not conform to our traditional fire service benchmarks for understanding fire behavior and fire travel in residential structures. The theory of fire remaining contained to a room or compartment within a residential structure due to the exterior walls may become antiquated. Fire will no longer need only a window to auto-expose and spread vertically. Nor will a window serve as the means of exit for firefighters inside the structure. Firefighters can remove the interior sheetrock and will only have to defeat one inch of foam secured by one inch nails to reach the outside via the exterior wall. The one constant in this process is that, with a high level of certainty, no fire department, fire chief, or firefighter ever got an email, call, or message that this new product was used. We must be attentive to changes in building construction industry trends and residential construction.

Urban or suburban, paid or volunteer, we have all made the commitment to serve our communities and provide a service that few would be willing to accept. To ensure we can maintain that professionalism, we must learn every facet of our job. As mentioned at the beginning of this chapter, the residential building is our battleground, and we must know it like the back of our hand. Unfortunately for us, our battleground is changing every day, so we must maintain our awareness and never lose that desire to learn.

That process can take a career to refine, but to start our matriculation we can adopt the basic premise that knowing whether a residential building is old or new construction is a start to developing sound tactical decisions. We must not get caught in merely "checking the box" that we took a building construction class. We must understand how a building will act when it is on fire. We must seize every opportunity to begin this process and our drills for building construction will open our eyes to the opportunity that is presented on every call.

Fortunately, we are provided with tools that can assist us each and every day that do not even require us to be a busy company running lots of emergency calls. We can start by using this chapter's drill that utilizes your local newspaper's real estate section (fig. 7–22).

Fig. 7–22. The real estate section of the local newspaper can be a great resource on what is being built in your area.

The Backstep

Review Your First Due

In every firehouse across America, it is common to find the day's newspaper scattered across the dining table with the sports page tattered and folded from the numerous readers. Yet, among this carnage is a valuable tool in starting the conversation on building construction. Pick up your local real estate section and you can be sure that a snapshot of your entire response area is right there in front of you.

Make it a weekly drill to review the real estate section and start the discussion with your members. Specifically, ask your personnel:

- Is this construction old or new? How can you tell?
- What hazards does this construction pose?
- How would this change your strategies and tactics?
- How would you communicate this hazard to your crew?
- How do you think fire will act when attacking this structure?
- Where will the fire go when it is attacking this structure?

This line of questions will initiate conversations about size-up, strategies, and tactics on the buildings you will fight fire within long before you ever respond to an incident. Now it is time to get out there!

8 Get Out There

A portion of the inscription found on the side of the FDNY's Firemen's Memorial, erected in 1913 on Riverside Drive and 100th Street in Manhattan, describes firefighters as "Soldiers in a war that never ends" (fig. 8–1). Unless a civilian has had a fire in his or her home, or had one occur to a family member or close friend, it is doubtful that the person will have a true understanding of the dramatic damage it leaves in its path. Even what we would consider a small fire can be mentally, physically, and emotionally draining for civilians to handle. As firefighters, we see the devastation created from fire routinely. We know that it happens at every fire to which we respond. Over time, we can become callous to witnessing this event, but we must never forget we are at war. We must stay in a constant state of readiness. We are fighting a battle that never ends.

Fig. 8–1. New York City's Firemen's Memorial

While we know that fire is a devastating force for a homeowner, it is our "war." In that nearly hundred-year-old statement from the FDNY Firemen's Memorial, there is recognition that battling fire is indeed warlike. It was fought in many of the same ways then as it is now. To fight today's war against fire, we must have the wherewithal to know that our war is conducted in the homes located on the streets of our response areas. Continually conducting reconnaissance of our areas and really getting out into our communities and taking inventory of the "battlefield" is paramount. In chapters 2 and 7 we talked about the benefit of reading the real estate section of your local newspaper, which can provide a great deal of information about the homes in your response area.

The buildings are our enemy. They are the office in which we do our dangerous work. Take into account the type and style of buildings where your company will be going to fires most often.

As is the case for most firefighters, the firehouse is our second home (fig. 8–2). Firefighters can spend up to one-third of their lives within the walls of their respective firehouses. Whether assigned or volunteering your time to be there, your firehouse provides a sense of pride, belonging, and fellowship. It can offer us a much-needed change of pace from the stressors of the real world, temporary shelter, and isolation from the hectic chaos that is life. We are among friends in the firehouse. There is a common bond and unstated understanding between us, as we coexist there. It's a comfortable place for us to be.

Fig. 8–2. Firefighters around the kitchen table in quarters

Our War

While the tendency to completely shut off the outside world might be appealing at times, we cannot simply close the doors to it (fig. 8–3). Firehouses are an integral part of the fabric of the communities we represent. We are tasked to serve the customers in our communities. We run fire safety lectures at open houses, put Santa on the rig, host birthday parties, etc., all in the name of doing the right thing for our community. But our true office is outside the red firehouse doors.

Fig. 8–3. Firehouse apparatus doors. (Courtesy of Nate Camfiord.)

As Francis Brannigan stated so eloquently, "We must know our enemy." The buildings in our response area are our enemy. Within them is where the true dangers lie. The best service we can provide our communities is excellent fire protection. Our community deserves nothing less. When they call us we must be ready. Getting out in your community is much more than getting in the rig and going for a ride for driver training. Assemble a crew and talk about the homes you see as you drive past. Knowing your first-due buildings is a huge part of being prepared for battle in your office.

Knowing the ins and outs of your first-due area is a great start to mastering the battlefield. But to stay on top of the changes within your community, and the new changes in the enemy which is your buildings, is to constantly stay aware of new trends in building construction, home building technology, and design. Contractors and building supply manufacturers will not be knocking on your firehouse door. They will not be showing up offering to show you new products that they are putting into the new homes in your community, some of which might kill you.

We have certainly come to know some such components in the past. Building components such as trusses, gusset plates, glued laminated timber, and pressboard plywood I-beams all fit the category. Take the plywood I-beam, which has been around for a while now. It is stronger in rated strength and can span greater distances than its dimensional ancestor. The problem lies when it is put under fire conditions. It has been proven over and over to fail more quickly than dimensional lumber.

No one came to the fire service and presented us with these now documented facts. We found out the hard way, many times with our own lives. We, the fire service, discovered and shared the information with little or no refutation of our facts from the building industry. No reputable building manufacturers are pontificating the potentially inherent fire service hazards in their new or existing products. They are not speaking at fire department training seminars, nor are they writing articles in fire service journals to make us aware.

It is up to those of us in the fire service to keep on top of the building industry as it puts out new products. We must research them as they relate to our profession. Making money is the end game for most in the building industry. Cheaper, faster, and using fewer resources—those are the dreams of the modern building industry. Builders want to make as much money as they can per home constructed, period. Yes, they want to make a good product, but they will continue to use materials (as long as they meet code) that may not necessarily be firefighter friendly. We must continue to relay new materials and their impact on our operational safety to one another. The Internet has certainly improved our ability to get this information shared quickly over email, blog, and community forums.

As the New York City terrorism slogan states, "If you see something, say something." Just because you find something new in your area does not mean it is an exclusive product to you. In fact, chances are while it may be new to you, it isn't new to someone else. Get the information out there and share your experiences. In doing so, you may just save the life of a fellow firefighter. We must continue to spread the word.

When it comes to the residential building fire and our attack plan, first-due knowledge is paramount. This is where our volunteer firefighters might gain a slight advantage over some career department members without residency restrictions. As volunteers, more than likely you live in the communities you serve. This gives you a distinct advantage over someone who travels 50 or more miles from another town to their place of employment with varying degrees of similarity in community layout.

You may have generations of family that have lived in your town, literally growing in town as the town has grown around you. You have watched subdivisions be constructed. You know the layouts and idiosyncrasies of the homes in those developments. You may also have an increased awareness of the infrastructure that surrounds those properties. You know where and what the pitfalls are of the water system. Street familiarization drills are pointless because you have intimate road knowledge. Perhaps most importantly, you and your fellow firefighters live in the buildings just like the ones where you will be fighting fires. Your house and your neighbor's house are your own enemies. You must use this knowledge to your advantage. You know the building types, the layouts, when they were built, and the components used to build them. As new members join your department, be sure to get them up to speed on the who's who and what's what of your town. Your knowledge of the information is great, but keeping it to yourself only benefits you.

Regardless of whether they are career or volunteer, firefighters must make good use of their time on every run. The "routine" emergencies, medical runs, etc., will give us an opportunity to be in the house. Once the emergency is mitigated, take a look around you. Is there a basement? Where are the service connections? What is the room layout? Where are the bedrooms? How are the stairs run? Note forcible entry concerns, access and egress points, the presence of room additions, and interior furnishings. Burn the building down, mentally, of course. Where will fire spread given the layout of the home? Where and what are the closest water supply options?

After every run, drill with your crew on their position and how they should respond given a fire in this building. We do not normally have the chance to get in and inspect most residential buildings, so this is our chance to respectfully take a look around. Many times, homes built on the same block and in the same time period have similar features. Ask the homeowner any questions that you may have regarding the home. You'd be amazed how happy they are to oblige and interact with the fire department.

As home burglaries are on the rise in some areas, occupants of residential buildings are becoming increasingly aware of home security. They are taking more and more actions to protect what they own. Residential building forcible entry is no longer the "mule kick" on a hollow core wooden door. Formidable forcible entry concerns are becoming prevalent in the single-family home (fig. 8–4).

Multiple locking points are becoming common in residential front doors

Fig. 8–4. Multi-point locking residential doors

I know when I was growing up in my parent's upstate New York home in the 1970s, we never locked the doors to our house. I can remember being told emphatically *not* to lock certain locks, ever, because we had locks on the doors that no one even had keys for! Obviously those days have long since passed. Forcible entry concerns originally found in the urban setting have made their way to the suburbs and rural settings, and security products once found only in cities are now available in big box home goods stores.

Building Relationships

Fire departments must maintain excellent working relationships with the local building department, code enforcement officials, and community planning boards (fig. 8–5). Any building code changes in the community will likely have some impact on our fire operations. Representatives from the fire department must stay abreast of these changes. Each fire department should have a representative or two who informs the members about what is going on in the community as it relates to building construction. Fire department members should be consulted and alerted by the planning board before new building complexes are constructed. This also applies to buildings undergoing significant renovations. We must strive to work with our local building and code enforcement officers to stay on top of renovations to existing homes and take part in preparations for new home developments.

and "burn the house down" in our minds. Where will the fire go? Where will we be tactically operating in the home? We need to be working hand-in-hand with our community boards to make our cities and towns the safest places to live and for us to do our work. Play a supporting role in the entire process and the benefits are truly our own. Again, once it's approved on the plan, it will be our problem down the road. It will be our responsibility to overcome any potential problems.

New Construction Assembly

New homebuilders are in many ways like the famous Johnny Appleseed. Even as many new construction projects slowed during the downturn in the economy, Johnny Appleseed continues to build residential homes. How can he afford to keep doing this? Who is this Johnny Appleseed Construction Company? Well, it operates in nearly every town, city, and state. Much akin to the literary figure, these builders seem to magically spread house "seeds" in any tract of open land or in your local farmer's back field (fig. 8–6).

Fig. 8–5. Building officials can help you forecast your department's needs. (Courtesy of FEMA.)

Fig. 8–6. Open land, ripe for new developments. (Courtesy of Google Earth.)

What are some ways to build and foster these relationships? Have these officials over for lunch. Extend invitations to them for your department functions. These are the people who approve plans for the homes and renovations going on in your community. Once the plans are approved and ground is broken, we will be the ones responsible to mitigate fires and emergencies thereafter. After the plans are approved is not the time to go to these boards and voice your displeasure on a particular design feature.

Many code officials were former contractors and builders themselves. Only those of us who fight the fires in these buildings really know what it is that we are looking for. Only firefighters are going to see the plans

Hopefully, since you have secured a voice on your local building committees, you have been privy to this pending development for some time now. Once the plans are all approved, firefighters need to get out there and take a look around. The best time to make your visit is when the Johnny Appleseed Construction Company breaks ground and is "sprinkling" the new home seeds on what was a farmer's field.

In order to save time and money, Johnny's residential construction plan usually has three or four houses under construction at a time, all with varying levels of completion. With this method, the company is able to keep many contractors working on-site at the same time. Most homes are completed in an assembly line fashion, where workers move from home to home, street to street, until finished. First the foundation, then the framing, plumbing, electrical work, and so on down the line.

This construction method can be an enormous asset for us as future responders. A fire today will be your responsibility, a fire tomorrow, the same. We need to know the enemy. Here is our chance to get in at various stages to see how fires and emergencies are going to affect these structures. Firefighters can see various construction components in position during differing stages of home completion. Due to the staggered completion of each home, we can see many phases of construction at the site in one visit to the development.

After a few months of construction, up sprouts a new community of 30-plus similar looking homes, all additions to your fire department's first-due response area (fig. 8–7). Most builders follow similar techniques for home construction when producing these multi-home developments.

Fig. 8–7. Up spring homes in your district. (Courtesy of Google Earth.)

New Materials

Statistically we know that most fires in the United States occur in residential buildings. As such, we *must* keep up with current trends in residential building construction. We can further our knowledge by understanding building construction and conducting site visits during the building phase.

We know that both old and new construction types pose well-documented hazards for firefighters and have specific fire travel concerns. Older homes, where dimensional lumber may afford us longer resistance to fire, often allow it to travel unobstructed through vertical voids, such as those found in a balloon frame home's exterior walls. Chapter 7 speaks at greater length of the hazards and the contrasts and comparisons of old vs. new construction materials and techniques.

Newly constructed homes are certainly not hazard-free. Some of the hazards we have known about for some time, including the truss, ply I-beams, and so on. Others are in the pipeline for residential application and are emerging daily. The construction industry, as mentioned earlier, wants to do things more cheaply, quickly, and using fewer resources. Building houses is not a static industry. It will be a constant challenge to keep our citizens and members safe from fires that occur in them by becoming students of the industry, as well.

The Field Visit

Get out there! Take your fire company over to the new addition in your first due and talk with the construction/on-site manager (fig. 8–8). While we prefer the company-level visits, some larger building developments may require that a higher-ranking officer set up an appointment with a representative of the construction company because of safety and security concerns. Either way, most site managers we have encountered are more than willing to allow a site visit. You are not there conducting an inspection per se, you are there for a familiarization drill.

Fig. 8–8. Members get out into a new development.

Before you even get out of the rig, what can you see? Are there design layout nuances such as flag lots and pipe-stem courts that may affect your water supply? Are there sufficient hydrants for the fire load of these homes based on their proximity to each other? Are the components of the structure "stick built" (2×8, 2×10), or are they all preassembled (e.g., truss floors, roofs, I-joists)? This may have a great impact on your operational times and the techniques employed. Look at the materials laying around the job site for clues about what may be inside the houses.

Take a walk around and do a 360-degree visual examination of the outside of the home (fig. 8–9). Note the presence and locations of windows and doors. What rooms lie behind or through them? Do the houses have basements? Basement fires are often some of the toughest and most dangerous fires we fight. Do you see variations in basement access points that may assist your operations (e.g., a walkout door)? Perhaps there are only small casement style windows (fig. 8–10). If the basement is pre-designed for living space, you may find larger egress windows with a window well in each home (fig. 8–11).

There is a large push of late to be "green." As Kermit the Frog knows, "It ain't easy being green." Many of the new green ideas are not going to be easy on us as firefighters, either. You can see it even at your local hardware store. If two products are side by side and one has "uses less energy" or has an Energy Star rating and the other does not, you would be tempted to buy the "greener" product. In newly constructed residential homes, you will see more green.

These new ideas will pose new hazards for us. Note any components you haven't seen before and determine how they will impact your operations. Energy efficient and green components such as solar panels are making their way into the housing market, especially since the government provides financial incentives to the homeowner and homebuilder for incorporating them in the construction. We are seeing more solar installations and greater R-values put in homes, insulating and making them more airtight than ever.

Fig. 8–9. Walk the site.

Fig. 8–10. Casement window not designed for egress. (Courtesy of Nate Camfiord.)

Fig. 8–11. Basement window style designed for egress.

How will these changes affect fire behavior? We have seen new structural insulation panels (SIPs) hit the market, which are essentially two pieces of OSB composite board with foam or polyurethane mixture innards glued in between (for more on SIPs, visit www.sips.org). This foam center core offers greater insulation qualities for the home but also can quickly add to the fuel load and increase the rate at which fire will travel in this void. With the foam support removed by fire, structural integrity will be compromised. SIPs style exterior sheeting is also becoming commonplace in residential homes.

Once inside in the building, start your recon of the house, beginning at the foundation (fig. 8–12). What is holding up these homes? Is it block or prefab insulated in-form? What is holding up the first floor? Are there laminated I-joists or steel beams over lally-columns? Identify the structural components throughout the home and the structural hierarchy of the dwelling, from the most important to the least important support members. How fire will attack them?

Once on the first floor, note any similarities to the layouts of each of these structures. Often there may be only two or three model floor plans for the entire development. Builders may vary the front facades to give the illusion of a completely different home, but the interior components and layout remain the same. They also commonly flip or mirror image the floor plans in homes next door to each other. Most importantly, mentally "burn the building down" and visualize the fire attacking the structural components of the home, floor by floor. What are the weaknesses and vulnerabilities of these components? How will they impact your operations? If the fire gets into the "guts," where can we anticipate the fire will travel next?

Fig. 8–12. Members inside a home

Be sure to thank the building representative, extending all the pleasantries that we normally do to the public. We are using this information to make ourselves and our citizens' lives safer. Many times the builder will have home layout plans that you can take back with you. Take the information to your members back at the firehouse and plan to incorporate your findings into your next company drill and your pre-plan books.

Back to the Firehouse

Quiz your crews during tactical discussions based on your discoveries. Use Johnny Appleseed's multi-home construction techniques to your advantage. Preparation starts the ball rolling for success and will make for a great drill. Develop fire attack plans and have suppression strategies in place before you respond to a fire in this new development. It is certainly wise to have these plans established long before you are advancing lines and crawling in with zero visibility and trapped victims. Proper pre-planning is just one facet in getting you, your company, and your department members combat ready.

The Renovation

In the wake of the housing crisis, many homeowners are looking for space in their existing homes rather than buying new, larger homes. This is causing many home remodels, many without required permits and building department inspections. Most of these renovations are taking spaces in the home that were not originally designed as living spaces and turning them into livable space. This happens in basement areas (addressed in detail in chapter 11), garages (attached or detached), and attic spaces. The hazard to firefighters is that most times we are unaware of the changes in the occupancy. This can cause us problems when we are expecting a typical situation.

Another home renovation on a larger scale is happening in older established residential neighborhoods, especially those that are in close proximity to major city hubs. The old adage in real estate of "location, location, location" has homeowners clamoring for these homes in desirable areas despite their smaller square footage. While the standing home is not necessarily what the new owners are interested in, the location is. Modifications can be grandiose. Figures 8–13 and 8–14 illustrate the transformation of an unassuming split foyer into a McMansion, nearly unrecognizable with the exception of the garage. The tie in between the new and old can cause us concern because there may be an increased number of void spaces and varying construction practices (lightweight vs. normal) throughout the addition.

Fig. 8–13. Pre-renovation

Fig. 8–14. Post-renovation. Note only the garage remained intact.

The McMansion is its own animal. They are large single-family dwellings with upwards of 5,000 square feet of living space. Most were constructed in the last 10–15 years, and they normally employ modern truss components and have large open spans for smoke and fire travel. Tactically, fire in a McMansion should remain largely similar to fires in other residential dwellings.

Even with these larger home sizes, we feel that the 1¾-inch hoseline is best suited for initial attack if compatible with fire conditions. However, secondary lines may need to be coordinated on the fire floor for quick extinguishment (fig. 8–15). While a 2½-inch hoseline is not normally thought of in residential buildings as an initial fire attack option, it can have its place. A company that takes a 2½ versus a 1¾ needs to be well practiced in its usage. Throw the normal complaint of short staffing in with ill-prepared, inexperienced crews moving a 2½-inch line in tight quarters and you are setting the stage for disaster. If you need extra gallons per minute (GPM), you need extra gallons per minute at the seat of the fire. If the team cannot get the 2½-inch hoseline to the seat of the fire, what use is the extra GPM?

Ladder company operations in larger area homes such as the McMansion will be stretched thin with a single company performing the required tasks. Later arriving ladder companies or rescue squads should check with the incident commander (IC) and first arriving unit to learn where they should focus their actions in the home.

As the IC, you need to have ample resources to properly and effectively respond to fires in these large buildings. Waiting to see if the first line "gets it" is not a sound tactical plan. Fires in McMansions will tax the resources of any department, so pre-plan accordingly and be quick to add additional alarms as needed.

Fig. 8–15. Fire in a McMansion will require extra resources.

Discussion

We must be students of the areas in which we serve. We must get out there into our community homes and stay abreast of changes to our response district. To some extent, we must also be students of the construction industry. Remember where we are fighting this "war that never ends." It is most often in the residential building. We must be prepared for battle at all times and in the best manner available. Having department members reporting to and gleaning information from your local building representatives about what is being planned will aid you in that quest. Invite them to lunch and make them welcome in your firehouse. The dividends they provide, in addition to your own building industry know-how, will do a great service to your citizens and your firefighters.

The Backstep

Get Out There!

As the chapter title clearly emphasizes, we must get out into our communities. We must have a good working knowledge of the "building stock" of the typical homes found in our response areas. Along with our knowledge of what currently exists, our fire departments must also be made aware of what is to come. Having access to the knowledge of what lies ahead may give you extra time to prepare for the changes. New building developments and their construction methods and materials can cause you to alter your fire operations, change the type of apparatus in your inventory, and cause a general alarm in quarters!

Fire departments must have a good working relationship with the local building zoning and code enforcement personnel working for the districts you protect. Whether the changes are due to large-scale renovations, new developments, or new approved building products, we must stay current with what is around us.

Find out the answers to the following questions, then discuss your findings with the rest of the crew.

1. Do you have a dedicated position within your fire department structure that acts as a liaison with your local government's building section? This person must create an avenue for communication with your local building representatives. Create this position and have the member chosen invite your local representatives to the firehouse for lunch or dinner. Explain to them your concerns with new and renovated structures and outline your response considerations, limitations, and capabilities.

2. Take an honest look at the inventory of structures in your community. Are your apparatus and tactics paired evenly with your findings? Are you providing the best service to the area to which you respond?

3. Go to the nearest local construction site and ask yourself the following questions. What has changed in the structural components and design layouts? What has changed in the finishes (both interior and exterior) and security? What policies can we adopt to best meet our needs with regard to changes in building construction?

4. Take your members through your local hardware store. Take a look around at the new products and spend some time in the door and window aisles. See how they are put together to determine the best way to take them apart. Look at the construction features of new products, especially those that pertain to home security. Do not be surprised when you see what is commonly available in your local hardware store at a residential dwelling fire near you.

9 On-Scene Reports: Nothing Showing Means Nothing

Do you or your fire department review the annual line-of-duty death reports that are issued by NIOSH and others that often state accountability as a contributing factor (fig. 9–1)? Do you happen to work or volunteer for a progressive department that institutes some form of accountability system on the fireground as a result of this information?

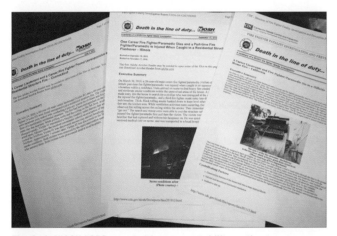

Fig. 9–1. NIOSH LODD report citing accountability on three different occasions

If you answered yes to either of these questions, then without even knowing it you and your fire department are supporting the use of on-scene reports. While it would seem that one has nothing to do with the other, the truth is that the on-scene report is the foundation of the accountability system on the fireground. It can serve as the benchmark for how your incident will proceed. We all have heard the "screamer" who barks pieces of miscellaneous and unrelated information across the radio with the end result of a chaotic fireground. Conversely, a calm and confident voice conveying relevant information on the fireground conditions usually starts and ends in an effective firefighting effort (fig. 9–2).

"Engine 1 on the scene, working fire."

Fig. 9–2. Officer giving an on-scene report via portable radio

What can we learn from this initial transmission, other than the fact the building to which we were dispatched that was reported to be on fire *is* actually on fire, and Engine 1 has declared it so? Not much, but this may be the system to which many firefighters have become accustomed and one they historically used with no detriment. To successfully implement the usage of on-scene reports we must first understand why we need to implement them and why are they important.

Paint the Picture

The on-scene report is the result of the first-arriving unit painting a detailed picture of what they witness upon arrival, along with confirming the information that was received from dispatch (fig. 9–3).

"Engine 1 on the scene with a working fire on the first floor of a single-family dwelling."

Fig. 9–3. Engine 1's report confirms fire on the first floor yet lacks additional information. (Courtesy of Kentland Volunteer Fire Department, www.kentland33.com.)

This time Engine 1 has given more information, confirming a working fire for which units were initially dispatched and giving the type of structure in which the fire is located. This information has confirmed the status of the incident for the incoming units and chief officers, setting in place their expected actions. Ultimately, the on-scene report is *not* for the first unit, it is for all of the other units responding to the call to assist. All the other units who, at best, may only see a plume of smoke in the sky. They are thirsting for more specific information in order to position apparatus and assume tasks based on conditions (fig. 9-4). Like soldiers entering a battle, responding firefighters are trained, motivated, and equipped with the tools to fight and win, but must wait for the information on their battleground and enemy to complete their strategy (fig. 9-5).

We must deliver more information, yet keep it clear, concise, and consistent. Screaming, trying to compress all of your sentences into one breath like you're preparing to dive underwater, or the perceived "cool" low talking all detract from the information you are trying to share. We want to be clear in what and how we say anything on the radio so that the focus is on the message and not the sender.

To be concise in your radio transmissions is to be prepared and calm. We are professional firefighters so we should not be surprised or overly excited by a building on fire. *Concise*, according to *Merriam-Webster's* means, "The giving of information clearly and in a few words; brief but comprehensive." To be concise on the radio and in the delivery of your orders takes training, experience, and quick thinking. We are faced with information overload on the fireground, so you must be able to decipher what is important and share it in a brief and comprehensive statement.

Fig. 9–4. A plume of smoke does little to describe building type or fire location, only confirmation of a working fire. (Courtesy of Kentland Volunteer Fire Department, www.kentland33.com.)

Fig. 9–5. Firefighters, like soldiers, must be prepared for the fight. (Courtesy of Sean O'Neill.)

Consistency can breed complacency, but in the hands of a professional it can breed confidence and remove some of the distractions on the fireground. If we provide firefighters with a template to commit to memory, a mental checklist that would serve as the structure to the on-scene report, we can then expect a clear and concise report. To build the consistent on-scene report we must think of what pertinent information we want to be shared. This is a vitally important step, as we want to remove as much of the non-essential information as possible while ensuring the essential information is included.

> *"Engine 1 on the scene, side Alpha, two-story colonial-style single-family dwelling, fire showing from two windows on second floor, we have our own water supply."*

We know that it is a working fire, what is on fire, where the fire is, and the water supply status, all in under 8 seconds. All this officer did when he arrived was apply his fireground knowledge to fill in the blanks of the on-scene template committed to memory. This is for some a return to childhood when you would complete a *Mad Libs* book for fun (fig. 9–6). The report has forced him and the other responding units to look for all of the points on this fireground and then deliver it in a clear, concise, and consistent on-scene report. Each of those points is essential information for initiating and completing a successful fire attack.

Develop and Execute a Template

Based upon the information needed for incoming units and the key information we want derive from the scene to make our attack, here is the basic template for the on-scene report (fig. 9–7):

- Restate address, if different from where dispatched
- Side of building
- Building size and type
- Conditions and location
- Actions for water supply (if engine)

This template is a simple five-point guide, but we should understand the relevance and importance of each point rather than merely accepting them as our template.

Fig. 9–6. A premade template can help to ensure consistent on-scene reports for officers. (Courtesy of Traditions Training, LLC, www.traditionstraining.com.)

Restate address, if different from where dispatched

How many fires are reported as, "Across the street from _____"? Most likely a civilian who is walking the dog stumbles upon the neighbor's house on fire, runs home, and calls 9-1-1. When asked the address, the civilian will recite his or her own. The reality is that while we are trained to handle buildings on fire and maintain

Fig. 9–6. The on-scene report is similar to filling in the blanks of a *Mad Libs* book.

a high level of calmness, the citizens we serve are not. This may be the first and only house on fire they will ever see, and it has caused them to lose any semblance of calm. At a minimum we must know where we are going, and having verification of the actual address sets in motion correct apparatus response routes, positioning, and water supply (fig. 9–8).

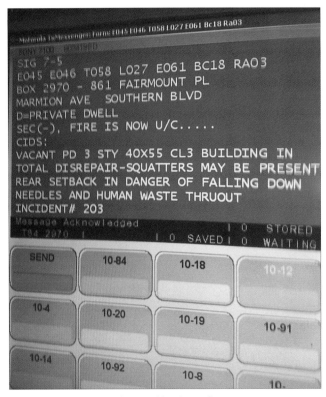

Fig. 9–8. An MCT screen for a working house fire

Side of the building

Once we have confirmed the address of the building, we want to know where the initial units are mounting their attack. Whether in an offensive or defensive operation, the assignment of a particular side by the initial arriving unit sets in place the response routes and positioning of incoming units. If units are dispatched for a basement fire and the first unit marks on-scene and gives a report confirming a basement fire with two stories in the front and three in the rear, it is essential that we know where they are positioning and mounting their attack. A two-pronged fire attack such as the basement fire demands coordination since hoselines will be entering from the front and rear. Companies may be performing VEIS on upper floors from specific sides. We cannot assume the address side of the building is always going to be the side on which we arrive and begin operations. Many times today, with the influx of population and need for more space, we see residential structures built in unconventional configurations. Let everyone know where you decide to start your firefight and the rest of the companies should plan accordingly.

Building size and type

We believe most firefighters would agree that tactical decisions would differ for a fire in the bedroom of single story rancher-style residential structure compared to a similar fire in an 18,000-square-foot, four-story McMansion. Both are residential structures, but each presents it own specific challenges and expected behavior when on fire. At a minimum, we should establish the structure in which we are going to fight fire so we can add to the mental image the incoming units are building. When we deliver this portion of the report, we are forcing our fire officer to visualize the number of stories of the structure so he or she can estimate hoseline stretches, determine potential exposure and extension travel, and identify immediate search areas along with other decisions. If we return to the basement fire scenario from the previous section, the identification of a two-story residential structure can quickly inform the next arriving engine that they may be able to deploy a line to the rear exterior walkout. Additionally, the incoming truck company will now know that they will either be able to use a porch roof or will need a 28-foot ladder to perform VEIS, if necessary (fig. 9–9).

Fig. 9–9. A house with a large porch roof is an excellent platform from which to work for search and ventilation.

How Does the Building Act When It Is on Fire?

The building type is the one area of the on-scene report that can either be concise or it can sound like a chapter out of a building official's book. Personnel may delve deep into the terminology by reciting building classifications types (Type I, Type II), often confusing incoming units. It is best to apply the KISS principle again here and relay only the most pertinent information that will affect tactics. When we discuss building construction we are ultimately concerned with how the building will act when it is on fire. What is the expected reaction of the building and where will the fire travel based upon the configuration and construction of the residential structure?

Given this data, we want to share information based upon the knowledge base of our personnel. If there is no requirement for learning building construction, then we cannot assume personnel will know what a balloon frame construction home is. What we can assume is that firefighters will know the style of homes in which they live, which are the identifying types such as ranchers, colonials, etc. With this information, incoming firefighters will start to visualize the interior layout of the residential structure much as you are probably doing now. A unit marks on the scene with a fire in the kitchen of a colonial. You know that if they enter the front door, they will most likely encounter a flight of stairs in front of them (fig. 9–10). If they advance and move slightly to the right or left of the stairs, they should move right into the kitchen where the fire was located. The entire incident planned and executed in a hypothetical fashion is based solely upon a quick sharing of information by the initial unit arriving on the scene.

Conditions and location

We have already established that merely confirming a working fire does not offer much more then what was known upon dispatch. We want to paint a clear, concise picture in a quick and informative manner.

> "Engine 1 on the scene with a working fire on the first floor of a single-family dwelling."

Fig. 9–10. View from the front door of a colonial, with the kitchen area directly behind

If we return to the previous report, we can confirm we have a fire, but how does the firefighter know it is a working fire? Can he see smoke? Fire? Operating on the guidelines of sharing relevant information that paints a picture, the on-scene report should include not only the presence of fire and smoke, but from where and how many windows. We are benchmarking what the fire was doing when we arrived, and we should be as specific as possible, but that does not require a doctoral dissertation on conditions.

The presence of smoke showing and not fire signifies to us the fire has not reached the perfect mixture to ignite. Depending upon the type of smoke (under pressure vs. lazily wafting out) we can tell how involved the fire is at that moment. If we note fire is showing, great! We now know where the firefight will most likely occur, and it has found a vent point and hopefully we keep the flow path centered on that location. Lastly, noting how many windows where conditions are showing is indicative of the location of the fire and where it is going. If Engine 1 marks on the scene of a two-story single-family dwelling with fire showing from four windows, we can instantly assume from our knowledge of these residential structures that this fire has extended from one room and may even have possession of the hallway and additional rooms (fig. 9–11). Conversely, if Engine 1 gave the report of light smoke showing from the open front door, we can assume that the fire may be small in size.

Fig. 9–11. Our inherent knowledge of homes we live in helps in determining the location and path of advancing fire.

Actions for water supply (if engine)

If you are the first-arriving engine on the scene of a working fire, you have one ultimate mission: *get water!* Fires are not extinguished with effective forcible entry and efficient primary searches, they are extinguished with the proper application of water to the seat of the fire. Therefore, regardless of your specific area's water supply issues (rural, urban, suburban, etc.), if you are the first-arriving engine you must start the water supply process. This does not mean you will "lay out," or forward lay to every fire call. We know that some areas do not have hydrants within miles. What it *does* mean is that you must identify the strategy for acquiring water and set in place the list of tactics to achieve this goal. This may be merely identifying a rural water supply setting and requesting the appropriate resources. Ultimately, we must never forget our mission on the engine company, which is to always get water.

Understand the Impact of Science

All of the information contained in the clear, concise, and consistent on-scene report is a trigger for the arriving officer to check the building in which the unit is going to fight fire for exactly what and where the fire is located. Secondly, it paints the picture for all the incoming units about what they are seeing and entering, benchmarking the incident at that specific time. This second point has become even more vitally important in today's fires. We hear often that fires burn hotter and go to flashover more quickly, but do we have actual data to prove this is true? Fortunately, we do through the work of NIST, UL, and other research-based organizations working to define fire progression from ignition to flashover and the impact it has on the residential structure.

One of particular interest and relevance to the on-scene report is the "Horizontal Fire Ventilation Experiments in Townhouses" report conducted in Chicago, Illinois. Essentially, NIST was able to use a strip of similar style two-story townhouses to conduct live fires with instrumentation to measure all facets of fire development. In one scenario they ignited a fire on the first floor in a couch and left only the window in the front bedroom of the second floor open along with the respective bedroom door. The fire progressed with smoke issuing around the front door and pushing with pressure from the only open window. At approximately the 3-minute mark, the smoke instantly ceased pushing from anywhere as if a switch had been turned off, and the fire disappeared. As occurs on the fireground, the truck and engine company arrived quickly and assumed their tasks. The truck instantly went to the front door while the engine stretched the hoseline. While the engine was stretching, the truck forced entry to the front and in an instant, the fire was back on (fig. 9–12)! Smoke was pushing out of the second-floor window, the first floor had smoke banked down to the floor, and soon fire was consuming the entire front room. This is great information about affirming a commitment to eradicating complacency, but what else does it mean and how do we use it?

Fig. 9–12. Forcible entry is ventilation!

For our purposes, it means that nothing showing means nothing! The on-scene report is not a safety umbrella that will keep us all safe and stop the fire from changing. It is a *benchmark of conditions when you arrive*, denoting how the fire is acting. We must always remain vigilant and eradicate complacency. If we return to the NIST scenario from above we can assume that when the fire is called in to 9-1-1, they are reporting fire in the house with smoke showing from everywhere and people trapped. When the first arriving engine arrives, the smoke has stopped, and they report nothing showing. The response of incoming units will most likely be, "How can it go from working fire and people trapped to nothing showing?" The speed of the apparatus slows down, the heart rates of the firefighters drop, and the high level of vigilance recedes as they wait for the "go in service" orders because it will be nothing. As we know from the example, that is not the case. A simple opening of a door reignites the fire.

From the perspective of the on-scene report we now know we must amend the report and offer the updated version to denote the current conditions. Given today's fire conditions and the dramatic and drastic changes they can undergo in a matter of seconds, we must develop and implement a system to start the accountability system on the fireground. We must offer clear, concise, and consistent information that paints a picture of what is occurring, where it is occurring, and what we plan to do. Pair the need for increasing our safety on the fireground with scientific information we are provided on the modern day fire, and the need for on-scene reports is evident. Aside from building the template that will sculpt your fire department's on-scene report, there must be an avenue, other than just emergency calls, to hone the skill. The "Cover Drill" is the perfect solution.

The Backstep

The Cover Drill

To conduct this chapter's drill, collect several of the trade journals in your firehouse with fire pictures of residential dwellings (fig. 9–13). Once you have a good collection assembled, all you will need is:

- Several willing participants
- Chalkboard or dry erase board (or anything on which to draw a picture)

Fig. 9–13. Trade journals, such as *Fire Engineering*, provide a training opportunity for all personnel.

Once both firefighters are ready, start your drill.

1. Pick one person to be the initial arriving officer and one person to serve as the scribe or "artist" (fig. 9–14).

2. Place the trade journal page face down in front of your initial arriving officer, and have your artist stand by at the drawing board.

3. Once all are assembled, have your initial arriving officer turn over the trade journal and give him or her 30 seconds to take in the entire scene. Once the 30 seconds is up, have the initial arriving officer deliver an on-scene report based upon what he or she sees in the photo.

4. While the initial arriving officer is delivering the on-scene report, have the scribe draw, to best of his or her ability, exactly the picture the initial arriving artist "painted" over the radio.

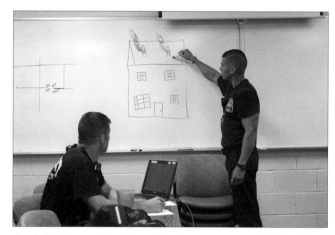

Fig. 9–14. We want the officer giving the on-scene report to literally paint the picture of the incident.

Do they match? Was the report a good depiction of the magazine photo? Rotate through each of your personnel with a different magazine cover allowing them to capture the scene and then deliver the report via radio.

Ultimately, we want the on-scene report to become not just a requirement of radio time for the initial arriving officer, but also a forced opportunity to truly evaluate the fire. The report should not just state that there is fire showing but should demonstrate an understanding of the impact this has on your tactical decisions. Essentially, you are removing the blinders and viewing all of the information presented to you on the fireground. This mentality has to be built-in training, so once the on-scene report is done ask your initial arriving officer the following seven questions relative to the fire seen in the photo:

1. What is on fire?
2. Where is it going?

3. How do you get to the fire? (Recall the layout of the structure based upon prior knowledge of floor plans and reported room on fire.)

4. List any of hazards or obstacles that you see that will hinder our operations. How will you overcome those obstacles?

5. Where would you place the first hoseline? Why?

6. Where would you initiate the primary search? Why?

7. Where would you assign your entire first alarm of units responding?

Not only does this type of drill ensure that you and your personnel are operating on the same page, but it also opens the conversation among all of the personnel who will respond to the fire. Include all of your firefighters in the training so the probie who has not run a fire yet will begin to understand what the commanding officer interprets on the fireground, along with his or her expectations.

10 Stairway to Heaven or Hell

An Honest Day's Work

It was a good day. My engine company had two jobs so far in this 24-hour tour, and we still had 6 hours to go. Oh, and did I mention I was fortunate enough to have had the nozzle for both of them? All firefighters love going to fires, right? One of the few things nearly as good as going to the job is being able to explain what happened to the firefighter who didn't. On this particular tour, in addition to having the nozzle at both fires, one thing set them apart. While having the "knob" at two jobs certainly is a great treat to experience, the best part was that I had a friend from another fire department visiting me, riding along with our company that day. He was looking for a few good stories to bring back, and he certainly got a few (fig. 10–1).

Fig. 10–1. Authors visiting at the firehouse

He was able to see firsthand the great engine company I found myself in and see the occupied structural work that we were doing. While two fires in a tour wasn't uncommon, it certainly wasn't as if it happened every day (but of course I didn't need to tell him that). The first fire of the tour was a kitchen fire in a six-story, occupied multiple dwelling (MD) on the fourth floor. While it was just the kitchen contents going, it was an enormous stretch (see fig. 10–2). It was a huge building, 200 feet by 200 feet, of ordinary construction. It featured an enormously deep recessed courtyard set back over 150 feet from the street.

Fig. 10–2. Buildings set back from the street make for long hose stretches.

Once inside, there were two narrow sets of stairs that allowed for transverse movements between each section (or wing) of the building. A key facet missing from this building was that there was no well hole to assist us in getting the line up the steps. Floor by floor, we hauled the 1¾ inch hoseline up the steps. Turn by turn, we eased the line around the spindles and newel posts. Once we finally reached the fire floor we kept pulling more line, kept humping. We needed a minimum one length of hose (50 feet) to be flaked out before we charged it to the ready outside of the fire apartment.

It required all members of the three first engines (including the help of my ride-along friend who jumped into the outside action) to help get the first line in position. To get the hose into the fire apartment and to reach the fire, we needed 15 lengths of hose (a combination of 1¾ inch and 2½ inch) from the rig. This was no easy, simple stretch. It was a coordinated symphony and no small feat to accomplish the task. As the Fire Department City of New York (FDNY) states in their radio communications, we truly used "all hands" and more at this fire.

As the ladder company held the fire in check with the 2½ gallon water extinguisher (the can) and started their searches, we moved the charged line into the apartment. How did the stairs play into the story? The stairs played an integral role in our quest for extinguishing the fire. They provided us the avenue get water to the fire. We had to get the line up to the fourth floor somehow, at the same time protecting the occupants inside. The stairs were the obvious choice.

We stated that there was no well hole at this fire. A well hole is the space formed between the upper and lower railing sections where they join together at the newel post. As a rough rule of thumb, a horizontal gloved fist in this space will be large enough to run the 1¾ inch line and couplings up through a narrow sized well (fig. 10–3).

Had there been a well hole, the company officer could have told us to use it since it would have certainly shortened the stretch. We could have traversed five stories with one length of hose pulled vertically up the well. Well holes can be great friends to engine company members (figs. 10–4 and 10–5). Commonly found in multistory multiple dwellings, you may find versions of these in larger residential applications. We must be sure that once we get the line into position, it is secured at the top railing with a rope hose tool or webbing so that it does not pull back down the well once the line is charged.

Fig. 10–4. Narrow well hole

Fig. 10–3. Measure the size of the well with a gloved fist; this well is too small.

Fig. 10–5. Large well hole

Fig. 10–6. Bottle with rope used to hoist hose

If the well hole is narrow, the lead firefighter can take the nozzle by itself in his hands as he goes up the stairs, leaving hose to be pulled up through the space created by the well. This leaves the rest of the hose at the bottom of the steps. It will pay up and out as the nozzle firefighter moves from landing to landing. The backup firefighter should manage this spot if at all possible. If the well is large, the nozzle firefighter can carry the entire lead length and nozzle into the well while transcending the stairs to flake out on either the floor below the fire or outside the fire door in the hall if the environment is not IDLH.

Also, if the staircase had windows in line with an accessible point at the ground level, had an extremely large open well, or even had a line of windows in line with the lobby on the first floor, we could have utilized a rope stretch (fig. 10–6). Instead of going around the steps on each turn, we could have effectively made an exterior well hole or standpipe running along the outside of the building. This technique is often utilized when the staircases wrap around elevators in buildings. Again, this is not a technique that you are going to use for the first time at a fire. It needs to be practiced before the bell sounds.

The engine company officer takes a gallon jug-sized bottle filled with rope (about 50 feet) and runs up the stairs to the floor below the reported fire. The member either removes or opens the lower sash of the window and sends the rope down by tossing the rope bottle to the street level below. The nozzle firefighter takes the lead length of hose to the point where the rope will be coming down. Usually, the nozzle firefighter then heads into the building to join with the officer on the floor below. As the nozzle firefighter heads upstairs, the backup firefighter grabs the deployed rope and ties up the nozzle and lead length, then gives the okay to begin the hoist.

The officer and the nozzle firefighter now hoist up the needed hose, enough to get it in place on the fire floor and reach the apartment (normally 50 feet or one length for the fire apartment, but also considering the line needed from the window to the fire apartment door). The backup firefighter ensures smooth hoisting of the hoseline and heads inside if needed. Members near the nozzle can begin to flake out the line in the public hall on the fire floor (if not IDLH) or on the floor below. Prior to calling for water, be sure that the line is tied off to the stairs with a rope hose tool or webbing so that it does not slip back out the window from which it came.

The Second Go Around

The second job of the 24 was another first-due job and, as in the first one of the day, also a kitchen fire. That is where the similarities ended. This was not the large multi-story multiple dwelling. It was a simple peaked roof, wood frame, two-story residential private dwelling (PD) with a basement. It was much easier in the hose stretch length department due to the size and proximity of the home to the street (fig. 10–7). Yet there was still a challenge: the stairs. How could the stairs in a two-story home be a challenge?

Fig. 10–7. Starting to stretch hose to the residential building fire

This home, while typical to the neighborhood on the outside, was far from it on the inside. We use the stairs at every fire in single-family residential buildings. What was the big deal? We know that we are going to use stairs anytime we operate at, above, or below grade. Because of this, we must be prepared to operate on all different types of stairs, typical or not. These stairs, while completely different, wound up being similar in their difficulty to manage. Hence the title of this chapter.

Originally, the run came in as "fire on the first floor." Fleeing occupants told us as we got off the rig on the block, "There is smoke everywhere in the house!" They weren't lying. As we looked at the two-story home, we saw smoke coming out every window frame, door, and crack in the place. We had beaten the truck into the box for whatever reason, and the engine company chauffeur had pulled past the house and was on a hydrant just beyond the building. Upon orders, we stretched a dry 1¾ inch line to the front door awaiting further communication from our lieutenant.

The officer radioed back that he hadn't yet located the seat of the fire. This was normally the truck company's job, finding the fire and starting their searches from the greatest danger (the fire area) to the least. Our engine boss had no thermal imaging camera to assist him to find its origin. He had no search team at his side. It wasn't going to be easy. He crawled back to the building entrance and saw us at the ready with the dry hoseline. He called the chauffeur and asked for water via the walkie-talkie. The engine company chauffeur told the boss he had good water (he had a working hydrant), that water was on the way, and charged our hose-line at once.

Once the line was charged and bled, the officer told us, "I've got an idea where this thing is, follow me." We only went a few feet into the dwelling when he told us to stand fast with the line. Smoke was thick and hot, and it wasn't lifting. The truck still wasn't in to effect ventilation tactics. In fact, the heat all around us was increasing, but there were no visible flames on the first floor. Where was this fire, and why wasn't it showing itself?

I thought that I kept feeling more intense heat coming from my right side. I put the hoseline nozzle under my left leg momentarily and moved toward that direction with my upper body. I only went about 4 or 5 feet, but I bumped into, saw, and felt something odd. It felt like a 3-foot-high fencelike dog cage about railing height, but there was a strange orange glow in the center. It looked as though it was a campfire glow, but it had a strangle gate type of device around it. It took a minute for the brain to drop the correct slide from my slide carousel into focus (for those of us who pre-date PowerPoint). It was a set of spiral stairs (fig. 10–8)! I shouted through my mask, "Hey Loo, it's over here, it's below us, in the basement! The basement steps are spiral stairs going down."

Fig. 10–8. Spiral stairs are not common in residential homes. Expect the unexpected.

Regaining the nozzle in my hands and moving the line in that direction, I tried to conjure up a plan in my head. I paused at the railing that surrounded the steps. We really only had one shot at this, and I knew what we had to do. The boss said to me, "We are going to make the steps, right? Make sure you have enough line," he continued. I made sure that the backup man had pulled in enough line behind me to make the bottom of the steps. In conference with the boss, we made the move. At the time, fire was beginning to lick up through the hole in the floor, exposing the metal staircase. I opened the line as I went down the steps, knocking the flames back as I went down the winding steel on my rear end.

The homeowner had literally cut through the floor from the first floor to the basement, recently adding these spiral stairs. Again, this may not seem like a big problem for anyone until fire breaks out and smoke begins to travel throughout the dwelling. Those stairs acted as a chimney for the smoke to travel to every open space in that home. The smoke was everywhere! Once the fire was out, the smoke had cleared, and the occupants were safe, it seemed rather innocent to them, a moot point. But we all know how easily it could have been a different outcome. Fire was ready to extend and consume that first floor.

Had that fire extended to combustible furnishings on the first floor prior to our arrival, once we knocked the first floor contents down with the line, we might have thought that was the end of it. We might have thought we had a contents fire held with just one line on the first floor. We did not realize that the entire time we were operating we were in one of the most dangerous spots to be, the floor above a fire! How long had this fire been eating away at those floor supports that we were crawling across and operating on? This incident illustrates the need to be combat ready.

In previous chapters we stated just how important it is to be combat ready, how you truly never know what you are going to find, even within the seemingly straightforward residential building. I certainly know that I didn't expect to find a set of spiral stairs in a 1940s balloon frame home. As often happens when the smoke clears, we see things that we must note, take back, and share with other members. We must tuck them away for the next fire we go to and add them to that mental slide carousel mentioned earlier.

As in the case of most incidents, the two toughest parts of both of these fires wasn't in the extinguishment itself. It only takes about 8 pounds of pressure to open the bale of the nozzle, and water is a very effective extinguishing agent when applied to the seat of the fire. The difficult part in both of these fires was the same: getting the line to the seat of the fire. This is consistently seen at many fires and is a common thread in multiple firefighter close calls and LODD reports.

The nozzle has to get water to the seat of the fire to extinguish it, period. Many times the stairs play a role in making that quest more difficult. Residential stairs will play a role in many fires that we respond to, especially in hoseline movement. Estimating and moving the hose stretch for our operations in and around stairs must be well rehearsed. Estimating the amount of hose needed is one thing, getting that hoseline to the seat of the fire to effect extinguishment is another. In both of the fires that were referenced above, the main challenges initially came from the stairs.

The Residential Staircase

Staircases in the residential single-family dwelling can be deadly for occupants and firefighters. Staircases can act as interior chimneys. They can provide unimpeded routes of travel for smoke and fire travel within dwellings. We must understand the variety of stairs that we may encounter when fighting fires in residential homes. Typical or atypical, we must be prepared. In the second incident, the spiral stairs were not expected for the style of home and the date of its construction. We must use the information on stairs and staircases found in this chapter as examples of what is to be expected based on the norms. We quickly see that we must neither learn with blinders nor be complacent in our study. Even though the norms exist in greater numbers, we must always anticipate the unexpected.

Occupants in residential buildings rely solely on their staircases for egress and access to various floors of their home. They are common necessities. As firefighters, we also put great emphasis on the stairs, both in locating their placement in the structure and controlling them during fire operations. Whether the fire is on an upper floor or in a below-grade area, both civilians and firefighters need to use stairs to get where they want to go. Unfortunately, under fire conditions, residents often attempt to use staircases the same way they normally do. While occupants are looking for a way out and utilizing them in their normal path of escape, firefighters are looking for entry, trying to find steps for the fastest possible avenue to the fire so they can extinguish it and protect civilian lives.

Even for a fire on the ground level where stairs will not necessarily affect the hose stretch, staircases can have an impact on our operations (fig. 10–9). While the initial line will find and extinguish the seat of the fire, other engine members should have a second line positioned at the base of the stairs going above. They should have enough hose to cover the fire floor and the floor above, as well all the other floors above the fire floor.

This is done primarily to address the case of fire extension, but this isn't their only task with that hoseline. That line may need to be repositioned to assist the first line if a problem arises. A sudden water loss, burst length, or nozzle malfunction may cause that second line to take over initial fire attack. The second line should be positioned through the same exterior opening as the first line. It should be brought to the position of the steps and held fast. In addition to effectively backing up the first line, that second hoseline is also now properly positioned to protect any life above the fire, civilian or firefighter.

Fig. 10–9. Residential fire on the first floor. (Courtesy of Kentland Volunteer Fire Department, www.kentland33.com.)

When operating at a residential building fire, the stairs must be located, protected, and controlled for our operational safety since fire and toxic smoke travel is largely aided by the presence of these interior structures. Controlling the stairs allows the greatest chance for civilian safety. Most civilian rescues are removed from the residential building via the main routes of egress, the stairs. While it does occasionally occur, it is not all that often that civilian victims are removed from second- and third-story windows via portable ladders.

We must recognize the importance that staircases have in all of our operations, from pre-fire control to overhaul. We must understand their structural building components, know the materials used, their limitations, and how to prevent and overcome problems impeding our goal of going up or down the stairs. Anytime we are going up or down steps, we must be extra alert. We must listen for the status of the first hoseline regarding its progress towards extinguishment. We must also recognize and plan to utilize a secondary means of egress from our location other than the steps we used to get there.

Staircase Design and Layout

There are two prevalent types of stairs in the residential building. Generally, and in its most simplistic terms, staircases can either be classified as open or enclosed (figs. 10-10 and 10-11). Their style however, can be varied to please the homeowner and best utilize the interior space of the structure. Most residential stairs are either *straight run*, which follow a straight line down from one floor to another floor (fig. 10-12), or *return*, where the stairs have a half-landing and then an angled turn, 45 or 90 degrees (fig. 10-13).

Fig. 10–11. Open staircase

Fig. 10–10. Enlosed staircase

Fig. 10–12. Straight run stairs

Fig. 10–13. Stairs with a landing between levels

The components that make up the great majority of residential building staircases are generally wood and occasionally light gauge steel. A general breakdown of the structure of a residential staircase is illustrated in figure 10–14.

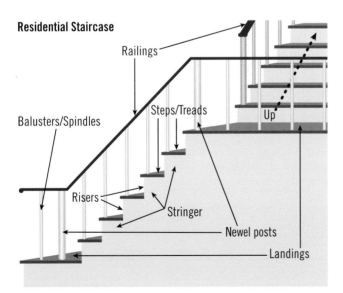

Fig. 10–14. Common staircase. (Illustration by Matthew Tamillow.)

In most residential homes, staircases from the basement to the first floor are usually the enclosed type with either a door at the top, bottom, or both. Sometimes these doors are required to have a fire rating, depending on building codes. As firefighters we know that doors are crucial limiters to fire and smoke travel within structures. Fire rating notwithstanding, simply closing a door may mean the difference between life and death for an occupant. We as firefighters know that doors can be life savers for us too, often allowing us to hold back fire momentarily or let a room lift from a smoke condition to aid us in search. For those reasons and other advantages, we must mentally landmark the doors we encounter. We must use doors to our advantage, as an additional tool to help us work smarter and more safely. Controlling stairway doors can be a vital factor in our operational success (fig. 10–15).

Fig. 10–15. Doors can help us contain and control fire.

Fire, heat, and smoke conditions at these doors must be communicated to operating members and consistently monitored. While we discuss basement fires specifically in chapter 11, it is certainly worth mentioning here. Once the door to the steps is located, announce over the radio that you have found it. A short message telling its status (intact/missing/burned through) should be communicated to all fireground personnel. In certain situations, this may be where we hold the fire in check, protecting the interior of the house and those in it. If the status of the door changes, operating forces must be immediately notified.

Open staircases in modern homes are a selling feature. Realtors and homebuyers alike are enchanted by grand staircases and large, open, chandelier-bedecked entrance halls. But when a fire breaks out, this open design can be deadly. What may be a small fire on a lower floor now has free reign to pollute the entire dwelling with its toxic by-products. Smoke from a lower floor can quickly travel to upper floor bedrooms, asphyxiating sleeping occupants. How many of us sleep with our upper floor bedroom doors open to hear the kids? The majority of fire deaths in residential buildings are from exposure to smoke, not fire. Additionally, some open interior staircases can have an open tread design (fig. 10–16). Stability may quickly become an issue under fire conditions.

Fig. 10–16. Open tread stair design. (Courtesy of Kentland Volunteer Fire Department, www.kentland33.com.)

Stairs tend to be a common congregating place in residential buildings for firefighters, but stairs are for us to go up or down. No firefighter should be camped out on the steps. Generally accepted building stairway widths from International Building Code (IBC) section 1009.1 list dwellings serving 50 or fewer occupants at 36 inches at minimum. At this width, fully dressed firefighters attempting to pass each other on the steps requires extreme coordination and can be nearly impossible in the best conditions. Add smoke, heat, and fire, and you can see how we have created a potential recipe for disaster.

Think about it like this: you are on the initial hoseline and have made the steps at a stubborn, smoldering, smoky fire in the basement of a residential building. While operating on the line, you suffer an SCBA emergency and need to get out of the fire area immediately. You have gone through all the mental and physical checklists to try to get it functioning again without success. Your mask has malfunctioned to the point that you cannot get good air out of it. You transmit the Mayday to the incident commander and move to seek fresh air, but what do you do? In two quick movements you follow the line back to the basement steps. You're almost there. You found the stairs!

You know that there is fresh air at the top because that's how you got to where you are. Only now, having found the stairs, you can't get up. You are face to face with another firefighter. You have no time to tell about your plight, you just have no air.

This firefighter is one of four members on the steps, looking for action. They have good air and you know it. You fight them, literally fight them, punching and clawing to get air from their masks. You can't speak, you can't think straight. You start to become overmatched by the effects of the carbon monoxide. All you want to do is get out. Firefighters looking for action are on your staircase, effectively blocking your way out. They have no idea why you are acting irrationally on the steps, shoving the first one out of the way until you collapse into the arms of the second, unconscious.

It takes that second firefighter 2 minutes to figure out what's wrong with you, then 5 more minutes to clear the stairs. Since the commotion you made has now gathered a crowd of onlookers on the steps, "the stuff" has effectively hit the fan. These firefighters are now flailing and failing as they try to haul you up the rest of the steps. This isn't a philosophical scenario. It has occurred, and it will occur again. You are now stuck on the steps, dying 8 feet below grade at what you thought was going to be a simple basement fire.

All firefighters at this fire want to get on the fire floor, and all units want to get a piece of the action. It takes both personal and company officer discipline, and perhaps a friendly reminder, to avoid the temptation to camp out on the steps. We have seen this phenomenon with greater frequency at fires on upper and lower floors. Clogging the stairs can be deadly. If you are on the steps, you had better be moving. You will impede egress if an emergency occurs on the fire floor. Either go up or go down. Stairs are not a place to hang out and wait for an assignment, or look over the shoulder of another member because you want to get a peek at the action.

Stair Design in Particular Building Home Styles

The stairs in certain building types, as mentioned in chapter 7, have predictable characteristics, such as style and location. We cannot stress enough that the building is our enemy, and as such we must take advantage of every opportunity at our disposal to get inside and look around. In many home styles we can often find telltale clues as to where the stairs may be located. Sometimes we can even see evidence of the stair location from the outside of the dwelling itself. We certainly know that these staircase locations should not be taken as a hard and fast rule for your local area, but they may provide a guide as to where to begin our search for the steps.

Colonial

The colonial home is often known for its grand staircases (fig. 10–17). In many areas they are referred to as center hall colonials. The front door of the structure normally leads to a central hallway that connects the front to the back of the home. In general, off to one side or the other of that front entrance hall are the stairs leading up to the second floor. Larger colonial style McMansion homes may have additional staircases, with at least one often found near the kitchen. A common construction trait in newer homes with basements is to have the basement steps located under the staircases going up, "stacked," if you will. This goes for straight run or return style staircases. A significant amount of square footage is often taken up by stairs in colonial homes.

Fig. 10–17. Center hall colonial home

Cape Cod

Cape Cod style home staircases are similar to those found in colonials in that most often they are centered off the main entrance. They commonly lead to bedrooms in the dormered upper floor (fig. 10–18). Basement stairs may be behind and under the steps to the second floor. Often, basement steps are in an area off the kitchen.

Fig. 10–18. Cape Cods normally have stairs to dormered bedrooms.

Split level

Stairs in split-level homes are most apparent when you open the front door (fig. 10–19). In this style home, we often climb a small stoop to gain access to the main entrance. Once inside, we find ourselves on a small landing. On either side of this staircase landing is a half set of stairs up and half set of stairs down. Be mindful that in this style home you are starting your operations on a set of steps. Larger versions of these homes may have additional steps in the rear in the vicinity of the kitchen, which is normally on the upper level of the house.

Fig. 10–20. Split-foyer homes have shorter run stairs.

Fig. 10–19. A split-level home's stairs go up and down.

Split foyer

As described in chapter 7, this style home is a combination of colonial and ranch with multilevel, short run staircases (fig. 10-20). Split foyers usually have some sort of up/down step arrangement as you enter the front door off the small landing. The variation of the home layout determines their placement. Some homes have the main entrance on the ranch section where you walk straight in at ground level, move to the middle of the home, and *then* find the short run steps up and down to the other levels. Kitchens are usually on the main ranch level with bedrooms above and the family room below.

Rancher/rambler

These home designs place all living space on one level. Stairs in these homes, if any, lead down. Depending on locale, these homes may or may not have basements. Ranchers and ramblers will likely have stairs to basements in either of two places, or both. They will be in the area where the garage attaches to the house and/or directly off the kitchen (fig. 10-21).

Fig. 10–21. Ranch homes may have basement staircases.

Variations and Clues

While we have given you a guideline for stair locations in a few residential building styles, the stairs can be found in many locations, sometimes where you least expect them (fig. 10-22). Nothing is more frustrating than searching for the stairs. While knowing the likely places for the stairs in your department's response area is key, you never know if a homeowner might have moved them in a renovation. Some building styles and designs can point you to the stair placement. Smaller style decorative windows along exterior walls (ones that could never be used for egress) may indicate a run of stairs.

Fig. 10–22. Renovations may move staircases to unusual places. This set of steps to the basement is accessed through a built-in cabinet.

When it comes to the steps, we must recognize their importance to us as firefighters. We must train and drill on maneuvering on them. Most fire academy burn buildings are not built to replicate the residential stair. They are much, much wider. Take some plywood and make stairs that are representative of what you will face on the street. Have your members practice passing each other and operate up and down the steps with hoselines. Check the drill that follows for tips to effectively and efficiently move hoselines through residential building stairways.

The Backstep

Moving Hoselines on Stairs

A flight of stairs shall not have a vertical rise greater than 12 feet (3,658 mm) between floor levels or landings.

–Extracted from IBC Section 1009.6

Why is the above quote important for firefighters to understand in relation to hoselines and residential buildings? No staircase in any residential dwelling shall have any straight run more than 12 feet. As such we know that we need greater than 12 feet of hose to reach the top or bottom of any staircase. How many times have we found ourselves in the precarious situation of getting hung up with not enough line in the middle of a set of steps? Two drill variations to remedy this are discussed and illustrated in photos below.

The first is rather simple and requires little thought. If we need to go 13 feet to make it up or down steps, advance the hoseline past the objective that distance or greater, then double the line back on itself and reposition yourself at the top/bottom of the steps (fig. 10–23). Doing this forms a large loop of hose in the area of the stairs, either at the top or bottom depending your direction of travel. Without much effort, you have got the hose you needed at the ready to make your move (fig. 10–24).

The second variation is the "push method." Once positioned at the bottom or top of the stairs, take the nozzle and place it between your legs. Push the line, either up or down the staircase, until you have reached resistance at the top or bottom (fig. 10–25). What you have effectively done is given yourself enough line. When you now pick up the nozzle and move on the stairs, you are bringing up or down that section of hose with you. If you have a stair design other than straight run to advance up, you may need to initially reposition yourself on the half-landing to ensure enough line makes it to the top.

Practice these two quick and easy residential stair maneuvers with a charged hoseline. Mastering these tips eliminates the situation of being stuck shouting for more line on the steps. Remember, the stairs are no place to hang out. If we make the move, we must *move*.

Figs. 10–23 and 10–24. Firefighter moves line past base of stairs approximately 13 feet and repositions at the base of the stairs.

Fig. 10–25. Firefighter pushing a line up stairs

11 The Fire Down Below

In chapter 10 we discussed the perils and pitfalls that firefighters can experience with residential staircases. Stairs, and their impact on our operations, reappear throughout this chapter as well. In this chapter's discussion of "the fire down below," we provide insights and tactical methodologies for attacking one of the most deadly and punishing fires that we encounter: fire in the residential basement (fig. 11–1).

Fig. 11–1. Basement fire in a residential building. (Courtesy of Kentland Volunteer Fire Department, www.kentland33.com.)

Limitations

Fires that occur in dwelling basements put instant limitations on fire department operations. The first instant limitation in most basement fires is the ventilation issue. Our options to vent basements are limited because windows serving basements are normally much smaller than those found throughout regular aboveground floors. Access and egress are the next deadly limitations. Basements may have only one set of stairs that lead into the space. This leaves us only one way in and one way out. Perhaps no other operation requires the same level of communication, coordination, and preestablished standard operating procedures or guidelines (SOPs/SOGs) for success.

Statistically, basement fires are always listed near the top for firefighter operational line-of-duty deaths and account for many injuries on the fireground. It seems that nearly daily we read reports of one of our own falling through a residential floor into the basement. Why is this? Why is it that the residential basement fire proves to be so difficult and dangerous for us as firefighters? There are nearly as many reasons as there are brothers and sisters who find themselves falling into their dark abyss.

It's 3:00 a.m. as you head back to sleep, interrupted for the third time tonight. No, it's not the firehouse alerting system sending you out on the next run. It's your third kid. You and your spouse always wanted to have four, but number three is trying the patience of you both. With the two older kids out of diapers and finished with the midnight feedings, you have forgotten how trying these first ten months can be. Between working at the busy firehouse and getting up three and four times a night at home, you're a train wreck.

Nestling back into the chilly winter sheets, trying to find that warm spot you recently vacated, you hear a muffled rumble coming from outside. Your wife rolls over and mumbles, "What was that?"

"Sounds like the kid next door needs to get that muffler fixed again," you mumble back. "Backfire, I think."

About an hour later, deep in well-deserved REM sleep, your oldest is now shaking your arm saying, "Daddy, daddy! There is someone knocking on the door!" You spring from the bed, grab a pair of pants and hit the stairs. Your neighbor and her kid are on your front porch.

"Kevin," she says, "The house is full of smoke!"

"Everyone out! Where's Frank?" you ask quickly, your wife now behind you on the phone calling 9-1-1.

"He's working tonight, we are all out." You run over to the neighbor's house, the same style as yours, a 1930s balloon frame colonial, as you begin to hear sirens in the distance. The entire house is obscured in a grey envelope of smoke. It's light, moving about, but has no real direction, nor is it moving with volume or pressure behind it. Having no PPE and knowing that no one is trapped inside, your options to assist in this situation are limited.

You decide to take a quick lap to take a look at the back. Perhaps you can see something to relay to the incoming companies. You open both outside gates. "This will help them when they bring the lines and help with access," you think aloud. As you head to the back to get a view of the rear, you note an orange glow and some black smoke coming through the small basement windows. "Ah, it's in the basement, at least I can tell them where the inside stairs are," you think.

You feel rather helpless as you watch the first-in engine company lay a line as they make the turn into the driveway. The pumper tanker rolls in right behind them, completing the initial water supply in your rural setting. You grab the engine boss, tell him you are "on the job," that you live next door, and let him know your initial findings. The crew pulls a preconnected 1¾ inch hoseline to the front door, and you help them flake it out. They call for water and move in. You help a few later-arriving firefighters with portable ladders, and head to the chief and check in at the command post, hoping to be able to listen to a CAN (conditions, actions, needs) report from one of the crews.

After the fire is extinguished, the engine officer finds you and thanks you again. He acknowledges that most of the time civilian information has to be taken with a grain of salt, but he thanks you for yours. He says it was truly the best he had ever gotten at a job. Beyond a doubt the preliminary information helped the team get quick water on the fire, and that in turn ultimately helped your neighbor. It turns out there was a malfunction with the boiler. That was the noise you had heard. The fire was contained to the basement with only minor extension to the first floor.

This basement fire had a moderately good outcome, no firefighters or civilians killed or injured, property loss contained at a minimum. Why is it that many times this isn't the case?

Access and Egress

For purposes of this book, we define a basement as a space in a building that has at least part of its room space (height) below ground level. This is not to be confused with a cellar. Cellars are normally found below the level of the basement. Some departments use the terms basement and cellar interchangeably. Have your department's SOPs spell out what spaces and what floors have what designations so that there is no confusion as to who is operating in which space. Communication will prove critical in the effective attack on basement fires.

Access to basement areas of the residential building is either internal (within the dwelling), external (commonly referred to as walkout basements), or a combination of both (fig. 11–2). It is critical to communicate to all members the type and location of these access points. The status of each of these points and their ease of use will be one of the main factors in our operational attack modes. Limited access and egress are just a few of the reasons why basement fires are so stubbornly difficult for fire operations.

Fig. 11–2. Residential home with walkout basement

What's in Your Basement?

Basements, like garages and sheds, are full of anything and everything that you don't necessarily want in the main living space of your home. When most homes are built, the basements are not normally finished off to be livable. The basement walls are normally stone or concrete, containing limited interior partition walls, open-tread stairs (most often with a door at the top), and open ceilings exposing the underside of the first floor. This poses a serious problem for us since there is no fire rating to the underside of what may be the entire first floor (fig. 11–3).

Fig. 11–3. The unfinished basement ceiling allows fire to immediately get into the support system of the first floor.

What else is in your basement? Answer the question for yourself. Basements can include everything from garden supplies to Grandma. Many homeowners with children convert that once unfinished basement space into living space. This can be additional storage, play areas for kids, craft rooms, and yes, even bedrooms for those pesky in-laws. In today's cash strapped economy, homeowners may create fully operational legal or illegal apartments in their basements. We have seen single room occupancies (SROs) in residential basements, sometimes in affluent communities where you would least expect it (fig. 11–4).

Fig. 11–4. Single room occupancy found in a single-family colonial style basement

Also, especially in the Northeast U.S., the building's mechanical systems are often located in the basement, including the residential building's heating/cooling units, fuel oil storage tanks, electrical service panels, and so on. A basic problem for us regarding all basement fires is access to and egress from the space.

If you follow the trend in basement fires, sometimes access isn't an issue at all. However, how firefighters access the basement isn't routine. It is not via a normal set of stairs or an outside door, and it's certainly not safe by any stretch of the imagination. Firefighters in every city and town are almost routinely falling through first floors into basements. Stop for a moment and think about why.

What has changed in construction or in our attack on the fire to increase the frequency of these events? When we respond and initially operate at basements fires by entering on the first floor, we are effectively crawling onto and operating directly on the floor above the fire, a very dangerous place. Think about this analogy: firefighters love to eat, and likewise, so does fire. Picture yourself crawling into a giant barbecue, only we are the meat and the grill rack is made of combustible wood. This is what we are doing when we enter the first floor headed to an involved basement fire.

But why the collapse? What is it about the construction? If we have a good fire in the basement, most of the time there is nothing to stop it from instantly burning into the structural components of the first floor. Most basements do not have finished ceilings made of fire-resistant material over the underside of the first floor's joists. This fire has access to all the supports and the underside of the subflooring. It's not like other levels of the home where we have sheetrock or plaster finishes to help keep the fire in one or two bays. The fire can instantly eat into the supports. Take into account the increasing amount of lightweight components in new homes, additions, and renovations. Several of these new products have been documented to fail earlier than their dimensional counterparts.

Please take time to check out an online course presented by UL called *Structural Stability of Engineered Lumber in Fire Conditions* on their website www.uluniversity.us. Another point of reference is NIOSH Publication Number 2009-114, *Preventing Deaths and Injuries of Fire Fighters Working Above Fire-Damaged Floors*. Fires in basements with increased usage of lightweight flooring systems are a recipe for disaster.

Recognition

When attempting to recognize a basement fire when it isn't obviously apparent on arrival can cause your fire senses to become clouded (this topic is discussed further in chapter 21). Many times, when the first floor is hot everywhere around you, it's hard to determine where to go. When you do move, have an outstretched tool sounding the floor and an outstretched leg in front of you. When heat is coming at you from all sides, the feeling that heat is oppressing you with no distinct direction can be a telltale sign that the fire is below you. Use a thermal imager (TI) to scan different sections of the floor, looking for changes in heat signatures (noting that carpet, tile, and hardwood will have different characteristics in their own right). While this may not prove to be an adequate indicator, it is certainly worth a quick scan.

This is also something that you should note when conducting your 360-degree check of the structure (see chapter 12 for more insight into the 160-degree check). This evaluation and/or having an outside firefighter check those basement windows can confirm the indications that you may have on the first floor (fig. 11–5). A quick check for smoke or fire in the basement may be noted by looking through windows. It is imperative that all fires receive an early basement check. If it's hot and smoky all around and you can't find any fire on your level, it's likely below you.

How can these fires grow to such intensity that they cause localized collapses? Basement fires have a built in delay, akin to a fuse on a bomb. There is often a delay in civilian recognition of a basement fires. Basements are closed off from living spaces, sometimes with two floors between them and the sleeping quarters. While the delay can occur at any time of the day, it is especially prevalent during the overnight hours, allowing fires to gain significant headway. Also, many basements that are used for storage only may not be equipped with smoke detectors, unlike our bedrooms and living spaces.

We can't always blame it on the homeowner, however. Oftentimes it is firefighters who do not initially recognize that the fire is in the basement or had originated there. Beware of the house with smoke from everywhere! An initial clue may be grayish-white smoke coming from every level you can see.

Sometimes we can be fooled into only looking at what is right in front of us. Let's say we locate a small fire in a sofa on the first floor. We think that is all there is to it, and many times it is. If presented with increased heat, more than should be detected for a couch fire, something more may be going on. There have been many occasions where there is a long delay in realizing that the simple couch on fire had ignited due to heat and fire travel from the basement level. Fire burns up and out in most instances. We need to check below us. We must be cognizant of this seemingly simple fire travel phenomenon. Make sure that basements are checked early in the operation.

Fig. 11–5. Outside team member checks basement windows for smoke or fire. (Courtesy of Nate Camfiord.)

Create a Plan

Basement checks are so critical that some departments have written SOPs/SOGs to mandate a check of them. The District of Columbia Fire Department is one such department that has a written policy on basement checks. Their policy is that in every building, regardless of height, construction, or reported floor location of fire, *every* basement must be checked. It is the explicit responsibility of the second arriving engine officer to check the basement and make a report to command with the findings.

If basement checks are as important to you as we feel that they should be, identify someone specific to check them. Write your departments SOP based on your resources to ensure that the basement is checked. It certainly does not have to be an officer who accomplishes this. In private dwelling fires within the city of New York, even with an officer on every apparatus, the incident commander (IC) routinely relies on reports that come back from the outside ventilation (OV) and roof firefighter for conditions that they cannot initially see. In responding to fires on the first floor or to the house with smoke coming from everywhere, I often ask the OV to quickly take a peek and report conditions found at the basement level. In Fairfax County, Virginia, the engine company officer when taking the 360-degree lap around the building will note and confirm basement findings to command.

With the knowledge that early, localized collapse of the first floor is increasingly common with today's basement fires, should we choose not to fight them once this has been verified and just back out and let it burn? Obviously, the answer is no. While basement fires are certainly dangerous, we can fight them head-on as long as we are prepared, practiced, and can anticipate what to expect. Recognition is the early key. Once we have determined and verified that this fire is in the basement, we can determine our operational actions.

When we have established that the fire is in the basement, it is just like a fire on any other floor. We need to confine, control, and extinguish it. Much like an established SOP/SOG that is written to assist us in recognizing a basement fire, your department should have a well thought out SOP/SOG to create a tactical plan in attacking them.

Before we mount an attack on the fire in the residential basement, we must make sure that we do our best to maintain the integrity and protection of the rest of the dwelling. This mainly comes from the control of the stairs, and perhaps more importantly, the control of the door at the top of the basement stairs. Once a basement fire has been recognized, our first thoughts must be to direct troops to first find and then confirm the status of this door.

Where are the basement stairs located in the majority of your department's dwellings? Only you can truly answer this question with any authority; that's why we recommend that you get out there and get into your buildings. We cannot obviously address the entire spectrum of buildings found around the country, but there are some particularly common places that they seem to appear.

One of the first places to check is directly behind the area under the stairs that lead to the second floor (fig. 11-6). It is a common practice for builders to stack stairs in the same position on each floor throughout the home to maximize space. Another commonality to the location of basement stairs is the proximity from the kitchen to the basement access. Also, rear and/or side entrances occasionally lead both into the first floor and split to provide access to the basement. The obvious Bilco style or walkout sliding glass door to a solid basement will lead you directly into it (fig. 11-7). Knowing your buildings will greatly increase your speed in locating the stairways and protecting the house and the members operating in it.

Fig. 11–6. Basement stairs are often located beneath the main stairs.

Fig. 11–7. Basement stairs may only be found on the exterior, sometimes shielded by Bilco style doors.

Ladder Company Operations at Basement Fires

Ladder companies normally precede the engine into the dwelling and locate the fire. This allows the engine company to stretch the initial hoseline to where it is needed instead of crawling blindly, dragging the hose behind them into the abyss. At the known basement fire, our first priority is to find the stairs to the basement, then notify the engine and command as to the status of that stair and the effectiveness of the door that is covering it (if applicable).

This transmission normally falls into one of two categories: the stairs are tenable or untenable. As the ladder company searches ahead of the line for the seat of the fire, the company officer has a decision to make. If conditions are untenable and fire has control of the basement, we must stay and hold the stairs. If conditions are favorable, with the line nearing our position, we may make access to the basement to further search for fire and victims. Let's break down this decision to stay or go further.

Fire at the top of the stairs

What if the searching ladder company finds the stairs before the engine moves in with the line and finds fire? How can we confine this? Door control is the first step to confinement. In certain situations, however, the homeowner may have removed this door for some reason. If the ladder company is searching ahead of the line while waiting for it to arrive and finds no door present with fire exhausting, what should they do? They should remove a nearby door and place it over the basement stair opening. Have a member use the 2½ gallon water can as needed to keep the fire at bay until the line arrives. At this point, the ladder company officer and IC must again take stock of their tactical options. Is there another basement opening?

Whoever finds the door to the basement stairs must announce the location and communicate the information to all members and the IC. If the ladder company first finds the basement steps, they must find the engine officer and ensure the accompanying hoseline is brought to that position. The location and status of the interior basement door must be announced early and often. If conditions at the door change, especially to the negative, all must be alerted immediately! If there is a breakdown in communication that starts at this point, we run the risk of losing the fire and injuring members in the process.

If an additional basement opening is known and the fire is making our basement access impossible from the interior, we must make an additional transmission to ensure a search of the basement. Once the initial engine company arrives at the top of the stairs, the inside team of the truck can move on. The initial ladder company now searches the first floor and all floors above for victims and possible fire spread.

If no additional access points for the basement are found, the ladder company moves behind the line to search as the engine extinguishes fire. Be aware of the possible use of a portable ladder to strengthen interior basement steps. Also, recognize that ventilation will be extremely limited and may need to be augmented by creating openings in the first floor (under windows).

No fire at the top of the stairs

If no fire is present at the top of the basement stairs and heat and smoke conditions are such that we can access the basement, what should we do? From the truck officer's perspective, there are a few options. Any and all perspective options have to be weighed against conditions found, then confirmed and verified with command prior to initiating them so that others know where you will be operating. Consultation and coordination with the engine officer and IC should take a part in your company's actions and decisions.

If you are going down the steps, *go down the steps*. The stairs are no place to stop and hang out. You are effectively in the chimney for the fire. Getting to the bottom of the steps quickly is important, so move with purpose.

The basement fire presents a myriad of challenges and has demonstrated it is one of the most dangerous kinds we encounter. The actions of the truck can certainly contain the fire along with providing essential information to the engine company, but no measure of spectacular truck work will extinguish the fire that is raging below the floor we enter. The engine company's mission does not change with the basement fire, but the manner in which it is completed must be rapid and coordinated. This rapid action allows the truck to complete their vital tasks and the engine company to extinguish a fire that is burning away the platform we are operating on. This coordination is essential to the isolation of the fire to the basement area of origin and the elimination of the spread of fire to upper floors, which can injure or kill our own personnel.

Engine Company Operations at Basement Fires

Through intensive and combat-ready training, the rapid deployment of hoselines is commonplace and one less obstacle the engine company, truck company, and incident commander have to consider on the fireground. Our job is to isolate and eliminate hazards and obstacles *before* the incident so we can focus on mitigating the issue that is the raging fire in the basement, so the next step in our engine company operations at the basement fire is coordination. This is not simply coordination between the engine company extinguishment efforts and truck company search and rescue efforts. It is the coordination of multiple hoselines from engine companies already operating and those that may be arriving later.

The goal we seek to achieve is isolation and extinguishment of the fire, and the coordination of this operation is similar to placing puzzle pieces correctly to achieve the finished product. Each hoseline is just one piece of the puzzle, and all companies operating must have clear and concise goals that are communicated before the incident. The front yard of a residential structure with fire issuing from three basement well windows and report of people trapped is not the correct place for the incident commander to explain a new tactic that involves coordination of four companies. The plans must be communicated and trained extensively, only to be executed when the correct situation is recognized.

The typical residential home in the United States with a basement may have no exterior access to the basement with only stairs on the interior of the home or may have various forms of exterior access (sliding glass door, wood frame door, Bilco style doors) along with the interior stairs. These two known characteristics set the tone for our pre-planned operations of hoseline deployment. First and foremost, the three options offered must be clearly described and mandated through standard operating procedures or guidelines. Secondly, proper recognition of the basement fire and structure access must be done, otherwise the tactics will not apply.

The Backstep

Basement Fire Engine Company Tactics

Paramount to initiating any of the tactics outlined in this section is the absolute need for the tactics to be plainly communicated to all operating units. Tactical plans for basement fires should be clearly spelled out and mandated in SOP, SOG, or operational manual format so that all incoming units have the same playbook. Much of our significant operational injuries and deaths at basement fires comes when the plan is not clearly stated and/or there is deviation from the initiated tactic. We must be all on the same page.

Option 1: Two-hoseline attack

After completion of the 360-degree check and the confirmation that this is a working basement fire with an exterior entrance, the two-hoseline attack should be completed (fig. 11–8). This demands two attack hoselines be deployed. Proper coordination of those hoselines in concert with the truck company is critical.

- The first hoseline is deployed to the exterior entrance. The hoseline and nozzle should be operational along with ensuring any forcible entry concerns through the exterior entrance are completed before confirmation that the hoseline is ready for advance on the fire.

- The second hoseline is deployed to the front door of the structure. Before entry, the hoseline and nozzle should be fully operational and any forcible entry concerns should already be addressed so access is not an issue. The second hoseline's primary job is to ensure no vertical fire spread via the interior stairs occurs and to protect the truck company completing a search. Depending upon the age of the structure, the intensity, location of the fire, and the location of the stairs, the second hoseline may either:

  — Stage at the threshold of the front door to protect the truck company that advanced to the upper floors via ground ladders or stairs. Additionally, this location may provide the best access to prevent vertical fire spread from the first floor to the second floor via the interior stairs.

  — Advance into the structure and stage at the top of the interior stairs leading to the basement. Recognition of floor conditions and the duration and intensity of the fire, among other concerns, should play into the decision of whether to advance to this position.

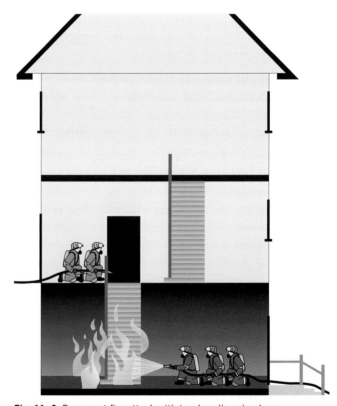

Fig. 11–8. Basement fire attack with two hoselines in place. (Illustration by Matthew Tamillow.)

Once both hoselines are confirmed in place and operating with no access issues, the incident commander should direct the hoseline at the exterior entrance to attack the fire. While the attack is occurring from the hoseline that is pushing to the seat of the fire at the exterior entrance, the hoseline at the top of the stairs or threshold of the entrance door should be utilizing a penciling technique to contain the fire to the basement while extinguishment is occurring.

Option 2: One-hoseline attack

In some situations the presence of an exterior entrance at the basement level may be not be present or accessible for fire department operations (fig. 11–9). In this situation, the engine company must implement the extremely dangerous tactic of advancing into the structure and down the interior stairs. If this tactic is employed the following should occur:

- The hoseline should be operational and charged to the appropriate pressure before entry.

- The location of the interior stairs must be known and communicated prior to entry. The truck company is most likely at the top of the stairs, clearing a path to their location, and communicating with the engine officer.

- Once the engine company reaches the top of the stairs, one length of hose is advanced into the area for movement down the steps. This must occur before the door is opened to the basement or an attempt is made to advance down the stairs. The stairs are a chimney and there must be an uninterrupted descent to the base. Not having a hoseline to make this advance can be fatal to companies operating in the basement or on the floor above.

- Once the door is opened to the basement stairs, confirmation of stairs that are stable is completed. This can be done with the leading foot of the nozzle firefighter, the extended tip of the hoseline and nozzle, or the thermal imaging camera.

- If the stairs are stable and present and enough hoseline is advanced into the room, the descent occurs. The nozzle firefighter advances swiftly down the stairs, extinguishing any fire along the path. At least one firefighter should be stationed in the room at the top of the stairs to ensure smooth flow of hose and to monitor conditions.

Option 3: No exterior basement entrance and interior stairs are untenable

- After the completion of the 360-degree check to confirm the presence of a working basement fire with no exterior basement entrance, and this information along with that from the truck company inside that the interiors stairs are untenable has been communicated, an exterior attack is appropriate.

- If a fire is well-advanced in the entire basement along with interior personnel in full personal protective equipment advising of untenable conditions, the survivability of civilians is diminished.

- Depending upon the type and design of the structure, well windows or escape windows may be present in the top portion of the basement foundation. Swiftly deploying handlines to these windows and quickly extinguishing the fire will drastically improve conditions in the basement. This may make it possible for hoselines to

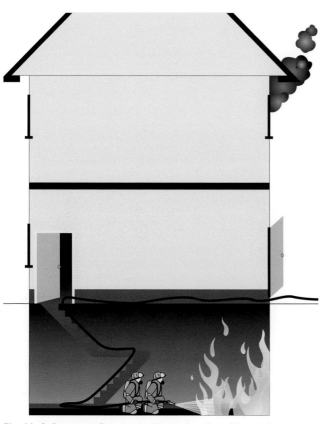

Fig. 11–9. Basement fire attack with one hoseline. (Illustration by Matthew Tamillow.)

redeploy to the interior to finish extinguishment and ensure no vertical extension.

- If no windows or access is available to the basement area, a cellar nozzle or Bresnan distributor may be used. Proper practice with these limited-use nozzles is essential as there can be no delay in deploying and operating them on a working basement fire.

- If either of the above tactics is used, the incident commander should ensure an attack hoseline is operational and stationed at the fire door. This action aids in preventing any vertical fire spread that may occur while the truck company is completing a search on upper floors after a VEIS tactic is employed.

Paramount to any of these tactics outlined above is the absolute need for them to be clearly communicated and mandated in SOPs/SOGs or operational manuals. There can be no deviation or combination of these attacks. On the working basement fire, the recognition of the fire should trigger the incident commander or unit officer to employ the correct tactic based upon the situation and/or conditions they witness during their 360-degree check. An additional benefit of these tactics is that, depending upon staffing levels, they can be implemented as staffing arrives.

If you are fortunate to work or volunteer in an area with the ability to deploy two handlines simultaneously, you are well prepared for this dangerous fire. If you have staffing shortages, then you must deploy the first attack line and ensure the next arriving personnel know where to deploy the next hoselines. In a short staffing situation with a two-hoseline attack, if the initial hoseline is charged and operational, the engine driver can deploy the second hoseline to the front door of the residential structure. In this situation, the next arriving personnel can immediately man the hoseline and prevent vertical fire spread.

The basement fire has demonstrated that it demands communication, coordination, speed, and accuracy in our operations for us to have a chance to defeat it. No other fire is attacking the platform on which we operate, seemingly waiting to trap us in its grip. To fight this enemy we must have a well-thought-out plan of attack that can not only recognize the fire in the basement but also employ the correct tactic. The foundation of success in the basement fire is in our training and coordination.

12 360-Degree Check

History Repeats Itself

"The building involved was a three-story, wood-frame structure with a basement and was constructed on a sloping grade that caused the building to have a different appearance depending on the side being viewed. Firefighters entering the building saw only one side and were not aware of the building's actual arrangement. The firefighters' distorted perception of the building may have impaired their ability to assess alternate escape routes."

This passage is the opening paragraph of the NFPA report on the fatal fire that occurred at Bricelyn Street in Pittsburgh, Pennsylvania on February 14, 1995, a fire that resulted in the deaths of Fire Captain Thomas Brooks and firefighters Patty Conroy and Mark Kolenda. When we are presented with information that specifically details factors contributing to the line-of-duty deaths, we owe it to the memory of those fallen to learn from them. We never want any of our fallen firefighters to die in vain with no change in our behavior or actions, as if we did not hear what they were trying to tell us. If we listen, specifically to the Bricelyn Street incident, we hear the importance of knowing what you are entering before you cross the threshold of the burning structure.

The 360-degree check performed by the initial arriving officer can provide valuable information that will complete the tactical decision-making equation. Regardless of whether you adopt the term "360-degree check" or call it a "lap" or "walkaround," we must take this step to ensure we are closing the loop on our tactical decisions. Specifically, where do we plan to deploy the most important resource we expel on the fireground: our personnel? Unfortunately, we have had many examples offered through line-of-duty death reports that cite lack of communication and tactical coordination. Many times, the information needed to avert these disasters could have been gained during a simple 360-degree check of the structure prior to entry. Pay homage to those lost in these fires and review the comprehensive reports compiled that offer insights into what we can adopt today in every department. Below is a short list of relevant line-of-duty death and close call reports:

- 3146 Cherry Road, N.E.: two firefighter LODDs
- 43238 Meadowood Court: seven firefighters critically injured
- 5708 Squirrelnest Lane: two firefighter LODDs

Each report details what appeared to be "routine" fires that ended in deaths or critical injuries to firefighters. While any fire we enter is inherently dangerous, we must eradicate complacency and develop a system that will offer us information to make sound decisions to extinguish the fire and save lives. Merely instituting a 360-degree check will not achieve this unless we first explain how and why we do this vitally important function.

"Engine 10 on the scene, side Alpha, single story rancher-style house, fire showing from side Alpha, one window on the basement level, quadrant Bravo, we have our own water."

The officer of Engine 10 has made his detailed initial on-scene report of the fire conditions, the driver is making his connection to the hydrant in front of the house, and firefighters have begun the deployment of a 250-foot crosslay to the front door. The conditions are all indicating that this is going to be a routine basement fire. The homeowner meets you in the front yard confirming the laundry room in the basement is on fire, adding to the belief that this will be a quick job. It should be simple. Enter the front door, down the steps, spray some water on the fire, and have it knocked down before the second engine even arrives. The officer jumps off the engine, quickly confirms from the homeowner that the structure is unoccupied, and joins his firefighters on the front step masking up to make entry.

Thick, black, pressurized smoke is pushing out the front door as they make their advance through the home to find the basement stairs. Moderate heat conditions are felt upon entry with a slight heat increase when the firefighters begin to crawl across the floor. After 5 minutes of stretching hoselines in blindness, yelling inaudible commands about needing more line, and building frustration, the engine company still can't find the steps. Just about the time the chief arrives on the scene he is greeted with a sudden fireball out the front door and calls of Mayday on the radio. The engine company has fallen through the floor and is trapped in the basement!

After the fire it is discovered that the basement was a series of illegal apartments or single room occupancies (SROs). The homeowner rented them out to illegal immigrants and sealed the basement door leading from his kitchen with a couple of deadbolt locks and padlocks to prevent them from entering his residence. To assist in dampening any sound migrating up from the basement, along with adding to the fire load, the homeowner haphazardly installed a layer of foam insulation between the basement rafters, too. The only entrance to the basement was the side Charlie door that led directly into the apartments and laundry room.

Could this disaster been avoided if the officer would have taken the extra time to perform a 360-degree check? At a minimum, he would have had the opportunity to size up his battleground and find the quickest way to the enemy. He would have noticed the single story home on side Alpha was actually two stories in the rear because it was built into a slight hill (fig. 12–1). He would have noticed a rear exterior entrance door with a single lock on it that would have allowed for unimpeded access directly to the seat of the fire. He would have been able to make sound tactical decisions based upon information presented at this specific fire and not assumed tactics based upon anecdotal fires. Experience allows us to build a plan, but actual information gathered allows us to execute the plan.

The burden of the firefighter to complete a 360-degree prior to entry is often complicated by the distractions presented as soon as they get off of the apparatus. We are sworn to protect the citizens in our community and have an overwhelming sense of duty to "get to work" when they greet us in the front yard (fig. 12–2).

Fig. 12–1. Topography can affect the levels of homes that are visible from the apparatus.

Fig. 12–2. Firefighters are greeted in the front yard of the house fire by the excited homeowners.

"In this incident, the first arriving engine crew was met by an excited homeowner who informed them that the fire was in the basement and directed them to the front door and down the stairs. Acting quickly, the engine crew prepared to enter the fire structure without conducting an initial size-up."

This is from the first recommendation cited in the July 29, 2009, NIOSH report on the line-of-duty death of Captain Broxterman and firefighter Schira in Colerain Township, Ohio. This fire initially was dispatched as an automatic fire alarm and was upgraded to a working fire. Upon being greeted in the front yard by the homeowner, Captain Broxterman and firefighter Schira entered the structure and never completed a 360-degree check. If a 360-degree had been completed, they would have found a basement level sliding glass door from which to mount an attack. This attack would have never placed them on the floor above the fire and possibly could have given them the opportunity to quickly knock the fire down. Ultimately, this is information that may have changed their tactics and the outcome of the incident.

The fire service sadly has many more examples that demonstrate the importance of completing the 360-degree check. Merely walking around a structure will not put the fire out, nor will it keep you out of harm's way. It is more important to understand why we take the 360-degree and what we are denoting. What will the information we gather offer that can define the tactics we plan to execute? If we do it correctly it will offer enough to make an informed fireground decision.

The Foundation for Success

The first step to executing an effective 360-degree check starts long before you ever run a working fire. It begins with knowing the area that you serve. Essentially this means knowing the battlefield on which we operate when we are called to war long before we ever engage in our fight against the fire. Once we do arrive at the scene, we want to scan the front yard to see if we can find any of the occupants. The intelligence information we receive from them can quickly supply vital information on the structure, the fire location, and the probability of victims still inside.

Quickly gather specific information from the occupant and make your lap around the structure. The direction you take does not matter unless you arrive with another unit, such as a truck company, then both officers can conduct a lap. In this case, each officer should initiate their lap on opposite sides of the structure and meet back on side Alpha to confirm observations and final tactics.

The Tactical Observation and Analysis

While your company is aggressively stretching a hoseline from the engine company, you should begin your process of information gathering. Regardless of the size of the structure or location, at a minimum, we should looking at the following items.

Topography

You may live in an area in which homes are routinely built into a hillside merely due to the topography of the land, such as in Pittsburgh and San Francisco. Two departments suffered line-of-duty deaths in which a contributing factor was the lack of being able to quickly identify the correct number of floors. If hillside construction is not a routine practice in your area, then you may soon be witnessing this new trend that is a reality of more people living in your community. With increased population, the need for housing increases and what was viewed previously as undesirable land before becomes more attractive (fig. 12–3).

Fig. 12–3. Topographical changes not only affect the rear of the structure, but also the front.

Regardless of the area, the need to identify any topography changes and associated changes in the stories is vital to making sound tactical decisions. While taking your lap, you must note:

- Visible and accessible changes in stories

- Any topography changes that may affect all four sides of the structure. For instance, is there a basement level door on side Delta but no windows or doors on side Bravo due to topography? Do you have a consistent and widely accepted term for the below grade levels (cellar, basement, etc.) so there is no confusion on the fireground (fig. 12–4)?

Fig. 12–4. Fire conditions from the rear of a residential structure. (Courtesy of Kentland Volunteer Fire Department, www.kentland33.com.)

- Conditions present from the additional stories visible and accessible. The location of the fire always has a dramatic impact on the tactics we employ at fires.

- Access concerns for apparatus and personnel due to the topography change. For instance, will an engine be able to stretch a line around to side Charlie due to the steep slope of the rear yard, or will the area be inaccessible for portable ladders?

Victims

Our first and foremost priority at every fire is to save lives. Every action we take is predicated on the successful completion of that mission. When we take the lap, we are scanning multiple areas of the structure and yard simultaneously. Scan every window of the home and ensure that there are no victims present.

- Consider the environmental conditions when you scan the windows. It is winter and 20°F outside, yet why is the top floor window open and issuing smoke? Did someone consider jumping and then fell unconscious, lying below the sill? Don't confuse the window that failed due to fire conditions with the window left open by an occupant.

- Take note of access to potential victims during the lap. Is the entire second floor of the home, which is most likely the bedrooms, accessible by a large porch roof? This is essential information to pass on to the incident commander and truck company. A porch roof can serve as an excellent platform for truck companies executing a VEIS technique.

- While the building on fire will attract your attention, you must also scan the yard. Fire and the by-products of combustion can cause the human mind to make split-second decisions that are unreasonable. Victims may choose to leap from a second- or third-story window instead of going back through the house on fire. When they make their leap, they may end up in bushes or shrubs near the windows, obscuring them from plain sight (fig. 12–5). If they do survive the fall, they also may walk a short distance from the area and collapse, so make sure you question anyone you find conscious. If a victim is unconscious, make notifications and request the appropriate resources, then finish your lap.

Fig. 12–5. Shrubs and bushes at the base of windows can hide victims who may have jumped from upper level windows.

Building status

We must be the resident experts on the structure that we plan to enter and risk our lives in protecting. First and foremost, we must identify the points of access and egress and the status of each.

- Windows—are they accessible? Are they casement with limited entry/access (fig. 12–6)?

Fig. 12–6. Windows seen on the Charlie side of a residential structure

Fig. 12–7. Bars on windows must be announced and removed.

- Denote the types and sizes of the windows when you view them. A bathroom window is smaller than a bedroom or living space window. A kitchen window that is located above the sink will be placed at a height above the counter.

- Are there objects that will impede your ability to gain access from the outside or limit your egress from the inside? Look for bars on the windows and how they are fastened, is the outside vent man able to quickly remove them with a Halligan or will it take a company with a saw (fig. 12–7)? Are there air conditioners in the windows (fig. 12–8)? Communicate these findings to the personnel going into the structure.

- Doors
  - Where are they? At fire level or below grade?
  - Easy to access or forcible entry concern?
  - Typical residential door or sliding glass that would introduce much more oxygen?

Fig. 12–8. A/C units can limit access and egress and should be removed.

Environmental conditions

Through the outstanding work of NIST, along with other partners in New York City and Chicago, the fire service was presented with tangible and accurate information confirming that wind-driven fires are dangerous. Unfortunately, we took the research to only apply to fires in high-rise structures. When Technician Kyle Wilson died on April 16, 2007, in Prince William County, Virginia while fighting a fire in a residential structure, we learned that wind-driven fires occur in the residential structure as well. No longer was this issue reserved for high-rise fires in big cities but it is also an issue in rural counties. High winds were pivotal in the instant promulgation of fire within the home that rapidly changed the incident from routine to tragic. Kyle's death is not the only incident in which we have seen such tragedy. A review of the historical line-of-duty deaths in the wildland interface will demonstrate wind is a major contributing factor. While we are not fighting wildland fires within the residential structure, the same wind that propels these fires fuels the fire in the residential structure through the vented window or other openings. Wind is just one of the environmental conditions we must contend with and note on our lap.

Ice and snow are another real concern we must note. For our brothers and sisters in the South, this is the white stuff that falls from the sky when it is really cold. For the rest of the country, snow and ice can be more than just an inconvenience. The presence of snow or ice can delay or prevent an uninterrupted lap of the structure. It can cover hazards such as pools, wells, and other obstacles. It can delay the completion of normal firefighting tasks on the fireground such as portable ladder placement, horizontal and vertical ventilation, and handline deployment. When you discover this environmental hurdle, communicate how it will affect the incident and adjust your tactics to reflect this issue. For instance, if it will take longer to get your 1¾-inch hoseline stretched to the front door due to 20 inches of snow, make sure the truck company is aware. The truck's normal operation may be to force entry and go find the fire, operating under the assumption that the engine company will be on their heels with the hoseline. Since forcing entry is ventilating an under-ventilated structure, we can expect a rapid increase in the volume of fire. This is not a positive sign for the truck searching without the protection of a hoseline due to a delayed hose team.

Fire location and conditions

We have learned from the innovative research conducted by NIST, UL, and other laboratories that fires have an increased heat release rate and reach flashover more quickly today. Thus, our need to correctly and quickly identify fire location, size, and intensity is paramount to making sound fireground decisions. Answering the following questions can help provide this information.

- Do conditions match on all four sides of the structure (fig. 12–9)? If we only have smoke issuing from the eaves of the roofline on side Alpha of the structure, we may assume we have an attic fire. Upon taking a lap, you find smoke pushing under pressure from the rear basement with an orange glow in the basement well windows. Conditions do not match, and we must adjust to the fire we now have in front of us. The strategies and tactics for an attic fire are vastly different than those employed at a basement fire with an exterior rear entrance. Being able to quickly diagnosis if the fire is above or below where we will be entering the structure is essential to our safety and survival.

Fig. 12–9. Conditions can be different from one side of a structure to the next. (Courtesy of Nate Camfiord.)

- Is only smoke showing? What is it doing? We must treat fire like a living and breathing animal that is on the hunt for oxygen. The fire issuing lazy smoke is not starving for oxygen (fig. 12–10). Conversely, the fire that is not showing any flames, pushing thick, acrid, billowing black smoke from every portion of the home, is deep-seated and on the hunt for more oxygen. Understand the difference and what the smoke is attempting to tell you. This decision-making process must be rapid. It is not an excuse for grabbing a chair and partaking in "observation" of smoke while the fire consumes more of the home. We discuss the behavior of fire in residential structures in chapter 13.

Fig. 12–10. Smoke must be interpreted before we enter.

- Is fire showing? Where and how many windows? Fire that is issuing from a window or door is merely confirming that all of the ingredients for the fire have come together at that singular point. The fire has gathered enough oxygen and is burning enough fuel to produce flames. You must note the intensity of the flames to help understand the concentration of the fuel load that is burning. The fire that is able to produce flames shooting out of an entire window and extending 10 feet up the structure is either aided by wind and/or ventilation or has a substantial fuel load powering the flame production.

While flames issuing out of a window are a great indicator of where we need to direct our initial hoseline and concentrate our search, we must also be mindful of where it is going. Is the window next to the fire in the same room or does a wall separate it? We need to diagnose where the fire will be traveling or has already traveled.

Hazards

According to the Humane Society of the United States, there are approximately 78.2 million pets owned in the United States. This statistic outlines just one of the many hazards we must be vigilant about when we take our lap. Not only does the presence of a large and mostly likely angry dog present a hazard to our operation on the fireground, but also it is likely the animal is contained in the yard by a fence (fig. 12–11). These two hazards hinder line advancement, portable ladder placement, and entry to the structure.

Fig. 12–11. A big dog can hinder our ability to complete a lap prior to entry.

Along with the hazards associated with pets, when you take your lap denote and communicate any of the following hazards:

- Imminent or confirmed collapse of the structure. You issue a clear and concise on-scene report from side Alpha of the structure upon your arrival. This may be based upon the volume of smoke you can see issuing over the house. As you approach side Charlie of the structure, you find that the smoke you saw from side Alpha was from the recently collapsed two-story porch attached to the rear of the structure.

- Swimming pools. While it would seem obvious that you should be able to avoid a large in-ground swimming pool, our attention is focused elsewhere. The simple lack of situational awareness as you pass side Bravo of the house viewing side Charlie could end in a splash. A swimming pool covered for the winter, a poorly lit backyard, and the possibility of getting entangled in the cover coupled with the water saturating your gear could spell your demise. Additionally, the large pool is likely in close proximity to the rear of home, which now eliminates valuable ground for properly positioned ladders to be thrown. Communicate and mark off the pool in the rear yard so personnel are aware and tactics can be adjusted as needed.

- Downed power lines. According NIOSH, between 1998 and 1999 four firefighters were killed due to electrocution, underscoring the importance of recognizing and respecting these hazards. The fire that is auto-exposing from the two rear windows could have easily burned through the residential drop line that carried power to the home. In accordance with Murphy's Law, they will be energized power lines that have dropped directly in the path of your lap of the structure. A lack of awareness coupled with being transfixed on the venting fire places us in the direct line of becoming another grim statistic. As you complete the lap, constantly scan the fire, the structure, the yard, and the path you plan to take for these hazards. If there is an electrical hazard, communicate the hazard and direct personnel to isolate the area. It is a simple task for the RIT, the outside team of the truck, or an EMS crew to isolate the area with clearly visible fire line tape. Additionally, ensure the truck is aware of the exact location of the electrical hazard. In the days of budget-restricted fire departments and the subsequent reduced staffing, it is very easy for a firefighter committed to getting portable ladders in place to come in contact with power lines with the ladder.

- Exposures. As discussed in chapter 7, the trend in building construction is to build larger residential structures on less land. This disturbing trend equates to the possibility of exposures becoming a hazard in completing our lap, aiding fire travel, and completing operational tactics. As pictured in figure 12–12, the placement of a ground ladder on side Bravo of this structure would be difficult, to say the least. Couple this with fire venting out of a window on side Bravo and we now have not only a room on fire in one structure but fire travelling to another structure and the inability to get ground ladders in place.

Fig. 12–12. Modern construction can place houses less than three feet apart, substantially increasing fire extension and decreasing our access.

Resources

A 1- or 2-minute brisk walk by the officer of the first arriving apparatus can provide the firefighters entering this deadly environment with information to successfully and safely complete their job. This miniscule amount of time expended taking the lap is the time your crew is advancing the hoseline and performing forcible entry. The behavior is ingrained through company training and strict policy, eliminating the possibility of the free-lancing firefighter entering the structure without you.

Information Gathering that Could Save Your Life

An argument against taking a lap is that the fire will build too quickly while the officer is "wasting time" running around the house. Once we understand fire behavior and that the majority of fires we face today are under-ventilated (until we get there, of course), we will understand that forcing the door and barreling in with no hoseline will do more harm than good. The fire that was contained to the kitchen is now consuming the hallway and rushing up the stairwell to the bedrooms, trapping the truck company and lessening the survivability of the occupants.

The bottom line is that we can perform these tasks of taking a lap, selecting and stretching hoselines, forcing entry, and conducting the primary search simultaneously if we train properly and eliminate complacency. Train each and every day to set the benchmark for how your fireground will run. Our actions and eventual selection of tactics should never be arbitrary; rather they should be founded in solid decision making based on experience and the information gathered at the fireground. The experience level may not only be fire duty, but also what you have gained from performing consistent and relevant training. One of the first steps in that training process is the completion of the 360-degree check.

Transmitting the information

When the assigned member completes and processes the information noted in the 360-degree check, any pertinent information must be transmitted to the IC and all operating forces. This should come in a radio transmission known as the "situation report." This radio report should not be confused with the initial on-scene report. The on-scene report sets the initial tone for what was noted as the members arrived. The situation report will solidify that initial information and/or make companies aware of any other potential problems or concerns that may affect the operation. The situation report is a radio report announcing the result of information gathered on the 360-degree walkabout.

The Backstep

The 360-Degree Check

Prior to conducting your 360-degree check, deliver a clear and concise on-scene report.

What to look for while conducting your 360-degree check:

- **Topography.** How many stories are in front and in the rear?

- **Victims.** Look for jumpers in the bushes or yard and identify VEIS locations and open windows.

- **Building status.** What size and where are the windows? Are there doors on alternate sides and levels? Any forcible entry concern? Can you quickly control the utilities? Are there obstacles to ingress/egress such as AC units, bars, etc.?

- **Environmental conditions.** Wind-driven fires do not occur only in high-rises. Be cognizant and adjust tactics as necessary. Subfreezing temperatures = frozen plugs and hoselines. High heat days = quicker time to fatigue for firefighters.

- **Fire conditions and location.** What is on fire? Where is it? How bad is it? Where is it going?

- **Hazards.** Be aware of any angry dogs, big fences, hidden swimming pools, or downed power lines.

- **Exposures.** Is there an exposure problem? How big or small is it? Is it occupied or vacant?

- **Resources.** You should know now whether what you have coming will take care of this fire or if you will need to get some additional equipment and firefighters to help out. Call now because the fire is not going to wait for you!

Once the 360-degree check is completed, transmit the information in the form of a situation report to the operating forces. Recommend any tactical changes to the IC.

13. Fire Behavior: Taking the Science to the Street

It's a living thing, Brian. It breathes, it eats, and it hates. The only way to beat it is to think like it. To know that this flame will spread this way across the door and up across the ceiling, not because of the physics of flammable liquids, but because it wants to. Some guys on this job, the fire owns them, makes 'em fight it on its level, but the only way to truly kill it is to love it a little.

—Donald Rimgale (Robert DeNiro), Backdraft, 1991

The movie *Backdraft* brought much deserved positive attention to the dangers of our occupation but, at its core, it was just another action-packed motion picture. Yet, Robert DeNiro's character shares a profound thought about fire with William Baldwin when explaining fire behavior. A thought that apparently Hollywood did not miss, but many firefighters who actually enter the firefight do not understand or appreciate. Fire is living and breathing, always looking for air like an oxygen-starved diver surfacing in the water. It's always looking for fuel like a man starved of food for weeks. Lastly, willing to dispense hate to anyone or anything in its path regardless of color, creed, religion, or affiliation like a dictator exercising iron-fisted rule on his path to power.

All of these attributes of fire are what make our job dangerous and rewarding at the same time. The sense of achievement at quickly extinguishing a fire or saving a civilian's property and possibly a life is paired with suffering the loss of one of our own in the line of duty. This relationship with fire demands, as Rimgale states, that we must "love it a little" to understand it and be able to extinguish it, thereby winning the firefight.

The first step to loving fire a little bit is to truly understand what fire is and why it acts the way it does (fig. 13-1). Due to the absence of firefighter re-certification training like we have with many other disciplines, the basics of fire behavior can become lost and undervalued. Yet, this information is paramount to understanding what fire will do when we arrive at the home on fire at 0200 hours with people trapped.

So, before we get into specific strategies and tactics of residential firefighting, let's go back to school.

Fig. 13–1. Firefighters routinely show their "love" for fire.

Back to Firefighter I

The fire triangle is the image that most of us have stamped into our firefighting consciousness as the ingredients necessary for fire to occur (fig. 13-2). The joining of forces of oxygen, fuel, and heat come together and create the epidemic that plagues the modern fire service each and every year. These three components must be present for fire to occur and each component must be of the appropriate proportion and quantity for the fire to continue. Given the fuel load in a house and the fact that the fire has started, the supply of oxygen is the linchpin that must be present in the right concentration for fire to grow. Heat is needed to bring the material up to a temperature that fire can occur. Once achieved, it is supplied by the oxygen source, giving it life. Lastly is the addition of fuel that was pivotal in the initiation of fire as it was the material first subjected to heat. Once the fire has been initiated, the level of fuel will decide the duration and, to some extent, the intensity of the fire. A fire that has an abundance of oxygen and fuel will burn until one of these key sources to fire's survival is removed, often by firefighters.

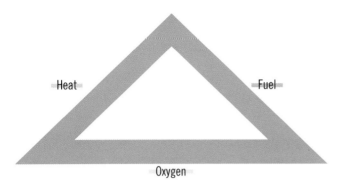

Fig. 13–2. The fire triangle. (Illustration by Matthew Tamillow.)

The fire tetrahedron welcomed a fourth member into the family with the addition of a chemical chain reaction (fig. 13–3). NFPA offered a very simplistic definition to this added component in a document titled, "A reporter's guide to fire and the NFPA," when it stated, "This chain reaction is the feedback of heat to the fuel to produce the gaseous fuel burned in the flame. In other words, the chain reaction provides the heat and gaseous fuel generation necessary to maintain the fire." This exothermic reaction, which means it produces light and heat, is sensationalized by the many materials contained in the residential structure that produce a wide array colorful flames along with rapid heat increase.

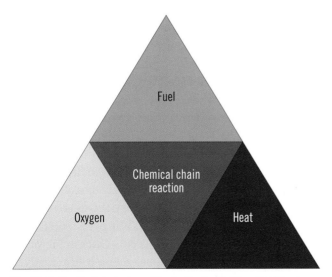

Fig. 13–3. The fire tetrahedron. (Illustration by Matthew Tamillow.)

Understanding the basic elements of how fire is created and sustained allows us to understand how fire will act. Under most circumstances, we can formulate a plan for how the fire will travel, but many times we are not privy to all of the information needed to make an educated guess. If fire would just wear a helmet cam, we could look to see where it is looming and quickly extinguish it, but fire is not so giving. Many times we introduce the final element that allows the fire to come to life. This trigger can range from simply fire projecting out of a window to some of the more dramatic fire events witnessed. An example of this phenomenon is offered in the incident below that was ultimately characterized as a low order CO explosion or cockloft explosion.

Engine 10 responds for a motorcycle on fire in the backyard of a residential dwelling. Upon arrival they find not only a motorcycle on fire but also that the motorcycle was next to the rear of the structure. Fire has spread up the side of the house into the soffit and the attic space. The residential home is your typical modern 2,000-square-foot home with lightweight wood components built with an ornate roof style that is aesthetically pleasing to eye from the curb but is nothing more than a concealed lumberyard.

The engine company is working diligently to put a knock on the fire on the exterior of the home while smoke is aggressively pushing out the gable end vent on the opposite side of the home (fig. 13–4). They knock the motorcycle and obvious fire down on the rear of the home and quickly move the handline back to the front of the house. They work to advance in the front door and up to the second floor to engage what now looks like an attic fire. Additional resources were requested but were not there in time to assist in this operation. The company works the line up to the second-floor landing over top of the grand open foyer. The officer looks back at his pipeman to ensure he is ready and then vigorously jams his tool into the ceiling to puncture a hole into the attic space for the nozzle to operate. BOOM! That is the last memory he had. Moments later he was in the front yard, surrounded by firefighters and paramedics trying to secure him to a backboard. His crew was there too, in various states of injury and disbelief. One firefighter was convinced he had an object impaled in his chest. Another was listless and had a look of shock on his face.

Fig. 13–4. Involved residential structure just prior to a low-order explosion

Unfortunately, the above incident is an actual event I was part of and the question you may have now is the question I had then. What had just happened? This incident was my department's first introduction to the low order CO explosion or cockloft explosion. Urban areas have historically experienced these types of fire phenomena in row homes with common cocklofts, but this was a first in a detached residential home for us.

The CO explosion had dislodged the roof and caused every room in the home to be engulfed in a fireball. The latter fact was only identified when it was observed that the curtains and any paintings in the rooms were melted due to the rapid nature of the heat. The paint on the walls looked almost undisturbed aside from the staining where wall studs were placed. The explosion had essentially seared the house like a steak placed on a 500°F grill, immediately imprinting its effect on the home.

Did this occur because of a new phenomenon in fire that would need to be vetted and researched through the fire service's elite scientific resources? No. It occurred because we did not have all of the information about the fire and where it was going. All we did was introduce the final element it needed to complete the fire triangle: oxygen. The small inspection hole coupled with the properly chocked front door created the oxygen flow that the starved fire was looking for. It had the heat from the ignited fire outside, which was now inside, and the fuel load of the lumberyard attic, but it was starved of oxygen.

The Impact of Oxygen

Fire behavior begins and ends with understanding the fire triangle and the effect that the building and firefighters have on it. "Coordinated ventilation" is a mantra that is tossed around the fire service with little emphasis on defining what and how (fig. 13-5). Rather, some would believe that just uttering it on the fireground will magically make it occur. The CO explosion incident was due to ventilation and we find today that every fire's outcome is impacted by ventilation and extinguishment, both positively and negatively. The fire service understands this demand for comprehension as our predecessors relied upon it to even make search, rescue, and extinguishment an option.

Fig. 13–5. Firefighters breaking out windows of a residential structure. (Courtesy of Nate Camfiord.)

Crawling into a flaming residential structure with only three-quarter boots and no SCBA, the perfect execution of tactical ventilation was essential (fig. 13–6). Allowing an avenue for the fire to ventilate away from the entrance they were crawling in allowed for survivability of firefighters and any trapped civilian. Since those days of limited personal protective gear, our fires have adopted a more vigorous progression to fully involved due to the contents of the structure and the design and components of the structure. The first step was to learn of the formation of the fire, but the next step is to define and understand the stages that lead to the progression of a fully involved residential structure.

Fig. 13–6. Firefighter with three-quarter boots. (Courtesy of Kentland Volunteer Fire Department, www.kentland33.com.)

"The fire was a routine, 'bread and butter' job but then it was like the fire became a blowtorch." How many times have you read, listened to a fellow firefighter, or viewed a video about a fire in which some variation of the above statement was made? It appears more and more common in today's fires and our responsibility is to understand why it is occurring. What is happening and why is it happening? All valid questions that can be answered with understanding the next step in fire behavior: the phases of fire.

The Phases of Fire

The tones ring in the firehouse for the reported kitchen on fire, your crew quickly dons their gear, mounts the apparatus, and heads out the door. Arriving on the scene you are greeted with smoking issuing from around a window on the first floor and the homeowner out front stating the kitchen is on fire. The engine is stretching their hoseline across the yard to the front door while you are preparing to force the locked front door. The door quickly pops opens and you control the door while awaiting the hoseline crew to get ready. A quick nod from the engine company officer that they are ready and you swing open the door. Before cresting the threshold and beginning your search for occupants and the seat of the fire, you stop and observe the fire conditions. Why? And what should it look like (fig. 13–7)?

Fig. 13–7. Firefighters dressing and preparing to enter. (Courtesy of Kentland Volunteer Fire Department, www.kentland33.com.)

We anticipate that the top half of the now open door will have billowing smoke pushing out while the lower portion of the door will be relatively clear. You hope it will be clear enough that you can see the seat of the fire, any obstacles, persons, and the room layout. This perfect scenario is occurring because the fire is entraining fresh air from the lower portion of the door we just opened, presenting the clear conditions. The top portion of the door is exhausting the products of the fire. Given the two-layer environment, hot gases at the top and cool, clean air below, we can see across the room. Unfortunately, the acrid, pressurized smoke pushing out the door is not just merely a by-product. It is a toxic mixture we must understand and respect or we will quickly suffer the negative effects of this oversight. This mixture is unburned fuel from the working fire. It is composed of flammable gases, and combustible droplets and particulate eager to ignite. The stratification and layering of these by-products are not merely impressive to view but also the initial phase of fires that occur in the residential structure that we must be aware of.

Rollover or flameover is a common phase of fire we see at fires in that it occurs at a higher frequency than other phases such as backdrafts and flashovers. Technically, it is defined by the NFPA as, "The condition where unburned fuel (pyrolysate) from the originating fire has accumulated in the ceiling layer to a sufficient concentration (i.e., at or above the lower flammable limit) that it ignites and burns; can occur without ignition and prior to the ignition of other fuels separate from the origin."

What does this mean to the firefighter in the residential structure? It means that we must understand that when we make that aggressive push with the hoseline to get to the seat of the fire, or initiate a search without a hoseline, we are operating under an explosive cloud of gases. The false sense of security we get from seeing where the "real fire" is and becoming complacent because it is just smoke above us is dangerous. A scan of YouTube will reveal many fireground videos in which you will see footage shot from the exterior of a burning structure where crews are operating inside and smoke is eagerly pushing out above them. The smoke will increase in intensity followed by small fingers of flames shooting out. Within seconds, the window frame or front door will be consumed with fire angrily projecting out followed by retreating firefighters.

The viewer is witnessing the progression of the fire to rollover, the spontaneous ignition of the toxic and flammable gases contained in the floating cloud above the firefighters. What makes the rollover dangerous is not just the increase in flame and heat production but the ability of the fire to hide (fig. 13–8). While personnel on the exterior of the structure may be able to easily see the buildup from smoke to fingers of flame to full involvement, firefighters inside cannot. When we are crawling in a low position performing a search or advancing a hoseline, we witness only turbulent and swirling smoke whisking in front of our masks. The fingers of flames are cloaked by the smoke, providing a false sense of security to advance further, searching with no hoseline or refraining from opening the nozzle until you get to the seat. Couple this with the protective ensemble we encapsulate ourselves in and you can understand why we advance further into the fire thinking we are safe.

Fig. 13–8. Fire routinely "hides" in the smoke billowing overhead of advancing firefighters. (Courtesy of Kentland Volunteer Fire Department, www.kentland33.com.)

The adage that we do not "pass fire" is maintained for a reason. It is dangerous to operate under a fire that is raging above you. Fortunately, we have tools, such as the thermal imager, that allow us to identify and extinguish this phase. If we are without these tools, we must be aware that this phase of fire can and will occur at the residential structure.

When rollover is witnessed from outside with a sudden explosion of fire from the top half of the opening that our personnel just entered or from a window on the upper floor, it is usually followed by shrieks of, "Flashover!" Unfortunately, while this distress signal will evoke immediate action from personnel on the fireground, it is incorrect. Rollover or flameover can be the precursor to flashover and while some may say the variation in definition is splitting hairs, the personnel inside that encountered the conditions would disagree.

The technical definition of flashover, according to Dan Madrzykowski of NIST is, "Transition phase in the development of a contained fire in which surfaces exposed to the thermal radiation, from fire gases in excess of 1,100°F, reach ignition temperature more or less simultaneously and fire spreads rapidly through the compartment. In other words, the thermal conditions within the fire compartment transition from a burning hot gas layer above a cooler air layer to burning throughout the compartment, floor to ceiling." In the simplest terms, this means everything, *everything*, ignites in fire at temperatures that will soar past 1,000°F.

Furthermore, when flashover occurs, we and any occupants trapped in the room will not survive. With exception of a firefighter being inches from an exit or window in full personal protective equipment and full of luck, the sudden influx of heat is not compatible with life. Flashover is not to be confused with a room that is completely on fire: this will occur at any fire that has an adequate fuel and oxygen source and no positive intervention from the fire service. The difference is in the one simple yet powerful word, *simultaneously*, meaning that radiation from the rollover causes the unburned contents and furnishings to ignite rapidly, causing the heat release rate and the amount of combustion products to increase.

Building upon our list of terms specific to fire behavior, the last we cover is the most powerful and least frequent: backdraft. A backdraft is an explosion that can have catastrophic effect on the fireground. When we encounter fires in our daily service, we typically arrive when the fire is in the growth stage. This occurs for the obvious reason that in this stage the fire is most visible, which then results in activation of alarm systems and multiple calls to 9-1-1. The fire service's rapid response allows us to usually reach the residential structure within minutes while it is most likely still in the growth stage. Since the fire is in the growth stage it is less frequent that we will encounter a backdraft, due to the conditions that must be present for it to occur.

A backdraft occurs during the decay stage of fire in which the fire has consumed all of the available oxygen within the room. The fuel and heat are still present and all the fire will need is a burning ember to sustain the fire. As we discussed in the beginning of this chapter, the fire may be pulsating and swirling in the direction of the smoke, much like breathing. This may be undetectable to us if we are entering a zero visibility environment.

Fortunately, the action of ventilation that we take at most fires prevents the occurrence of backdraft; unfortunately it can also cause this deadly event. The keys to remember when discussing backdraft is that it is an explosion, not just a rapid progression of fire like a flashover. Secondly, backdraft is missing only one element—oxygen—so we must be diligent in recognizing the signs of backdraft and how we introduce oxygen (such as ventilation). Recognizing a fire that is in the decay stage paired with the unusual actions of the smoke can all be signs that backdraft conditions are evident. It is important to remember that while it may seem the separation is trivial, the explosion is caused by oxygen being introduced and not an ignition of smoke like in rollover or flameover. To be a professional at this occupation, regardless of whether you receive a paycheck or not, is to understand that you must be a mix of a structural engineer, hydraulic engineer, and a fire protection engineer amongst other skills. Fire will not wait for you to recognize the need for professional development, but it will definitely motivate you to learn.

Flameover/rollover, flashover, and backdraft are terms that have been in the fire service jargon for years yet there is a need to revisit and re-educate ourselves on these terms. Fires in today's residential structures are vastly different from when these terms were initially defined. All of these terms are related to the growth of fire in the structure, some more rapid than others. We must understand that fires are growing more rapidly in the residential structure now than in the past. Simply offering that statement is not a solution or an education to why or what is causing this change.

Four major issues affect the change we see in the rapid progression of fire in the residential structure: construction methods, fuel composition, firefighter actions, and science. Each has a significant impact and, ultimately, causes the same end result: the propagation of fire in the residential structure. We spend a little bit of time on each issue and start first with construction methods and practices in the residential structure and their affect on fire behavior. Many of our peers have penned journals, books, and manuals on building construction and it would be redundant to repeat their work here. What we can offer are the big offenders when it comes fire behavior and, more specifically, fire travel in the residential home.

Construction Methods

The advent of the modern home brought about changes in the configuration of the "typical" floorplan to become more spacious, achieve Feng Shui, or just to be artistic. No change has aided fire more in its mission to spread rapidly then the open stair (fig. 13–9). The stairwell, which channels heat, smoke, and fire throughout a residential structure like a chimney, was initially fully enclosed with a door in a remote portion of the home. Now a staircase can become a central focal point of the home with wide-open elaborate space and no enclosure to enhance an open floor plan. While this may be appealing to the eye and the social engagement aspect, it allows fire to travel unimpeded from the bottom of a home to the top with little or no resistance. In the homes with the enclosure aspect being limited to vertical walls, it does not limit the fire spread but rather channels it at a higher velocity.

As we denoted in the after action report by NIST on the Cherry Road incident in Washington, D.C., on May 30, 1999, the fire raced up the basement stairwell to the firefighters at 18 mph! Those fallen firefighters in full protective gear did not make it and neither will any citizen. When we are operating in a residential structure we must be cognizant of the location and type of stairwell. There are several tactics we can employ to aid us once the fire does reach the stairwell.

- If operating above the fire floor, all windows must be ground laddered. If the conditions warrant and the command is given, completely remove the entire window, making the window into a door. This may be the only means of access for the truck performing the search.

- A common theme throughout this book is the paramount need for coordinated ventilation and consideration for the actions that make that possible. The need for coordination is vital since when we break a window we can be completing the flow path for the fire that is lying dormant. This simple oversight can take a small fire on the first floor and instantly charge it up the stairs and out the second-floor window we just removed. A fire will race up this "chimney," dispensing havoc on anything in its way. Fortunately, we can offer a measure of control to hamper this rapid fire spread and it is as simple as closing the door. When the decision is made to break a window and enter a structure to perform VEIS or a truck is searching ahead of an engine company, the door to the room must be controlled. Notice "closed" was not used but "controlled" since we do not want to arbitrarily close every door and run the risk of it locking behind us. Control the door means to verify if the door will lock behind you and then proceed with either closing it if it does not lock or moving it to a closed position with the ability to stop it from closing and locking (fig. 13–10).

Fig. 13–9. Open staircase

Fig. 13–10. A Halligan can be used to control a door that could potentially lock behind you.

your advance the entire length of the stairs and make your move. Aside from the fact this is the fire channel and you run the risk of being incinerated, this is also the means of access and egress. If the truck is bringing a victim down, you are hindering the citizen's survivability. If one of your personnel is being chased by fire on that upper or lower floor and needs to get to the stairs to get out, he or she will not stop for you in the stairwell.

The second facet of building construction that has an impact on fire behavior and travel is the age of the home. Due to renovations and modifications, not every home is still the same building construction type as when it was built. However, it does hold true most of the time. In the simplest terms, the residential homes in which we fight fire can be characterized as old or new. The older construction homes would be balloon frame construction homes, wood frame construction built with larger diameter wood such as 2×6 and 2×12 lumber (fig. 13–11). The new construction would be the homes built today with lightweight wood components, cheap fasteners, and engineered wood that is comprised of wood chips and glue.

- Immediately identify the location of the stairs and notify the engine company. Whether "holding the stairs" is done with a door removed from another room, a 2½ gallon extinguisher, or a hoseline, it must be done to keep the fire in check and/or extinguished. A special note about the 2½ gallon extinguisher that all firefighters must be aware of: the current fire load of a residential structure today may quickly overwhelm the capabilities of a water can. There is a fine line between becoming complacent and relying upon 2½ gallons of water to suppress a well-advanced fire versus not even bringing the water can to the fire because you believe it is useless when it could, under the right conditions, provide knockdown of a small fire. Understand modern fire behavior and the conditions present, and apply the right tool!

- Do not operate on the stairwell! If on the truck or rescue, either get up or get down. The stairs are not the place to stage. If on the engine, make sure you have enough hoseline to make

Fig. 13–11. Old style home

The older homes are not resistant to fire, yet many firefighters hold them in that regard compared to the volume of fire and damage we see in new homes. The older homes present two dynamics that aid in fire travel that we must be aware of and deal with tactically. The presence of balloon frame construction, with unimpeded access from the basement to attic, is a serious fire travel issue. This superhighway with no tolls is a fire's dream. With little to no fire stopping, a small fire in the basement can reach a jackpot of fuel load (walls and attic) in seconds. All the while it can mask itself behind the walls, only showing once it has a firm grasp on the structure.

The presence of knee walls in the structure can serve as the melting pot for the fire that was small in the basement (fig. 13–12). In older homes, along with newer structures, the knee walls can provide large void spaces that children love to play hide and seek in but are usually chock full of belongings and provide limited access even in perfect visibility.

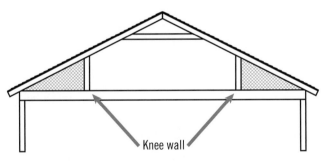

Fig. 13–12. Knee walls. (Illustration by Matthew Tamillow.)

Fuel Composition

We are not talking about unleaded or diesel fuels or even hybrids. What we are talking about are the items within modern homes that compose the fuel that stimulates and sustains the fire. The UL study on legacy and contemporary furnishings clearly illustrates the disparity of a fire in your grandmother's house with natural furnishings versus the modern home with synthetic furnishings. The video shows a fire that progresses to flashover in minutes in the modern home whereas the same fire in the legacy home may not even be detected in the same amount of time due to it being so small.

What does this mean to the modern professional firefighter and his knowledge of fire behavior? Enough that it could change your tactics, predict the travel and intensity of the fire, and even save a life! The fuel composition of the modern residential structure is nothing more than stored energy. Much like a fantasy movie in which a child finds the magic box and, upon opening it, beams of light and energy shoot out blinding him with its awesome power, the fuel load in a modern home is just waiting for us to open that "box" and release that awesome power.

The modern furnishings, wall coverings, carpets, just about everything has an awesome amount of energy that is just waiting for us to release. An overstuffed couch, fancy draperies, or a decorative floor rug seem so unassuming when you encounter them in your day-to-day life. Add a fire to that structure and they become like the running of the bulls in Pamplona, Spain, releasing their stored energy to the fire and allowing it to charge through the home.

It is very important to remember that many of our fires are ventilation limited to start. We arrive and perform our normal fireground operations of forcible entry (i.e., ventilation tactic) and provide the ingredient of oxygen the fire was seeking. Once this flow path is created, the fire has the current exchange in place and all of the stored energy from the furnishings can charge directly at you. The modern day residential structure's fuel composition as provided by interior furnishings creates the dramatic uptick in the rate of heat release that can easily trap or kill firefighters and the citizens we are sworn to protect.

Firefighter Actions

The expectation when the fire department arrives at the scene should be similar to that of a doctor's Hippocratic oath: *do no harm*. We should arrive at the scene of one of our citizen's worst days with their home ablaze and take actions that improve the situation, not worsen it. Yet, when we exercise uncoordinated ventilation, we can cause harm and damage to the home, occupants, and our own personnel. Uncoordinated ventilation can be simply explained as the random and haphazard introduction of oxygen to an underventilated fire. This is often done by the overzealous but well-intentioned firefighter who breaks out windows of a home on fire before any means of suppression is in place.

In a small, rural fire department Engine 8 and Truck 10 arrive at the scene of the house on fire with five total people on both pieces of apparatus. In a few more moments, some additional volunteers will arrive to render some assistance. With their five personnel, the initial companies will need to secure a water supply, force entry, initiate a primary search, deploy a hoseline, and begin ventilation. The tasks outnumber the personnel so they will make tough tactical decisions that require some tasks wait and others be paired together. This scenario creates one situation where we see uncoordinated ventilation.

The truck driver needs to get ground ladders thrown and ventilation initiated on top of setting the aerial ladder up for operations. His time is limited and he makes the decision to throw ladders to the rear of the structure. While there he will use the ladder to clear out each window, saving him a trip back or up the ladder to clear with a tool. Unfortunately, when he performs this task, the engine is still deploying their line and the truck officer just forced and made entry. The simple action of clearing the window earlier than tactically needed has now caused the under-ventilated fire that was starving for oxygen to erupt. The fire spontaneously consumes the kitchen and fire gases are racing to the two exhaust vents: the upper portion of the open front door and the two second-floor windows that are now pathways to the outside. At the same time, fresh air containing oxygen is being drawn into the fire from the lower portion of the front door. This additional oxygen increases the fire size and energy. The windows did need to be ventilated, just not when there was no hoseline in place ready to extinguish the fire. This scenario results in the truck officer becoming trapped on the stairwell he was climbing to initiate his primary search and calling a Mayday. All due to one well-intentioned, yet haphazard decision that lead to uncoordinated ventilation.

This dramatic effect on the fireground, our battleground, is not limited to small rural departments. It occurs from the largest career departments to the smallest volunteer departments, emphasizing the need for all firefighters to understand the impact it has and how to correctly perform coordinated ventilation. Understanding fire behavior is the first step in deciding when to ventilate. That begins with understanding many of our fires are under-ventilated and are just waiting for us to arrive and introduce oxygen.

Through training and actual firefighting experience, personnel will understand and demonstrate the correct tactics. A perfect scenario is for the truck driver to recognize fire is in the rear kitchen but has not yet vented out the window. The truck driver uses a portable radio and calls the inside crews to determine if they are ready for ventilation. The engine officer verifies there is proper amount of pressure in his hoseline and that it is advancing through the front door to seat of the fire. Citing the perfect opportunity for ventilation that will allow fire and heat to escape opposite the advancing hoseline, they direct the truck driver to vent the window (fig. 13–13).

Fig. 13–13. Truck driver removing a window with fire venting. (Courtesy of Nate Camfiord.)

Perfect execution of ventilation allowed for an exit point of the byproducts of extinguishment, allowing the engine to push in and finish the job. The truck crew quickly advanced to upper floor and completed a thorough primary search. The linchpin of the operation was the coordinated tactics of all personnel on the fireground, all working toward the goal of preservation of life and property. A deep understanding of fire behavior is the foundation that directs our tactical decisions. Fortunately for the fire service, we have excellent resources that supply accurate and relevant information.

Science

Whether probationary firefighters know it or not, they are fire scientists the day they take the oath to serve and protect (fig. 13–14). What is even better is that they don't have to do any actual research. Someone has done it for us, and we just see the effects. Truly incredible agencies like the National Institute of Standards and Technology (NIST) and Underwriter's Laboratory, Inc. (UL) have been diligent in their research and information sharing on the modern residential fire. They've taken to heart the Santayana saying, "Those who cannot remember the past are condemned to repeat it." These two agencies have identified the exact issues that are injuring and killing us at residential fires and use science to explain, teach, and help firefighters develop tactics to prevent it from occurring again. Our job is to take the science to the street! We must not let those that have died in the line of duty die in vain, nor let the valuable information NIST and UL share go unused.

Fig. 13–14. Firefighter taking his oath

While reports from fatal fire recreations have been incredibly beneficial to our learning, www.fire.gov offers two of the more recent documents that provide information that directly deals with tactics at the residential structure. Not in any order of importance, as each of these reports offers facets of relevant and potentially live saving information, we will discuss the UL study, *Impact on Ventilation on Fire Behavior in Legacy and Contemporary Residential Construction* first. A copy can be found at the following URL: http://www.ul.com/global/eng/pages/offerings/industries/buildingmaterials/fire/fireservice/ventilation/.

UL study

UL constructed two residential homes in their large fire facility in Northbrook, Illinois. The first of the two houses constructed was a one-story, 1,200-square-foot home with three bedrooms and one bathroom and eight total rooms. The second was a two-story, 3,200-square-foot home with four bedrooms and two-and-a-half bathrooms, for 12 total rooms. The second house mirrored many of the homes we encounter today with a modern open floor plan, two-story great room, and open foyer. Multiple experiments were conducted utilizing different ventilation options to see the impact on fire behavior. These varied from ventilating the front door only, the front door and a window near and remote from the seat of the fire, opening a window only, and ventilating a higher opening in the two-story house.

Each scenario was repeated three times to examine the consistency of the results. For some in the fire service the results of their study will confirm what they have witnessed first hand pushing down a hallway to fight a fire. To others, it will be an eye opening, evidence-based red flag to evaluate their tactics as they may be antiquated. For all, it should be a catalyst to focus on fire behavior training and the need to be a pro every day! We must learn about our enemy, fire. When the enemy decides to change it does not send us an email, text, or publish a Facebook post. It uses complacency and the element of surprise to dispense its damage.

While the UL study offers various recommendations and data, four specific points are directly related to this chapter's mantra of fire behavior (a very important caveat: while we mention four points, every point in the study is extremely relevant and should be reviewed, not just these four).

- ***Stages of fire development:*** *The stages of fire development change when a fire becomes ventilation-limited. It is common with today's fire environment to have a decay period prior to flashover which emphasizes the importance of ventilation.*
  Our traditional thought process that fire will follow a specific game plan that we have taught to our firefighters for more than 60 years is antiquated and dangerous. The elements must exist for fire to initiate and sustain, but the

manner in which the fire travels from growth to extinguishment is different. The path the fire will take is directly impacted by what we do! We do not arbitrarily perform a task on the fireground to appease the chief, we do it because it is coordinated and strategically relevant. We must understand how fire will act and how each phase of our operation will affect the operational success. The foundation of this recommendation is cemented in company and department training. Rehearsing your strategy and tactics prior to the fire is paramount to success.

- **Coordination:** *If you add air to the fire and don't apply water in the appropriate time frame the fire gets larger and safety decreases. Examining the times to untenability gives the best-case scenario of how coordinated the attack needs to be. Taking the average time for every experiment from the time of ventilation to the time of the onset of firefighter untenability conditions yields 100 seconds for the one-story house and 200 seconds for the two-story house. In many of the experiments from the onset of firefighter untenability until flashover was less than 10 seconds. These times should be treated as being very conservative. If a vent location already exists because the homeowner left a window or door open then the fire is going to respond faster to additional ventilation opening because the temperatures in the house are going to be higher. Coordination of fire attack crew is essential for a positive outcome in today's fire environment.*

This is such an important recommendation, it is stated again in specific terms. We affect the maturation of the fire, specifically how quickly it will mature into a flashover consuming everything in its path. We know the fire will need to be ventilated but *when* is the key element. We must coordinate the operation of the attack with the vent to achieve success. Much like a well orchestrated football running play, the offensive lineman must block at the right time and right place, the running back must patiently wait for the hole to develop, and the quarterback has to hand the ball off at the right time. If any position involved in the play does not follow the plan, then the running back loses ten yards, sacked in the backfield. In our profession, the running back is you, and we don't get sacked, we get burned or killed.

- **Smoke tunneling and rapid air movement through the front door:** *Once the front door is opened, attention should be given to the flow through the front door. A rapid in-rush of air or a tunneling effect could indicate a ventilation limited fire.*

The job of the firefighter is to build a strong knowledge base of fire behavior and then take that knowledge to street for application. The greatest hurdle may be the applying of knowledge in the stressful and instantaneous environment that is the fireground. We have to remove the blinders when we arrive and not focus solely on the fire issuing from the one window or the panic-stricken parent in the front yard. Observe the conditions from when you arrive until you enter zero visibility. What is each action we take, or the environment invokes, having on the fire? In the 5 minutes since you have been on the scene has the smoke changed in color, intensity, or direction? Has the fire self-vented and changed the scope of the incident? Have environmental conditions, such as wind, affected the stage of the fire? All these factors emphasize the importance of what this study recommends, that smoke needs to observed once we do a tactical action, such as opening the front door. We can't blindly rush into a zero visibility, high heat situation without taking a second to observe our surroundings. Many names are given to this operation such as crew resource management, situational awareness, etc. Let's just make it common sense and make it part of what we do every day.

- **No smoke showing:** *A common event during the experiments was that once the fire became ventilation limited the smoke being forced out of the gaps of the houses greatly diminished or stopped altogether. No smoke showing during size-up should increase awareness of the potential conditions inside.*

Simply put, nothing showing means *nothing*. The dispatch rings out, "House on fire at 1st and Main Street," followed by the typical reports of "next to," and, "Behind the house at . . ." All are telltale signs that you are responding to a working fire. You arrive as the first unit on the scene and the frantic neighbor is pulling your sleeve and pointing to the house on fire that does not have one iota of smoke coming from it. Your first indication is she must be crazy or blind, what could she possibly be referring to? Since you are a company that trains and acts like it is the biggest fire of your career until you get there and determine otherwise, you proceed to front door ready for the worst. You open the front door, and whoosh! Thick black smoke billows out the front door and top floor window and the couch just inside the front door erupts in fire. This scenario has been played out on the Internet by fire departments responding to actual incidents, in training fires conducted by NIST in Chicago, and in the aforementioned UL study. No longer is it an anomaly but becoming more commonplace for the residential fire. As professional firefighters, we must treat every alarm for a reported house fire as the biggest fire of our career until we get to the incident and determine otherwise. In the process, we must not become complacent and let our guard down. We cannot slip into the attitude that nothing showing on arrival means, "Don't stretch the line because I don't feel like racking it," or, "Don't layout coming in because I don't see any smoke."

The modern fire is working diligently to hide and bite us when we become complacent, so we must understand modern fire behavior and adjust our tactics to match it. Couple all of the data offered in this report and you can see a connection between our actions and fire growth. Relying upon smoke showing or fire showing as confirmation of a fire is not reliable anymore for all fires. Be prepared for the fire to awaken by having sound skills and practices that focus on scaling the incident down once you confirm there is no fire.

NIST study

A second study, soon to be published by NIST, is the *Horizontal Fire Ventilation Experiments in Townhouses* report. These experiments were conducted in suburban Chicago in attached two-story townhouses, recreating a fire that starts in a couch on the first floor with a window open on side Alpha of the second floor. Many aspects of the fire were evaluated and recorded including temperatures, smoke flow and fire path, and fire growth. The unintended data that presented itself may be more remarkable than some of the data that was intentionally recorded.

As the fire was ignited in the couch the smoke was following a typical path to the oxygen source, in this case an open second-floor window. Unexpectedly, at the 3 minutes, 45 seconds after ignition the smoke that was steadily issuing out of the 2nd floor window completely ceased. The testing staff quickly determined that another window or door was not opened or failed while the evaluation continued. As the fire department personnel began their typical operation of forcing entry and stretching hoselines, the smoke, like a light switch being flipped, immediately begin pushing out the window and front door. The fire had appeared to be progressing through the typical stages of fire, consuming the oxygen that was present within the closed structure. Although the townhouse still contained fuel and heat, without the oxygen the heat release rate decreased. The decrease in energy production caused a decrease in temperature, which resulted in a decrease in pressure within the townhouse; hence the smoke could not overcome the pressure outside the window and stopped flowing. That is, until the fire department arrived and opened the front door. As fresh air carried oxygen to the seat of the fire, with gusto, the smoke and fire was back on and with a sense of urgency, like it was making up for lost time. The fire transitioned to flashover within 80 seconds of the officer opening the front door.

Much like the UL study, the NIST study offers a plethora of data that every firefighter and fire officer needs to review. Specific to fire behavior, two glaring points are critical enough that we included them in this chapter:

- Even if a window is open, the fire can still be under-ventilated until we arrive and force open the front door. This creation of a continuous flow path is the pipeline the fire was looking for. Firefighters and fire officers must understand that forcible entry is a form of ventilation and, like all ventilation, it must be coordinated.

- Nothing showing means nothing! This is not a misprint. You are reading the same mantra from the UL study, thus demonstrating the importance of this statement. We must assume every reported fire is a fire until we get there and determine otherwise. This does not mean when you enter the block or hit the on-scene button, but when you enter the home and confirm it is or is not a fire. Fire is very welcoming of complacency. Our profession should not tolerate it.

Much like many other occupations and trades, the ability to have continued success hinges on the capacity to adapt to current trends. For the fire service, the current trend is fires that will reach flashover quicker with the right supply of oxygen and fuel. Hotter fires coupled with personal protective gear buffers this heat and increases a false sense of security and complacency. Pair this with the decrease nationally of "fire duty," personnel fighting fire on a daily basis, and the true need for firefighters to learn and study is paramount.

Firefighting's 2,000-year mind

The United States Marine Corps have a belief in a "5,000-year mind." This belief is that in a trade with over 5,000 years of documented warfare, there is no reason why a solider should not study every day. Study to learn about their enemy, their battleground, and the positives and negatives of warfare. The fire service is not without its history either, from the Roman Empire to the first organized fire departments in the United States, we have 2,000 years of history. Couple the history from our past with the new scientific data we are provided, and there is no excuse for why a firefighter should not intimately know fire behavior. We must respect and know our enemy to defeat it, and the foundation of that successful battle plan is knowledge.

The Backstep

Fire Behavior: Taking Science to the Streets

Fire behavior begins and ends with understanding the fire triangle and the effect that the building and firefighters have on it. "Coordinated ventilation" is a mantra that is tossed around the fire service with little emphasis on defining what and how. Rather, some would believe that just uttering it on the fireground will magically make it occur. Coordinated ventilation is an orchestrated series of events between personnel inside a burning structure and personnel outside performing tasks.

For this relationship to work successfully and the outcome to have a positive impact on the fireground, it must be planned and practiced long before the event. To facilitate this effort, this chapter's drill focuses on a company level drill to hone the tactic.

- A strong relationship between the engine company officer and the truck/rescue officer, or the assigned firefighter who will perform ventilation must exist.

- Gather all personnel who will be operating inside the structure and those who will perform ventilation and watch live fire videos on the Internet. While watching these videos, discuss the progression of fire and the point that you, as the company officer or incident commander, would want the building ventilated. Discuss why you would deem that specific time is correct for coordinated ventilation. Stress the findings from the scientific studies that when we apply air sooner and in larger amounts than water to the fire, the fire increases and the safety decreases.

- Determine among personnel operating on the fireground the common terminology that will be used between the inside and outside personnel to execute coordinated ventilation. Communication is listed on an all-too-regular basis as a contributing factor in line-of-duty deaths, so working out this terminology beforehand is extremely important. This can be as simple as a two-word transmission such as "Take it" or a more one detailed such as "Remove the window on side Charlie now." Develop and implement the terminology and distribute it to all members.

- Now that the terminology is defined, it is time to practice the operation. If you can secure a live fire training facility, that is ideal. If not, the training drill can continue, either with smoke machines or an actual live fire training. Have the engine company advance an attack hoseline to the fire. Have the outside ventilation team communicate with the inside operating forces in order to time the ventilation when needed. The outside personnel should swiftly, and in a position that is out of the direct line of by-products, vent the window.

- Upon completing the operation, assemble all of the personnel for a critique of the operation and perform a roundtable discussion, gathering each person's input. This is an invaluable tool for every officer and firefighter to understand the level of knowledge we have on fire behavior and ventilation.

- Once the critique is completed, repeat until each person can serve in each role and perform the assigned tasks. Having physically performed the task helps each firefighter to develop mastery at their position.

- The final portion of this training drill takes place back in the firehouse and is just as important as the above drill. Our ability to perform our jobs at a high level is hinged on us being students of the profession. A key step to this is to routinely review the scientific data that is offered to the fire service and learn how to implement it into your daily operations. Perform a monthly drill on the newest data out and discuss how your department plans to implement it. The following websites provide an excellent introduction, but half of the battle is finding intelligent sources from which to learn:
  - www.fire.gov
  - http://www.ul.com/global/eng/pages/offerings/industries/buildingmaterials/fire/fireservice/ventilation/

PART 3

Engine Company Operations

14 Preplanned Engine Company Riding Assignments

In more than 2,000 years of firefighting history, there remains a constant tactic that has had the most impact on the development and extension of fire: the application of water! When in the right volume and properly applied, water extinguishes almost every fire. The implementation of ventilation, positive pressure fans, or effective searches will not extinguish a fire. Only water in the hands of a competent operator will do the job (fig. 14–1).

Fig. 14–1. Getting water on the fire will have the greatest impact on the outcome of the incident.

Understanding Our Mission

The above statement may seem to be common sense and a very elementary operation in fire science, yet it is often the most overlooked task on the fireground. The key to preventing the omission of the proper application of water from happening on your next fireground is to lay the groundwork for the process, and that is accomplished through riding assignments. Predetermined assignments provide guidance on the tasks to be completed and the manner in which they will be executed. The theme we want to continue to emphasize is combat readiness. In this case it means nothing more than preparation.

Consider the analogy to a football team again. The professional players learn and practice their positions for endless hours. Once the fundamentals are learned, the specific plays are defined and practiced, with each professional performing a specific task that is pivotal to overall team success. Game day arrives and the plays are communicated to the team via their leader, the quarterback. When the whistle blows, they are executed perfectly. Yet, what happens when one player goes left when everyone else goes right? A 10-yard loss and failure to complete the mission is the result.

This sounds eerily similar to what we do in our fire service careers. We learn and practice our trade in recruit training. We learn the "plays" from our officers for what we do on the residential house with fire in the basement, the kitchen, the attic, or maybe even the attached garage. When the alarm sounds we put the plan into action, with each professional performing a job that is pivotal to the success on the fireground. The same result is achieved when one of the professionals pulls the wrong hoseline, does not chock the door, does not remove a kink in the yard, etc. We suffer a failure in our system that leads to the inability to get the right amount of water to the seat of the fire to extinguish it, thus preventing all the other tasks from occurring (search, ventilation, etc.).

Start the Accountability Now

Bearing in mind the importance of this task, the development and implementation of engine company riding assignments is not only justified, but essential. The design and implementation of riding assignments offers several notable benefits, the first being the vital measure of accountability. In almost every close call and line-of-duty death report in the last 20 years, some measure of lack of accountability has been denoted as a contributing factor (fig. 14–2). Accountability measures are often overlooked or discounted in the spirit of the "things need to get done now and we don't have time for it" mentality. Yet they provide an initial safety blanket for personnel in the IDLH so that if trouble does occur, rapid intervention teams can deploy to the correct location to assist.

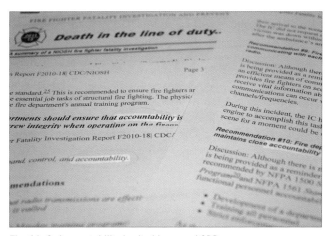

Fig. 14–2. Accountability is cited in many LODDs.

By using a simple and effective riding assignment system for the engine company, we can convey in one short sentence where all the personnel on the engine company will be operating and the tasks they will be performing. The first-arriving engine officer voices the final radio report before entering the structure fire.

"Engine 25 will be advancing a 1¾-inch line through the front door to attack the fire on the first floor."

Engine 25 has their standardized engine company riding assignment set, and now the incoming chief, RIT, and all other incoming units know where the engine officer, nozzle operator, and backup will be operating and what tasks they should be performing.

Time is of the essence on the fireground with fire growth and victim survivability affected every second, so any measure we can implement to increase our deployment is essential. Engine company riding assignments, once taught and honed, will eliminate any conversation about what the engine needs and where. A simple, "250 foot 1¾ inch hoseline to the front door," will suffice to explain who is doing what and where they need to be now.

There is no, "I want you to be the nozzle operator. I want you to back him up and push hose in from the front door. I am going to take a lap." This continued commitment to excellence on the fireground not only breeds professionalism and competence, but also gets the fire out more quickly.

Train Like You Plan to Combat Our Enemy

A question that firefighters, fire officers, and fire chiefs often ask themselves is, "What should we train on tonight?" In the career system with a consistent schedule it is often easy to plan a comprehensive training plan. In small volunteer departments it is more difficult with their flexible staffing and unknown personnel skill level. Regardless of affiliation, the implementation of riding assignments can provide the foundation for your training program. Clearly defining the scope of what each position performs and then applying that to specific fireground situations (basement fire, second-floor bedroom, etc.) helps provide individual training programs. For instance, one training session can be focused on nozzle operator evolutions. This can include topics such as:

- Nozzle selection and operation
- Nozzle inspection (fig. 14–3)
- Stream selection principles
- Hoseline selection based upon fire load and fire involvement
- Hoseline stretch estimation techniques
- Hoseline entry procedures (quick scan of floor before entry, sweep of floor prior to entry, etc.)

These six topics can be expanded to cover one night each or, depending upon time constraints, can be one night of practical exercises. Ultimately, the key is that a clear and consistent policy is developed and implemented and the opportunity is provided to hone and perfect those skills in training.

Fig. 14–3. Firefighter conducting a nozzle inspection

Hopefully the benefits listed above demonstrate the importance of riding positions for the engine company, but they do not address the recurring question of staffing shortages. How can an engine company that may arrive with only three firefighters perform riding assignments? First and foremost, to successfully and safely perform engine company operations on the fireground you must have at least three personnel! Two people is not acceptable, and we must develop means to supplement the shortage. As is a common theme in this book, we do not offer a blanket statement without an explanation and option of how to fix the problem.

Essential Versus Eventual

Whether you have the luxury of large urban area staffing (five to six personnel) or are faced with a dwindling volunteer shortage (two to three personnel), you can still implement riding assignments. The key to being successful is to understand and communicate that three is the starting point for deployment and operation of hoselines. The following sections describe the essential positions of the engine company.

Chauffeur

There is no way around this position. Someone has got to get us to fireground and operate the pump so that we get water to the seat of the fire. One advantage of well-trained and professional chauffeurs is that they can supplement other positions later in the incident. Once a positive water supply is established and water is directed to the personnel on the hoseline, the chauffeur can assist in flaking hose in the front yard, throwing portable ladders, and other fireground necessities (fig. 14-4).

Fig. 14–4. The engine company chauffeur can perform many jobs on the fireground. The first and most important is getting water!

Engine officer

Whether it is because every close call or LODD report mentions command and control as a contributing factor, or simply that you understand the importance of someone being in charge, someone has to be the boss, the director, the quarterback of this operation. The engine officer needs to:

- Decide on the strategy and communicate the tactic to his nozzle operator.
- Select the correct length and flow for the fire conditions.
- Decide the fastest and most efficient route for the hoseline selected.
- Watch the changing conditions.
- Maintain situational awareness.
- Monitor the crew.
- Constantly assess if the tactic is working.

While all members should demonstrate these traits, the engine officer has the overall responsibility for this operation. Depending on the staffing, the engine officer may need to operate in dual roles. This is the burden of being the boss. Not only may you have to monitor conditions, but you may also have to do this while you serve as the backup position. We discuss the intricacies of the backup position later, but officers must be ready to recognize and handle multiple tasks in a hazardous environment.

Nozzle operator

This firefighter holds the last line of defense in his or her hands and can have the greatest impact, negatively or positively, on the fireground (fig. 14–5). Nozzle operators must be professional in every sense of the word, from preparation to execution of the task. While officers may be ultimately responsible for the selection of hoseline, a nozzle operator should be their greatest asset. If an officer has done a good job and the nozzle operator is prepared, the selection of hoseline is done by the nozzle operator. This person recognizes the fire conditions, the extent of fire, and the layout of the structure, and confirms it with the officer upon descent out of the engine company. Once the line is deployed to the point of entry, the nozzle operator must:

- Assess the fire conditions and acknowledge any changes.

- Quickly assess the layout of the structure as the door is forced open. Once the truck "pops" the door, a quick peek down low can reveal the location of the fire, obstacles along the way, victims, and the presence of a floor.

- Assume the firefighting position (fig. 14–6). Today's firefighter faces a multitude of hazards offered by current building construction that demands a more upright body position. By simply assuming a body position in which your leg is out in front of you, head up, facing your obstacle, and keeping your body low you can ensure:
    - You are constantly probing for obstacles or lack of a floor with your foot. No more banging your head into an unseen wall!
    - You can monitor fire conditions in front and above you.
    - You can quickly sit back onto your back leg to stop forward momentum that would normally send you into a hole.
    - You can direct your nozzle reaction into the ground or obstacle to lighten your reaction burden.

Fig. 14–5. The nozzle operator is charged with delivering water to the seat of the fire.

Fig. 14–6. The correct firefighting position is a heads up, leg out position.

- Maintain enough hose in front of your body to be able to direct the stream in an arc of 180 degrees in front of you. The nozzle should not end up looking like a shoulder purse you are carrying underneath your arm (fig. 14–7). Keep at least 2 feet of the nozzle and hose in front you as you advance. In this position, you can quickly drop to direct nozzle reaction into the ground or object and then direct your stream, with ease, to any position in front, overhead, or below you. Additionally, if you need to quickly pass the nozzle to your backup position to knock fire that has spread around or behind you, it only takes a quick handoff.

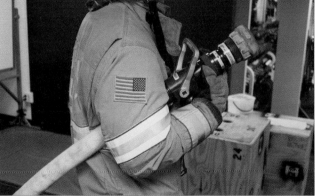

Fig. 14–7. If you advance the nozzle incorrectly, it looks more like a shoulder purse than a firefighting nozzle.

- Push in! Fires are not successfully defeated by a retreating mentality or tactic. Do not spray water and wait until the fire burns to you. We attack the fire to eliminate the problem and make the building safe again. Conversely, a nozzle operator must be able to assess when his actions are not having a positive outcome on the incident and recognize that it is time to change tactics.

- Don't forget the 200-foot tail you are dragging in with you. Proper deployment means that we do not simply focus on all that is in front of us and hope for the best behind us. You move with purpose to the seat of the fire but must recognize that after a turn, you may have to take a momentary pause in your offensive advance. Use the wall or static piece of large furniture to serve as your backup firefighter while more hose is advanced around the obstacle. You can make a knock, or keep the fire in check, from this offensive position until enough hose is up to advance. Then we push in and extinguish the fire.

- Know your nozzles, their capabilities, and their limitations. This is a vital tool and you must be able to select and operate your nozzle. Recognize the fire conditions and know if your flow is working or not, and fix it if necessary.

- Be creative in finding a backup person. You may not have the luxury of having another firefighter directly behind you to absorb the nozzle reaction. This should not force you into the bad habit of allowing the nozzle to slide up under your armpit so you are fighting fire with crocodile arms. Direct the nozzle reaction into the ground to absorb the back pressure, find a large piece of furniture, or even use a corner or wall (fig. 14–8).

Fig. 14–8. Firefighter using a piece of furniture for nozzle reaction

Hopefully you can add more personnel, whether it is another arriving engine or personnel arriving in other vehicles. When these personnel arrive, they will be the "eventual" positions, those that are vitally important but can't be filled immediately due to staffing shortages. The eventual positions of the engine company are backup and control.

Backup

While this task is listed in the eventual position section, it must be filled quickly to facilitate effective hoseline movement. It is listed here is because it takes more than one person. It can be filled by any number of incoming personnel. At nearly every fireground, the need for three, four, five or more backup positions may be necessary to complete the stretch and get the line on the fire. The backup position is not a spectator who crowds inches behind the nozzle operator to get a front row seat for the fire (fig. 14–9). In reality, the person in the backup position doing the job correctly may never see the fire.

Fig. 14–9. If done incorrectly, firefighters performing the backup position will all crowd up behind the nozzle firefighter.

The backup position must, at a minimum:

- Eliminate kinks and pinch points from the engine company to the last obstacle. This may require removing furniture or other objects.

- Stay one obstacle back from the nozzle operator while the advance is occurring. This ensures that a smooth flow of hose occurs around the last pinch point.

- Maintain 10 feet of hose to assist in the final push. This may require pushing a small section of the charged hoseline into an adjoining room, forming a loop, or maintaining a "snake" up a hallway (fig. 14–10). The last technique maintains the advancing hose in a snakelike fashion up a hallway to the nozzle operator and not a taut line. If the line is taut, it is stuck on a pinch point or unable to advance further. By maintaining the snake, you are ensured that the hose is not taut, and if a small amount is needed quickly it can be pulled from this slack.

- Monitor conditions in front and back of your position. The nozzle operator and officer will be only a couple of feet or even inches from the seat of the fire when they are working to extinguish it. This close encounter along with the immediate and turbulent impact of the hoseline on the fire will create instant zero visibility conditions for them. The advantage of being one obstacle back from the firefight is that you can see the changing conditions, hopefully denoting a change from free burning to extinguishment. You can also determine if the firefight is going well or if the fire is traveling overhead or around the nozzle operator. As the Department of Homeland Security slogan goes, "If you see something, say something."

Fig. 14–10. Hoseline snaked up a hallway can help determine if the nozzle operator is advancing.

Control

In larger departments, this is a luxury that allows for the depth of experience that can be pivotal to successful hoseline selection. A single person who can quickly assess and estimate the proper stretch of hose to reach the fire is an asset. For most fire departments, this position will be filled by later arriving personnel. After the hoseline has been stretched, they can move on to another task. The control can serve as the bouncer, and everyone wants to get to the big fire. One of the greatest hindrances to successful engine company operations, or really any portion of a firefight, is overcrowding in the fire room. There is no need for an entire engine company, truck company, and rescue squad to crowd into a 10×12-foot bedroom to watch one firefighter apply water to the base of the fire. What is sure to occur in this situation is ineffective hose movement, inaudible yelling for more line, and a fire that continues to grow. As the last person before the fire room or at the point of entry, you can suggest to the overzealous crew wanting to get in and see the "red devil" that there is no room and help is needed along the hoseline behind you. This allows you to also evaluate the fire attack, monitor conditions in the common hallway, and assess ventilation needs.

Additional positions

The primary job of any later arriving engine company is to ensure that the first line is in place and advancing to the fire. There is no need for two lines that come up short when it would only take one well-positioned hoseline to accomplish the task. If the first line is operational, later arriving personnel should prepare not only for stretching an additional line, but also assess if more personnel are needed at pinch points. It only takes one turn without a firefighter taking time to "feed" line to stop a fire attack in its tracks. Additionally, it only takes one firefighter to deploy an additional line to the front door while the rest of crew eliminates kinks in the front yard or aids at pinch points.

Understanding the "Why"

The misconception that riding assignments only apply to large, well-staffed fire departments is dispelled by explaining the purpose and scope of the work the engine company performs. Once the entire game plan for an engine company is completed, the need for riding assignments is demonstrated and can be implemented. Aside from providing a defined and efficient fireground, it also breeds the professional and competent behavior we seek from our personnel. By building a clear set of goals and expectations for each position you can create the foundation for a training program that works to accomplishing the primary goal: getting water on the fire!

The Backstep

The 2½-Inch Hoseline

The 2½-inch hoseline is one of the most powerful firefighting tools we can utilize on the fireground. It is also one of the most intimidating because of the high nozzle reaction and difficulty in maneuvering. If your recruit school or firefighter training was anything like ours, the 2½-inch was used as a discipline device (fig. 14–11). The class was not performing well on a task, so the punishment was to deploy and flow the 2½-inch to teach us a lesson. Of course, this does nothing but condition firefighters to fear the tool and not appreciate how effective it can be.

To practice this drill you will need the following equipment:

- Two personnel (three is optimal) to fill the nozzle and backup positions
- 2½-inch hoseline of a length you typically have pre-connected or deploy
- One play pipe style stack tip nozzle (fig. 14–12)
- A commercially sold hose strap or 3-foot loop of webbing or rope (fig. 14–13)

Fig. 14–11. The 2½-inch is often used as a measure of discipline, conditioning behavior to never use this functional tool.

This chapter's drill should dispel the fear of the 2½-inch hoseline and teach you how to effectively deploy, maneuver, and advance the 2½-inch with limited staffing. The classic way to deploy the 2½-inch in recruit training was across a parking lot in a standing position. Once you reached your objective or defensive position you would form a loop and man the line with one firefighter. This is a complete underutilization of an effective firefighting tool. While the 2½-inch is not going to replace the speed and efficiency of a 1¾-inch line, it can fill a role other than defensive firefighting. The key is to work smarter and not harder.

Fig. 14–12. A play pipe style nozzle with side handles

Chapter 14 Preplanned Engine Company Riding Assignments

Fig. 14–14. Correct nozzle position with knee into the hoseline, directing the nozzle reaction into the ground

Fig. 14–13. A hose strap made from spare 8 mm rope or commercially sold hose strap is essential for proper and rapid advancement.

The key to effective movement and use of the 2½-inch line is to control and direct the nozzle reaction. The typical nozzle reaction for a 2½-inch hoseline ranges from 90–120 pounds of force that you must accommodate. Notice that the word accommodate, not fight, was used. You don't need to fight the nozzle reaction, you just need to put it somewhere other than your chest. The use of the play pipe nozzle with side handles allows for the firefighter to assume a firefighting position in which they use one hand on the nozzle and place their knee into the hose about 2 feet back from the nozzle (fig. 14–14). This position directs the nozzle reaction into the ground and not into the firefighter. The use of the play pipe allows the firefighter to easily direct the stream up and down in a 45–60 degree range and 180 degrees left and right. Additionally, this keeps the firefighter's line of sight straight ahead, able to observe fire conditions and to assess when to advance.

Our objective with the 2½-inch line is to not remain in a static position. We must be able to not only negate the nozzle reaction, but also quickly advance the line. Gripping and holding onto the 2½-inch hoseline can be a challenging feat, with the backup firefighter often becoming very fatigued due to the weight and size of the hoseline. To overcome this obstacle and advance the line, we utilize a hose strap. Strap the hose strap, webbing, or rope around the hose utilizing a girth hitch about 4 to 5 feet behind the nozzle position (fig. 14–15). Kneel next to the line and monitor the nozzle position's firefight. If more line is needed, grab your strap and advance. If the nozzle operator is in an offensive stance and darkening the fire down, you can slide your strap farther back to bring more hose into a room or hallway to prepare for the advance. As more personnel arrive, they can fill in along the line with hose straps and work to advance the line.

Fig. 14–15. Correct backup firefighter position utilizing hose straps for advancement

This technique in no way makes the 2½-inch an effective tool for advancing through a Collyer's mansion condition in a small single-family dwelling. It does, however, allow a short-staffed crew to advance and darken down a large volume of fire. Once darkened down with the large stream and effective push, they can quickly switch to a 1¾-inch hoseline and push in to the seat of the fire to complete the extinguishment.

No manner of innovative techniques or tools will overcome ineffective communication of a hoseline crew. For this technique or any technique to be effective, all of the personnel on the hoseline need to communicate. Advance and retreat as one unit and not as a disjointed group that looks like a derailed train.

15 Backup Is Not a Direction of Travel, It Is a Position

As you can read from the title, this chapter deals with the position of the backup person on the hoseline. The importance of the position is so vital to the overall success of the incident that it requires its own chapter to explain the intricacies of the skill.

The Backbone of the Hoseline

First and foremost, operating as the backup on a hoseline is not a direction of travel or merely a spectator, it is a position much like the officer or nozzle operator. The value of the backup position is not any less than the aforementioned positions. On many incidents it can be the linchpin to the success and can prevent the confluence of water supply errors that allow a fire to grow beyond our immediate suppression ability. The backup is the backbone of the operation, with the nozzle operator being the eyes leading to seat of the fire and the officer the brain that directs the operation. Without the spine, the body cannot move or operate and without a professional, well-trained, and competent backup, the hoseline cannot perform its job. All three systems must work in concert to achieve success in our objective of putting the fire out (fig. 15–1). While the backup may not be as big a hero as the nozzle operator who charges into the toxic and heated environment, the actions of the backup make those heroic acts possible.

The backup position is also not a position to be filled when the last available person arrives. This could be likened to when the chubby, non-athletic kid is picked last at the playground football game. We want a superstar and we want him early in the position. In the fire service, we learn day in and day out that the success of any firefight hinges on how the first line goes. Run that one fire in your career that demands several turns and elevation changes of the initial hoseline without a good backup position and you will never forget it. A chaotic event will most likely ensue with no water getting on the fire and bewildering looks as to why you could not put the fire out, all because of an ineffective or complete lack of a backup position being filled.

Fig. 15–1. A three-person engine company properly advancing a hoseline

Protecting the Blind Side

The backup position can be likened to the offensive line of a pro football team with a superstar quarterback. Everyone knows the quarterback. He is the face of organization and the first to be adorned with all the praise when the outcome is successful. Conversely, you never see the offensive line being hoisted up on shoulders and carried across the field after a long touchdown drive to win the game. Yet, if they didn't do their job of protecting the quarterback and allowing him to demonstrate his skills, the ball would have never left his hands and no TD would have been scored. The backup position is the offensive line, providing the avenue for the nozzle operator to demonstrate his or her skills in extinguishing the fire. Backups anticipate what the nozzle operator needs and exhaust themselves to make sure there is ample supply of hose and support to apply water to the fire. Without them, the nozzle operator is left screaming at the first turn for more line and operating in a retreating position.

Now that the demand is clear for the backup position and that there is a need for a competent and professional to operate in the role, what do they do? As with many topics in the fire service, we are very poor sometimes at explaining the why and the what. Why do we have to hold the hose in this body position? Why do we need to carry smooth bore and fog nozzles? What is the flow range of the initial hoseline on the engine company? What is the job of a backup firefighter?

All these are questions that should be asked and answered correctly and quickly. Our goal should be to not have any policy in place for arbitrary reasons ("We have always done it this way!"). "We do it this way because of the reasons x, y, and z," is what our explanation should begin with anytime we are asked. When we expect firefighters to enter environments to risk their lives, at a minimum we should provide a clear reason why we do what we are doing. Colin Powell expressed this very statement in his speech "U.S. Forces: The Challenges Ahead," when he stated, "We owe it to the men and women who go in harm's way to make sure that this is always the case and that their lives are not squandered for unclear purposes." We owe it to firefighters who risk their lives that it is for a clear purpose. That begins with understanding what the backup position is and what its expectations are.

The Intricacies of the Backup Position

The manner in which a firefighter executes this position will vary from fire to fire based the specific conditions encountered. Some fires will require more hose handling, others may require fire behavior observation and room control. Regardless of the level of the involvement on each incident, all the expected tasks should be known and clearly communicated. Let's start at the beginning and break down each part of the backup position.

1. **Chase all kinks.** Nearly every hoseline that is preconnected or dead load racked on an apparatus is a single person deployed hoseline. Many will become two person lines out of our wanton desire to help out, even though it may hinder the operation. Once we recognize we have another job, the obligation to help pull hose onto your shoulder will lessen. The backup firefighter's job begins at the engine company and works up to the last obstacle (fig. 15–2).

   - If the hoseline is pre-connected, make sure it is completely clear of the hosebed. This can be your first and greatest kink, essentially strangling the water supply.

   - Ensure that the nozzle operator does not drop any of the shoulder load until the entire trailing end of hose is stretched. If the nozzle operator meets resistance due to the line becoming snagged on a bumper or wheel of a car, he or she will think it is time to start deploying. This will only make a mess of spaghetti in the front yard.

Fig. 15–2. Two firefighters deploying a one-person line becomes more of a hindrance than a benefit.

2. **Assess and decide on the shortest and most efficient path of the hoseline and make sure the hoseline gets there.** The nozzle operator may have to run 30 feet around the 4-foot decorative fence in the front yard to get to the entrance gate since one person can't jump over it with a shoulder load (fig. 15–3). Fortunately, as the backup position you are travelling light and can grab the section of hose, hop the fence, and save 30 feet of hose that could become pinched at the turn or may be needed to make the push to the fire room.

Fig. 15–3. Houses with long fences can affect hoseline selection and placement

3. **Turns and excessive bundles of hose must be eliminated.** The residential structure offers a plethora of hoseline obstacles including rock gardens, lawn ornaments, vehicles, and porches. All these obstacles seem to grow arms and grab hoselines when we deploy. The backup must track the path of the nozzle operator and negate these hazards. Whether it is a call to "stop" the nozzle operator so the backup can pull the hoseline out from the obstacle before it is pulled taut or to make a simple flip of the hose, it must be done. If a gross estimation of the distance is made and you end up with 100 feet of extra hose at the door, decide quickly where to put it. Water is coming and it will be almost impossible to unravel this mess when it is charged. If there is 100 feet left on the shoulder of the nozzle firefighter, grab the coupling before the load drops, ensure the nozzle is being held firmly, and perform an alley stretch. This action quickly places the extra hoseline in an elongated "S" design leading directly into the residential structure (fig. 15–4).

Fig. 15–4. The alley stretch is just one of many ways to overcome obstacles to hoseline advancement.

4. **Determine the number of turns to the fire room and work the hose around them.** The greatest advantage we have as firefighters in residential structures is that we live in the same type of buildings. Your subconscious kicks in when you arrive at the colonial-style residential dwelling with a fire in the kitchen. Most likely you are doing it now as you are reading this. Go through the front door, step left to go around the stairs, and the kitchen is in the back, right? You have imprinted the routes of hoseline travel throughout all of the residential structures you serve. When you are operating as the backup, you must be this visionary and decide quickly how many turns you think the nozzle operator will encounter on the attack. Once we have this path nailed down we can then anticipate the number of backup positions we will need to fill and how you will manage the advance. This may require the singular backup position to push a loop into the house, move up to the next turn, and repeat, pushing a loop in front of the nozzle position up until the last turn. Ultimately the backup must be proactive and not merely reactive as it may be too late to save an ineffective firefight when you are hoping to catch up.

5. **Maintain the correct body position on each turn.** Proprioception is a term used in the medical field that essentially means "one's own" or the position of one's body. This sense of where and what your body is doing is essential in serving as a successful backup. You must not only know

where and what your body is doing, but you must also anticipate imminent obstacles. The backup must be able to recognize the obstacle of a turn and position in the correct location. The difference between being on the outside of a hose that is instantly pulled taut against a turn versus being pinned on the inside of this movement is the difference between being an asset or liability. The backup must anticipate the movement of the hose and put him- or herself in the position of success by knowing exactly where to be to effect positive movement. When a turn is expected and realized, the backup must get into a crouched or a kneeling position in which the body weight is centered. This will allow the operator to utilize leverage and strength to make the push around the corner. The backup must be on the outside of the hose with a hand always on the hose (fig. 15–5). The simple gesture of keeping a hand on the hose provides the backup with instant feedback on whether the line is flowing water, advancing, or in a static position. The feeling of water flowing, stopping, and a line being advanced is very easy to feel with a gloved hand; it allows the backup to perform the next task seamlessly.

beckoning them to come closer. They are not thinking about anything behind them until that is what stops them. To perform the backup well is to understand and facilitate the pace of the fire attack. If the backup is operating in a hallway, the focus is to push the hose up the hallway so it snakes in an "S" formation against the two interior walls. As the line advances, you will see the "S" return to a straight line (fig. 15–6). If another turn is not made and there is no need for the backup to slide up the line, maintain the "S" in front of you until the conditions demonstrate a knock on the fire. *Do not just push!* Hopefully that is clear enough. This is not a technique to see how quickly you can push the nozzle operator into the fire. Nor do you want the one firefighter who will be in the most dangerous position, the seat of the fire, to lose the lifeline (the nozzle) because you pushed it out of his or her hands. It can be difficult to relocate even a charged hoseline once you become separated from it, and the guilt of causing an injury or death is not worth the display of your strength. Facilitated and honed through training and demonstrated on the fireground, the backup and nozzle operator must know each other's pace and have that firm trust between one another.

Fig. 15–5. Correct body position when operating as the backup

6. **Facilitate the pace of the fire attack.** The backup must be so good that no one ever thinks about him or her. The nozzle operator and officer are focusing on the task at hand, the fire that is

Fig. 15–6. Maintaining an "S" formation up the hallway is useful in advancing a hoseline.

7. **Keep eyes up and ahead.** The nozzle operator enters the hazardous environment of the fire room and immediately interjects water at the base of fire and begins bouncing the stream off of every wall. This action is incredibly successful in putting out fire but it also changes the conditions in the room from somewhat visible with a line of stratification to utter and complete chaos. Fortunately, the nozzle operator and officer can rely upon their backup to maintain a vigilant eye on the conditions, confirm the successful steam conversion, and alert them to advancing fire behind them. This safety blanket can only be provided if the backup does not operate like an elephant being led into the circus. The image of this large animal, head down, latched to the rear of the elephant in front of him and oblivious to any of its surroundings, should not be the way we enter fires or advance to the seat of a fire. We must maintain a low position by being in a duckwalk or offensive shuffle. This not only allows you to stop suddenly if an obstacle is encountered, but it also allows for the helmet to not strike the SCBA that has risen up the back of the firefighter. In this position, the backup can stay low but also watch for change in smoke color and intensity or a fire that has wrapped around behind the attack crew.

8. **Know nozzle reaction.** If the fire is a short distance from the engine company or the building is not full of obstacles, the backup may end up directly behind the nozzle operator. This is not an excuse to become a spectator, merely watching over the shoulder and observing the nozzle operator fighting the nozzle reaction. The backup position is even more important now because it allows the nozzle operator to have unimpeded firefighting range (table 15–1). The key to this advantage is to understand nozzle reaction. Combating nozzle reaction does not mean I nuzzle my hip and shoulder into your back and apply as much force I can to you to overcome the reaction of the nozzle kickback. This may work for a 2½-inch hoseline, but it will push a nozzle operator right into a fire with a 1¾. Plus, it lacks the professional demeanor that an engine must possess. As a backup you must be part scientist, observing and understanding the change of water through its various forms, but also a mathematician. You must know what the nozzle reaction is for each of the nozzles that you have on your apparatus.

Table 15–1. Nozzle reaction for the most commonly used nozzles

Fog Nozzle Reaction			
	Nozzle pressure		
gpm	50 psi	75 psi	100 psi
100	36	44	51
150	54	66	76
175	62	77	88
200	71	87	101
Smooth Bore Nozzle Reaction			
	Nozzle pressure		
gpm	50 psi		
7/8 in. (161 gpm)	60		
15/16 in. (185 gpm)	69		
1 in. (210 gpm)	79		

The recognized formula published in the *National Fire Protection Handbook, 17th Edition*, for fog and smooth-bore are as follows:

- Smooth bore: 1.57 × bore diameter squared × nozzle pressure

- Fog nozzle: 0.0505 × rated flow × square root of the nozzle pressure

At this point your head may be spinning, and it would seem it is well outside the scope of your challenging freshman high school math, but do not despair. The basic fundamental fact we want to remember is that you probably only have two to three different types of nozzles on your engine company, so you would only have to compute the figures for two or three lines and commit them to memory. Fortunately, most of the nozzle manufacturers do the math for you and provide a helpful chart in their documents.

Regardless of the level of mathematical difficulty, as a professional you must understand that when you are backing up a nozzle operator you are there to apply the proper amount of force. The generally accepted rule in the fire service is that one firefighter can handle 60–70 pounds of nozzle reaction. When the 70-pound threshold is crossed, the need for a static object or another firefighter to assist in countering the nozzle reaction is needed. Our recommendation is to find something (wall, couch, floor) or someone (backup firefighter) to take this force. This extends the time you can perform the duty and lessens the energy you expend. This reduction in stress and exertion means you can preserve your SCBA because, as we know, air is not free in the IDLH so we want every pound of it we can have.

Lieutenant Andrew Fredericks of the FDNY offered a quote that encompasses the importance of the backup position in the engine company operation when he stated, "Disciplined engine companies are deliberate, patient, and professional." Nowhere within the quote does he recommend rushing in without purpose or blindly operating with reckless abandon as an option. The backup position is the epitome of this statement. Backups are deliberate in their actions and purpose, patient in watching and supplying the nozzle operator, and professional in their anticipation and delivery on the fireground. The foundation to this position, and any position on the fireground, is to be knowledgeable. That starts with understanding the why. This chapter's drill not only explains the why, but also builds esprit de corps by breeding the professionalism we seek in our trade.

The Backstep

Nozzle Reaction Drill

With the assembly of some tools and equipment that are readily available in nearly every firehouse in the United States, you can perform this highly effective and educational drill. If the tools are not in your firehouse, the cost is minor in comparison to the knowledge gained. The result of this drill will provide you the true, empirical nozzle reaction for each of your hoselines at the pressures you pump with your nozzles.

Tools needed:

- Two 5- to 10-foot loops of small millimeter rope or webbing (fig. 15–7)

- Digital or standard scale that can measure up to 150 pounds. The preferred scale is a hunting scale with a hook or fastener at the end to secure your rope or webbing (fig. 15-8).

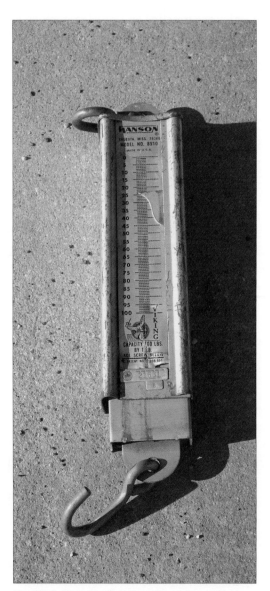

Fig. 15–8. Digital scale or fire extinguisher scale

- A column, pole, or other object to serve as the static anchor

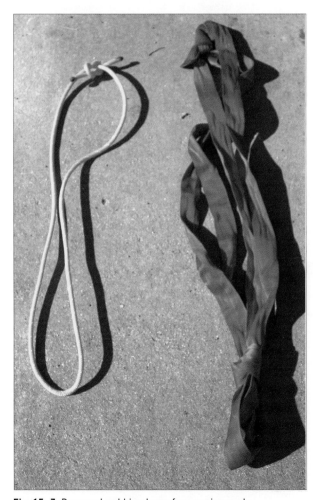

Fig. 15–7. Rope and webbing loops for securing scale

Procedure:

- Identify a sturdy anchor that you can wrap your rope or webbing around and that can sustain weights up to 150 pounds being pulled against it.

- Wrap one 5- to 10-foot rope or webbing loop around your anchor with a girth hitch.

- Attach the top end of the scale to the loop provided by your girth hitch.

- Deploy the hoseline and nozzle for which you plan to record the nozzle reaction to the end of the scale.

- Wrap the second loop of rope or webbing on the hoseline immediately behind the last coupling leading into the nozzle. The girth hitch is an effective hitch to use in this step also.

- Attach the loop provided by the second girth hitch to the bottom of the scale.

- Ensure your hoseline is in a straight line with the scale and the webbing is in line with the nozzle. The nozzle should be pulled back until the loop around the anchor, the scale, and the loop around the hoseline are taut.

- Throttle your hoseline up to the recommended pump pressure and observe the line retracting, pulling the two loops and scale taut.

- Once the recommended pump pressure is achieved, open the bale slowly until it is fully open.

- Once the nozzle is fully open and you have ensured the hoseline is in a straight line leading back from the nozzle, allow the nozzle to be held completely by the two loops of rope or webbing and scale (fig. 15–9).

- Record the readings on the scale when the line is at the recommended pressure, fully supported by the scale, and the hoseline is not kinked or turned.

Fig. 15–9. With the hoseline charged to the correct pressure, note the pounds of force recorded for the nozzle reactions.

The numbers you record are the true and accurate nozzle reactions that your nozzle operator is facing each and every time a hoseline is deployed and operated. This test can and should be performed for every hoseline you use so that your engine company can be deliberate, patient, and professional.

16. Where Is My Water?

The common mantra we have espoused throughout this book is the combat-ready behavior coined by our dear friend and mentor, Lt. Peter Lund, FDNY, who passed away in 2005 (fig. 16–1). The foundation of this behavior is a sense of preparedness and sense of respect. The respect is for fire and the damage it can unleash. Ultimately every run could be the biggest fire of our career and the most challenging to overcome. Regardless of the scope or size of the incident, we must overcome these obstacles and do the job we took an oath to complete in the face of any adversity. A key foundation to doing our job since the inception of the fire service is the establishment and delivery of *water*.

a static water source. Regardless of the means, the job does not change. The engine company is relied upon to complete this task on every fireground and we must not falter, we must not fail (fig. 16–2).

Fig. 16–2. While much has changed in the fire service, the goal of the engine company has not.

Fig. 16–1. Pete Lund was a mentor to many. (Courtesy of Traditions Training, LLC, www.traditionstraining.com.)

One Mission and One Mission Only

An engine company has one initial purpose and that is to bring water to the fireground. This may be done with an onboard pump, tapping into a system of hydrants, or establishing a draft and positive water supply from

In the suburban setting with hydrants, engine companies must determine the water supply system that is consistent and will provide water to the fireground. This can be the act of performing a forward lay from the hydrant to the fire, a reverse lay from the fire to hydrant located in close proximity, or a split lay entering a reported fire location (figs. 16-3, 16-4, and 16–5). To be effective, each of these types of lays requires planning and communication. The lay that is selected is dependent upon the location. For instance, a residential structure fire on a dead end court or cul-de-sac may require a split lay entering the block. What is consistent is that it does not matter what the specific situation is, hose will be laid in the street and we will be shortly filling it. There is a no valid excuse in hydranted areas why an engine company on a reported residential structure fire would not lay out. The following are some excuses that may be used:

- Complacency: "We have been to this place two times today," or "We never run fires."

- Ignorance: "I thought the second due would get the water," or a complete lack of defined procedure within the organization.

- Time consuming: "No one wants to rack 4-inch large diameter hose at 3 a.m. in the rain."

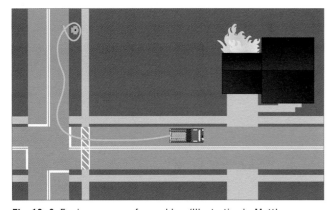

Fig. 16–3. Engine company forward lay. (Illustration by Matthew Tamillow.)

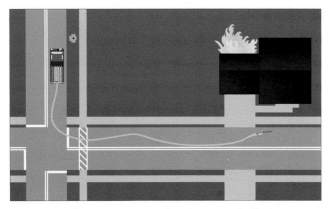

Fig. 16–4. Engine company reverse lay. (Illustration by Matthew Tamillow.)

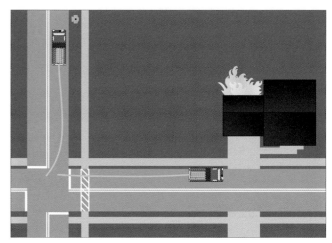

Fig. 16–5. Engine company split lay. (Illustration by Matthew Tamillow.)

Would any of these excuses serve as a consoling statement to the family that lost their children because the fire that was burning in their home was merely waiting for you to arrive and pop the door to show itself? Would the children of the firefighters who died in the line of duty understand why the room flashed over and killed them as they ran out of the 500 gallons of water in the onboard tank? Absolutely not!

Laying Out

Nationally, fire duty is down and EMS calls are on the rise, which means there are fewer opportunities to hone our craft. Additionally, it means we should seize every opportunity to hone this craft when we *do* run the possibility of a fire. Take a moment and break down the run numbers for your company and it may offer a peek into how few times you actually get the opportunity to lay out. Notice we did not say the amount of times you have to layout and, unfortunately, rack hose. Why? Because as professionals who love our craft we look at every situation as an opportunity. It is an opportunity to evaluate how well the fundamentals are being demonstrated, an opportunity to evaluate how well your fireground runs, heck, even an opportunity to get the wrinkles out of the hose!

For instance, say that on average about 20% of the calls a company runs are suppression oriented. Out of that 20%, about 10% are for reported or actual house fires. In a company that runs 500 calls a year, this would equate to 50 times a year that hose would be dropped in the street, fifty times a year that you would get the opportunity to see how well the water supply system can and will operate. This is about once a week that you have the opportunity to demonstrate this skill, or for the naysayers, about once a week you would have to rack hose (fig. 16–6).

Fig. 16–6. Firefighters racking hose. (Courtesy of Roger Steger.)

In the rural setting, the same theory applies because the time it takes to establish a positive water supply can be lengthy. The intricate nature of rural water supply operations first demands an established procedure so that decisions are not made at the most critical time. The scene of the working fire is not the time or place for conversation on appropriate tactics for water supply. Anytime that we run a reported house fire, we must take the opportunity to set in motion the plan we have for water supply. Pair this with the decisive fire officer who quickly investigates and determines the presence of an emergency and we can erase the conclusion that we lay thousands of feet of hose for nothing.

The key for the engine company to bring water to each and every reported and actual residential fire is several fold and begins with correct behavior. Every single engine company, from the first arriving to the last one on the final alarm, must come into the game with finding and delivering water as a priority. This is not advocating for the third engine on the fourth alarm to forward lay from a hydrant 4 miles away. It is advocating for this engine to ensure that the first engine has an established water supply, the firefight is well supplied, and future water supply needs are being met. This can be accomplished by simply listening to radio traffic of the incident. Listen closely to the first engine's on-scene report, the reports from the incident commander on the success of the firefight, and how the incident is progressing. This information will quickly tell the professional engine company if additional water or assistance is needed in completing a water supply.

Once the engine company is on the road and en route to a house fire, the next step in successful water supply operations begins. If you have four engine companies responding to a reported residential structure fire, do all four of the engine companies respond by the same route? If so, how do you overcome the helpful, yet over-zealous police officer who a parks his cruiser directly in front of the hydrant and runs to "help" (fig. 16–7)? As a result, the first engine company is required to park in the middle of the tight street with a parade of engine companies parked behind them. While this may be a once in a lifetime incident, the benefit of having the third-due engine respond in a different direction is beneficial. We all know we are only as good as the last fire we have run.

Fig. 16–7. Police can hinder our operations in their attempt to help.

In a typical forward lay, the first and second engine company will respond the same direction because the second must pick up and secure the first engine company's water supply. If the third-due engine takes a different route, they can view additional sides of the structure, lay out or secure water from a different source, and offer a report of side Charlie conditions to arriving companies. The sequential order is unimportant. The fact that later arriving engine companies are responding from different routes is vitally important.

The action of the third engine company also aids in another function of the initial engine, which is to always leave room for the truck at the front of building. This may not seem like it has anything to do with water supply, but engine companies tend to focus on the idea that the closest hydrant is the best hydrant. Many times this is correct, but when that beautiful beacon of water supply is right square in front of the burning structure and would completely block the use of the aerial ladder or movement of portable ladders, it is not. Consider all the hose we bring to the fire. Forward laying a few hundred feet will leave the front open for the truck, provide an additional water supply directly in front of the structure if needed, and allow for free flow of equipment to the fire (fig. 16–8).

In the fire departments around the country that either forward lay or split lay to many of their residential structure fires, the act of laying out provides some additional benefits. Often the location of the hydrant is within eyesight of the reported structure fire. When your driver or layout man gets out to wrap around the hydrant or perform the split, the officer is provided a "time out." The officer and the crew are provided a 10- to 30-second pause before exiting the cab and beginning the firefight to view the structure and deliver an on-scene report (fig. 16–9). This report is vital, and one of the common reasons for it being eliminated is because of the need to go to work as soon as the brake is pulled. If it is performed 300 feet away and updated with any additional pertinent information in the situation report, then we capitalized on this time out.

In the firehouses around the country where a truck or tower is either responding with or in close proximity to the initial engine company, the act of laying out can provide an excellent tactical advantage. Once the location of the hydrant is identified and communicated, the driver can pull the engine company to the curbside of the street to perform the layout. While this act is being completed, the truck can pass and gain their tactical position. Additionally, it will now allow the engine the opportunity to establish a sound tactical position that will not hinder the truck operations.

Lastly, laying out gives the engine company a moment to cement their plan of attack in the last cone of silence they will have before arriving. The officer views the structure, confirms that there is a working fire on the second floor, and communicates it to the crew. The seconds while the engine company lays out is last moment the firefighters in the back can confirm their assignments and the officer can communicate last second orders. Once arriving on the scene, the chaos begins with screaming family members pleading for you to save their house, fellow firefighters barking orders, and other fireground obstacles.

Fig. 16–8. An engine should pull past or stop short of the structure, depending upon the truck response, to allow for truck placement. (Courtesy of Kentland Volunteer Fire Department, www.kentland33.com.)

Fig. 16–9. While your chauffeur is laying out is a great opportunity to deliver an accurate on-scene report. (Courtesy of Kentland Volunteer Fire Department, www.kentland33.com.)

Since the days of the Roman Empire, firefighters have performed one singular function to put out fires. We deliver water! While the modern fireground poses many hazards the Romans did not face, it does not change the fact that putting the fire out is the greatest lifesaving measure we perform. This act can be challenging at times with fires in which people are reported trapped and the engine company is the only unit on the scene. Regardless of this, science, common sense, and our basic caveman roots tell us that we will need water to eliminate the hazard that is causing this problem. We must never forget this mission and always be prepared to perform it on every reported and actual residential structure fire.

The Backstep

Developing a Water Supply Plan

The delivery of water is the most important function the engine company performs on the fireground and it has not changed even though our environment has. The fires, the buildings, the apparatus, and the personnel and staffing have changed, but to mitigate fires we still provide the same agent: water. Paramount to completing this task is understanding, developing, and implementing a water supply plan for every reported fire. Note that this is not just confirmed fires but all reported fires as we must be prepared and be proactive, not reactive, to an emergency incident.

Analyze the number of emergency incidents you run on a yearly basis. From that figure, determine the number that are suppression related. From that figure, determine the number of incidents that are for reported fires and we believe you clearly see that you have a limited amount of time to hone your supply line skills. Make it a mandatory action anytime you are dispatched for a reported building fire that the engine company will lay out or simply lay supply hose from a static water source. If not feasible, mandate that a water supply plan be established. This is accomplished by first having detailed maps of your response area denoting the location of water sources (hydrants, ponds, wells, etc.). Once you have water supply sources established, institute common terminology for the three types of supply hose deployments:

- **Forward lay.** The engine company will lay a supply line from a hydrant to the fire.

- **Reverse lay.** The engine company will position at the fire and personnel will deploy dead loaded hoselines from that position to the fire. Once the attack hoseline has been selected and deployed, the engine company will then proceed to the water source. Another variation of the reverse lay is for the attack engine (first arriving) to position at the fire and notify the next arriving engine of a reverse lay from their location. The next arriving engine will stop at the attack engine's position, deploy their supply line to the attack engine, and then the second arriving engine will proceed to the static water source. This tactic is often dependent upon arrival order, as the path to the static water source can quickly become obstructed by arriving units such as trucks or rescue companies.

- **Split lay.** The engine company will identify an area where a static water source is in a location that is not in direct line of the fire location. This could be a pipe stem, flag lot, or a dead end court or cul-de-sac. The first arriving engine will notify the second arriving engine of the location that they will deploy their supply line from and place it around an immovable object (tree, wall, etc.). The second arriving engine will position at the denoted location, complete the connection of the supply lines (connect the first engine's hoseline to second engine's hoseline), and proceed to the static water source. It is absolutely crucial that the second arriving engine driver, or another firefighter who is clearly delegated the task, hook the first and second engine hoselines together. While it may seem elementary, it can be overlooked in the chaotic environment of a house fire and will have a disastrous impact on the outcome.

17 One More Room: Making the Stretch

Recently, a new building code came out that stated, "All residential structures must ensure that any fire located in the structure will be within the distance of their respective fire company's pre-connected hoseline."

Helper or Handcuff?

Ponder that thought for a moment. The building industry has finally come to its senses and will now make our job easier. Of course, the above statement is an absolute bold-faced lie! Yet, if you travelled across the United States and reviewed engine company operations you would think that we operate in this manner. It is safe to assume that you would witness engine companies that are ill-prepared for deploying an attack line to a fire 400 feet away from where they have positioned, when all they have are 200- to 300-foot pre-connected hoselines.

The tactical issue of overcoming long setbacks due to large yards, alleys, and other man-made and geographical obstacles is difficult. Yet this does not mean we can let this challenge stop us, as we have to complete our task and get the hoseline in place (fig. 17–1). Conversely, not every fire we go to is going to become a logistical and tactical nightmare and may, instead, be quickly extinguished. Some fires will be 50 feet from where we position and will be extinguished within minutes by your highly competent crew. The overriding element of our profession that has remained consistent is that we deal with something new every day. There is no such thing as routine.

One of the most innovative, and at the same time crippling inventions for the fire service was the pre-connected hoseload, often called crosslays or mattydales. The invention of this hoseload allows for quick racking and deploying and can greatly assist in limited staffing departments. But the implementation of this hoseload has led to a measure of the dumbing down of the fire service.

Fig. 17–1. Firefighter performing a long line stretch. (Courtesy of Kentland Volunteer Fire Department, www.kentland33.com.)

Specifically, the implementation of the pre-connected hoseline provided a safety blanket that firefighters could rest under, assuming that all fires would be within this distance. For many fires and many years, it worked seamlessly and provided the correct tool. In modern residential structures we face obstacles that were not present 30 years ago. The size of the residential home has increased exponentially, the construction methods increase the fuel load and fire path intensity, and the contents burn at much higher heat release rate. These factors present setbacks and obstacles, resulting in larger and more intense fires, which require more hoselines and more water. Additionally, fires that were historically in compartmented rooms are now travelling rapidly across a wide open floor plan (fig. 17–2).

Fig 17–2. A McMansion on fire can present more like a commercial fire.

The first step in eliminating a complacent engine company and ensuring that when you arrive at the working house fire you can seamlessly extinguish it is preparation. Is your piece of apparatus ready for a parade or ready for a residential structure fire? Dissect this statement, and you understand that being a combat-ready engine company is having a piece of apparatus with hoselines racked appropriately. The statement, "Looks pretty, pulls pretty" rings true in this application (fig. 17–3). The tools that are needed, based upon experience and defined tasks, are mounted and ready for rapid deployment. These are all traits of an engine company that is ready to overcome any obstacle or hindrance faced on the modern fireground.

Engine 242 responds as the third arriving engine on a working residential structure fire. Engine 230 has laid 600 feet of supply line in a forward lay to a dead end court and is aggressively fighting fire. Engine 136 secures the water supply for Engine 230. The first tower ladder has positioned on side Alpha and is performing rescues. The tower ladder has also blocked any future responding apparatus from getting into the block. Engine 242 arrives, checks in with command, and is given the assignment to stretch a hoseline off of Engine 230 on the front of the structure. They trot down the court to Engine 230 and quickly scan the engine for an available hoseline. The crosslays were deployed by the first engine. The last 1¾-inch off the rear was taken by Engine 136. The only line left is a 3-inch supply line. You have to get the line in place and complete your assignment, but how?

This scenario happens on a daily basis across the country. Sometimes it is negated when the fire is quickly extinguished. Other times it is an exercise in futility. All members who ride and operate on an engine company must remember the mission. We deliver water. This is not limited to when we are first or second arriving. No matter when we arrive, this is our core mission. This is not to say that an incident commander may have more than enough water on the fire and will deviate and assign a different task to the engine company. This tactical move does not change the focus and mission of the engine company as they train, prepare, and respond to emergency incidents.

Fig. 17–3. Hoselines racked correctly will deploy correctly. (Courtesy of Kentland Volunteer Fire Department, www.kentland33.com.)

Preparation, Preparation, Preparation!

When we return to our scenario above, what would be *your* solution? There is not one simple answer, but rather there should be several options that you can offer. This is based on your preplanning and preparation for such an incident. A commonality among the fire service is our level of preparation and ability to adapt and overcome. Much like forcible entry training preaches to have not just one plan for getting through a door, but have a plan A, B, C, and D, we must have the same approach to getting water on the fire. This ability to recognize, adapt, and overcome will not occur by simply hoping for the best when are confronted with the situation. Nor will it occur by having a board meeting at the backstep of the engine company to discuss options while the building is burning and your fellow firefighters are dependant upon the additional flow you are expected to provide.

The foundation of overcoming these challenges that are omnipresent on today's fireground is in your preparation and knowledge. Preparation for these engine company challenges begins with knowing the area that you serve. This means not simply knowing the address of the local convenience store or your favorite watering hole. It is knowing response routes that aid in positioning, the buildings that present challenges to access, or residential structures that will demand a higher fire flow (fig. 17–4). Once we recognize these situations we can begin to build our plans for overcoming the challenges and getting water on the fire. The next step is to actually put the plans in place. The reality is that to accomplish this task it will take your short staffed crew greater effort to accomplish multiple tasks simultaneously. Each member of your crew will have a specific task that will need to be executed to perfection and in a timely fashion. This only occurs through practice.

In the scenario presented previously, the tactic selected may be to have one member grab the high-rise rack, another grab a gated wye, and the third member grab sections of the 3-inch hoseline deadloaded and stretched to the front door of the house. This is a scenario that would quickly put a line in place *if* all of your members are prepared. One hiccup, such as the firefighter not knowing how to deploy a deadload line, another firefighter having no idea where or what a gated wye is, and the last member not knowing how to remove the straps from the high-rise rack, and our formerly seamless operation has become cumbersome. Instead of an asset to the fireground, we are suddenly a liability.

Let's change the scenario to not only be an issue with limited availability of hoselines but also overcoming the obstacle of a long setback. In the engine company, the setback is the distance from where the engine company has positioned to where the seat of the fire is located. It is safe to assume in today's society of putting more on less that you could be confronted with a home that has large expanse of a front yard, or a large square footage that may take you 200 feet inside the front door to reach the seat of the fire. Regardless of the specific scenario, the 250-foot pre-connected crosslay you have on your engine company will not reach your tactical objective. A plan must be in place for the engine company that arrives first, second, and third to reach that objective.

The long line debate may be as old as the fog versus smooth bore nozzle debate, and it will have as many advocates and naysayers, too. Ultimately, regardless of your position, the mission remains the same in that you must get a line in place. The fire does not care if you think throttling your pump above 250 psi is bad, it does not care if you think 400 feet is too long for a 1¾-inch hoseline. We must get water on the fire. Hopefully this consistent message is getting through.

Fig. 17–4. Firefighters participating in a street drill

Under Pressure

The long line conversation demonstrates yet another hat we must wear on the engine company. Every member of the engine company must be a hydraulic engineer, well versed in the why and what of the water delivery system. They must not arbitrarily accept what is passed on through the years but rather, through practical testing and mathematical calculation, determine what is valid. For instance, a common argument in the fire service against long lines is that the pump pressures will be too high. They could exceed 250, 260, 275 psi and that is not good for the pump. Why? Some fire departments conduct a yearly hose test by raising to these pressures or higher. Couple that with the fact that most 1¾-inch hose issued before 1979 is service tested from 300–600 psi! These numbers are well above the pressure we need for a fire fight, water that is needed for firefighters in peril, and for citizens that may be gasping their last breath of fresh air waiting for the fire to be suppressed.

The difference lies in that one number represents testing and the other number represents the life and death situations we face every day. We should not accept any of these numbers without understanding the why, understanding what these numbers mean, and more importantly, understanding our hose. To accept what one manufacturer says when you purchase a piece of equipment is naïve and shortsighted at a minimum. Take the time to determine the capabilities and limitations of your particular engine company, hoselines, and nozzles. A company performing flow testing on their engine company and flowing their hoselines can quickly dispel or prove what their vendor espoused upon purchase. Regardless of the outcome, they will have accurate information for their engine company (fig. 17–5).

With modern hose construction that allows water flow through hoselines with less turbulence and essentially makes the water slicker, we can achieve long line success. Pair this with a nozzle capable of higher flows (150, 175, 200 gpm) at lower nozzle pressure, and we have the perfect water delivery system. A water delivery system is needed that can handle both the bread and butter residential fire that is 100 feet from where the engine company is positioned and the residential structure that has 4,000 feet of living space and is 400 feet from the street. Once the water delivery system is ready for your firefight, we need to have the plan or plans in place for overcoming fireground obstacles.

Fig. 17–5. Engine company flow testing is a definitive measure we can take to have accurate flow capabilities.

The Backstep

Achieving Long Lines with Simplicity

When presented with a long line challenge—a situation in which you must stretch a hoseline greater than distances of 300 feet—we must have multiple options in place. Since you are combat ready you have already established a company of competent professionals. The engine company is set up for success and ready to overcome anything from the common to the most unorthodox situations faced. Now we must have our tactics ready to handle the challenge. The solutions are limited to your imagination and determination to test and validate. The following are few options we have vetted that work and can work on your next fire.

- **Dead load lines.** The preconnected hoseline does offer a fast and effective solution to getting water on the fire for many fires we run, but it also can be a limiting factor. Keeping one attack hoseline (1¾-inch is our preferred choice) in a deadload rack off the back of your apparatus is practical and can overcome most long line issues. 500 feet of 1¾-inch in a flat load with last 100 feet in a horseshoe shoulder is our preferred rack. This will allow the nozzleman to take the 100-foot horseshoe on his shoulder and advance to the fire while the remaining members can effect the stretch (fig. 17–6).

- **High-rise rack.** It does not matter if you have a high-rise or not, having this 100-foot bundle of hose racked in an accordion load secured with straps works! Change the name to apartment rack, setback rack, it does not matter. In this scenario the pre-connected hoseline is stretched and another 50 feet is needed for the firefight. The backup position can grab this pack and advance to end of the preconnected hoseline, remove the nozzle or breakaway tip, attach the 100-foot rack, and then attach the nozzle to the end of the new long line. In under a minute the line is in place and water is on the fire. If the distance was underestimated and the rack was not brought with the attack crew, the solution is simple. The advantage of this pack is that your engine chauffeur can complete the long line addition for the attack crew. The chauffeur will disconnect the preconnected line at the first 50-foot coupling. Attach the 100 feet of the high-rise rack and then reconnect the hose to the preconnected hoseline. The backup position will need to run back and ensure the new 100 feet of hose does not get caught on any obstacles between his position and where the chauffeur made the connection (fig. 17–7).

Fig. 17–6. A dead loaded hosebed may be useful. (Courtesy of Kentland Volunteer Fire Department, www.kentland33.com.)

Fig. 17–7. High-rise packs can useful in many applications other than just a high-rise fire.

- **Pony sleeves.** One of the preconnected hoseline's hindrances is that it is preconnected. Think for a moment when you have to disconnect the preconnect. The difficulty is in climbing up on the engine company, reaching down into the tray where the crosslay is located, and trying to uncouple the hoseline from the swivel valve. Most likely it is a coupling that has not been exercised in years and feels as though it is welded together. None of these factors make the process of disconnecting efficient or expedient. If we return to the scenario with Engine 242, the easy and rapid removal of their own pre-connected cross-lay which they could bring to Engine 230 would have solved their issue. While we advocate for never relying upon a single water supply and fire pump, some

situations will dictate that we have to. Engine 242 could have brought their 200-foot cross-lay on their shoulders to the chauffeur of Engine 230, commandeer a discharge, and deploy their line within minutes. The key to this technique is the addition of one small section of 1¾-inch hose! All pre-connected hoselines should have a 6- to 8-foot section of 1¾-inch hoseline that will run from the swivel valve to the first 50-foot section of the attack line. In this configuration, the officer can voice to the crew that they will be removing the entire crosslay for a long line. The nozzle operator can pull the entire cross-lay onto his shoulder and step away from the engine. The backup man can simply disconnect the 6-foot pony sleeve as pictured. The hoseline can now be carried to another engine or location to be deployed. Additionally, if you have two side-by-side cross-lays, the first crosslay can be deployed, and the second disconnected and taken to the end of the first. Once at that location, the two lines can be connected and advanced. With a crew of three, a 400- to 600-foot line can be constructed, advanced, and flowing in minutes (fig. 17–8).

Every day in every fire department across the country, firefighters are facing more and more obstacles to performing our job. This can range from building construction practices to street designs to fire load characteristics. All of these challenges must be overcome so that we can do what has always been required of us, get water on the fire. The tips offered in this chapter should be the catalyst to start the design and implementation of your combat-ready water delivery system.

Fig. 17–8. Pony sleeves can take a time-consuming task and make it a rapid process to deploy a hoseline to a remote location.

18 The Last Line of Defense: The Firefighting Nozzle

Let's start this chapter with a very easy test that has simple yes or no answers.

1. Do you have a standard or mandate for the daily checking and inspection of your SCBA?

2. Do you have a standard or mandate for the inspection of your personal protective equipment (PPE)?

3. Do you have a standard or mandate for the daily inspection of your respective piece of apparatus?

4. Do you have a standard or mandate for the daily inspection of your firefighting nozzle?

We would venture to guess that you answered an unequivocal yes to the first three questions. The fourth question you probably had to ponder for a moment. Ask yourself, "Do we have a policy?" This is probably a no because if you have to ask yourself that question, it means you don't have one. You can take solace in the fact that you are not alone. This practice is lacking in many fire departments across the country.

Take a moment and think about how ridiculous it is to have a standard or mandate for the first three and not one for your firefighting nozzles. A policy is in place for:

- The inspection of the piece of apparatus that delivers you to the fireground and will either deliver water to the fire or assist in ascending to upper floors (fig. 18–1).

- The inspection of the personal protective gear that allows us to work in extreme temperatures.

- The inspection of the SCBA that allows us to cross the threshold of the residential structure that is involved in fire and battle in a toxic, zero visibility environment.

Fig. 18–1. The engine company doing what it does best—getting water! (Courtesy of Kentland Volunteer Fire Department, www.kentland33.com.)

Yet we do not have a policy in effect for the inspection of the one piece of equipment that aids us in the most dangerous position on the fireground. The one tool we rely upon to operate seamlessly when we are inches away from the enemy. We overlook this vital tool when we have inspection mandates for all of the other items. The firefighting nozzle is truly the last line of defense for the firefighter. It is the tool that eliminates the one hazard that causes civilian and firefighter injuries and deaths. If it does not operate, our solution (i.e., another hoseline) is minutes away at best and those are valuable minutes we do not have when advancing into a room involved in fire, moments from flashover.

A Day in the Life of a Firefighting Nozzle

To put it in greater perspective, think if you treated your nozzle as you do your SCBA. We store our SCBA in protective brackets inside of the apparatus along and after every use have mandatory inspections, hydrostatic testing, and thorough inspection for any damage (fig. 18–2). Most likely the firefighting nozzle hangs off the back of an engine company, beating back and forth against the apparatus as you respond from emergency to emergency (fig. 18–3).

Fig. 18–3. Nozzle covered in grit and vital parts broken from hanging off the back of an engine company

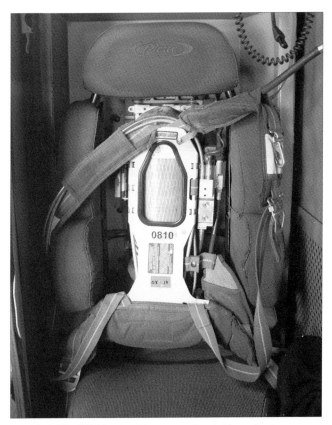

Fig. 18–2. The SCBA is stored in a secure bracket in a climate-controlled cab to ensure it is ready to work.

It is subjected to severe temperature fluctuations several times a day going from an apparatus bay that may be heated to 70°F to the outside that can be as low as 20°F. It is exposed to road grit and grime and constant UV rays that all work to degrade the components. A simple tip to prolong the life your firefighting nozzle during the winter months and ensure it performs on the fireground is to simply wrap the tip of the nozzle with a plastic grocery store bag. Place the bag over the tip and secure it with a rubber band just prior to the bale (fig. 18–4). With this in place you can provide protection for the moving parts of the nozzle from road grit, grime, salt, and whatever else you drive over. When you check your stream prior to entry, which is mandatory, the bag and rubber band will shoot off from the pressure of the water, leaving the tip nozzle ready for the firefight.

In many cases the nozzle is never inspected, cleaned, or maintained, but we still expect it to perform perfectly every time. It would seem our expectations are bit far-fetched and need to be re-evaluated.

Fig. 18-4. Plastic grocery bag over a nozzle for protection

There is hope! The first step is change the behavior of our firefighters. That happens through proper training. In the classic war movie, *Full Metal Jacket*, the drill instructor played by R. Lee Ermey is conducting his evening routine of lights out by having the soldiers recite the rifleman's creed while sleeping with their weapon. The purpose being that each solider must learn that his weapon, his last line of defense, is a part of him and must be as valued as the life of one of his crew. Novel idea, huh? We are not advocating you make all of your personnel recite a creed and sleep with their nozzle, we are advocating for developing behavioral change. Consider for a moment implementing the nozzle position's creed.

> *This is my nozzle. There are many like it, but this one is mine. It is my life. I must master it as I must master my life. Without me my nozzle is useless. Without my nozzle, I am useless. I must operate my nozzle truly. I must extinguish fire quicker than the enemy who is trying to kill me. I must kill him before he kills me. I will. My nozzle and I know that what counts in the firefight are not the number of lines we deploy, the length of hose we use, or the smoke we eat. We know that it is the hits that count. We will hit.*
>
> *My nozzle is human, even as I am human, because it is my life. Thus, I will learn it as a brother. I will learn its weaknesses, its strengths, its parts, its accessories, its gallonage, and its bale. I will keep my nozzle clean and ready, even as I am clean and ready. We will become part of each other.*
>
> *Before God I swear this creed. My nozzle and I are the defenders of my community and my company. We are the masters of our enemy. We are the saviors of lives.*
>
> *So be it, until my brothers all come home and there is no fire.*

That about covers it, right? A roomful of new firefighters eager to jump into this new opportunity to serve their community and this is what you present them. The message is clear that not only is this important, but the person who is going to lead me is committed to it. There is not a part of the creed that does not hold true to what we do, day in and day out.

The first step of a behavioral change is addressed, so the next step is to understand why this has not occurred in the past. *NFPA 1964* provides a comprehensive checklist that can be implemented in any fire department. Unfortunately, many departments either do not have, or do not know how to access NFPA standards. This adds to the belief that nozzle inspections do not occur due to ignorance and not malice. Fortunately, this is a problem that is easily corrected with proper training and the information provided below.

The Nozzle Inspection

To be able properly inspect a nozzle, some general information must be consistent among all personnel. This information will not only assist in properly checking the nozzle, but also aid in your knowledge of this vital tool.

- Know all of the gallonage settings on your fog nozzles or flow of the smooth bore tip (fig. 18-5).

- Do you have a flush function on the fog nozzle gallonage selector? Know that the flush function will allow for small amounts of debris to pass through the nozzle (fig. 18-6).

- Know how many clicks in a zero visibility environment it is until you are on the flush setting.

- Be familiar with which way to rotate the nozzle in zero visibility to ensure you are using a straight stream (right to fight, left for lobster). If the nozzle is in the incorrect stream when you enter the fire room, the steam will endanger your crew and any potential civilian victims.

- Do you have a personal or departmental procedure for a nozzle malfunction? We must have a quick mental checklist to run through in order to diagnose and mitigate the situation (i.e., flush function or use of breakaway nozzles). Decide if this is a Mayday trigger in your department.

Fig. 18-5. All members must know the settings on your gallonage ring.

Fig. 18-6. The flush function enables a nozzle to pass a small obstruction.

For a fog nozzle, regardless of the manufacturer, the minimum shall be checked:

- Overall condition/appearance
  - Evaluate the extent of damage that the environmental conditions have had.
  - Take it off and flush it with clean water to remove any grit or grime.
  - If the appearance is tattered or worn, is it cosmetic or is it structural?

If firefighters are ever unsure of what is considered cosmetic, structural, or worn, they should always consult with the manufacturers. They can provide invaluable expertise on the nozzle.

- Condition of the washers (fig. 18-7)
  - Expose every washer in the nozzle assembly. There are washers at each connection point.
  - Evaluate the condition of each washer, as they can dramatically affect the flow of the nozzle. Remove the washer and squeeze it between your thumb and forefinger. If the washer quickly returns to normal shape and is not cracked then it is okay. If not, replace it immediately.

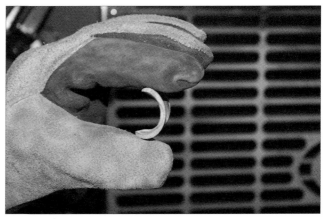

Fig. 18–7. All washers should be removed and inspected.

- Free movement of any movable parts (bale, couplings, gallonage selector, etc.)
  - Ensure anything that moves and is essential to the delivery of water operates freely.
  - The bale should open with a reasonable amount of resistance, not so hard that it

can't open and not so loose that it could open accidently. Check any setscrews for tightness.

- Evaluate the condition of the bale and the attachment nut. Many of the bales utilized today are plastic and can be easily damaged.

• The gallonage selector is on the correct setting, if applicable.

- The amount of water you will typically flow should be decided prior to a fire. Discuss with your personnel based upon the fire load in your first-due area and stream capability. If you have all single-family dwellings, your fire load and nozzle compliment will differ from a company situated among nothing but commercial high-rises.

You may need to change the gallonage in the middle of the firefight, so you must ensure the selector moves freely.

While the smooth bore nozzle has fewer movable parts and the gallonage is limited to the interior bore diameter of the tip on the nozzle, the same basic tenets of the check apply to the fog nozzle. You must ensure that the movable parts operate freely and that the tips are securely fastened, not lost or loose due to wear and tear. Additionally, the quality of all the washers in the couplings and the removable tips must be assessed.

The smooth bore nozzle is very effective but demands that the operator and officer have a firm knowledge of the capabilities of the flow and assessment of fire load. While the fog can have the versatility of selectable gallonage, the smooth bore is limited to the bale and the tip on the nozzle, but is also always set to flow a solid stream. If you are committed to the smooth bore then you should carry spare tips in your turnout pocket. This will allow the experienced operator to evaluate the effectiveness of the current tip and adjust to a smaller or larger tip, if necessary. As with the fog nozzle, the change in flow must be communicated back to the pump operator.

The nozzle is truly the last line of defense when we make the decision to mount an interior attack. That tactical decision demands that personnel are aware of the operation, capabilities, and limitations of their respective nozzles. To be truly combat ready, all firefighters must be vigilant in performing a thorough check of the nozzle and ensuring its effective operation.

Planning For the Unplanned

It is incumbent upon the engine company officers to foster dedication to professionalism by discussing hypothetical tactical situations with their company. Develop a plan for when the nozzle doesn't work and propose solutions to ensure that fire suppression continues. Discuss the strengths and limitations of the fog and smooth bore nozzles based upon your first due, staffing levels, and experience. Lastly, as with all training, conduct drills with personnel in the environment that we work in normally: zero visibility (fig. 18–8). Methods to simulate real life situations include blacking out the facepieces, wearing firefighting gloves, and describing each part of the nozzle. In zero visibility, firefighters should be able to determine:

• Which direction to rotate to straight stream

• Location of the grooved band on the breakaway nozzle so they can extend the line or remove to clear an obstruction

• Location of the raised lugs, which denotes the selectable gallonage

• Which way to rotate to get less gpm or more gpm and the flush function, if present

Fig. 18–8. Drills should be conducted in zero visibility.

A commitment to completing our job is demonstrated through ensuring you and your equipment are ready to go when the bell rings. Develop and enforce a policy on checking your nozzles. Learn about the capabilities and limitations of the nozzles. Finally, train for the unexpected with your nozzles in the environment in which we operate.

The Backstep

Nozzle Obstruction Drill

The possibility of an obstruction entering the pump of the engine company, flowing through the hoseline, and lodging into the tip of your firefighting nozzle exists every time we operate. The solution many will offer is to simply back out and call for another line, but if you have a truck company operating with you, this is not an option. If you leave, they are left vulnerable and could be injured or killed. Prior to retreating with a broken stream, we have to:

1. Immediately call for a second line to advance to our position.

2. Immediately begin your troubleshooting checklist for overcoming nozzle obstructions.

The second point demonstrates the need for all engine companies to have a troubleshooting checklist for nozzle obstructions. This can be accomplished through a two-fold process, the first of which is the knowledge gained through nozzle inspection. The inspection ensures personnel will know the capabilities and operation of all parts of the nozzle. This knowledge makes it possible for personnel to overcome an obstruction in zero visibility. The second step is to implement a realistic drill to recreate these conditions.

Equipment list:

- 200 feet of 1¾-inch hoseline
- Firefighting nozzle, preferably a fog nozzle
- Engine company with an available 1½-inch discharge
- Obstruction

The obstruction can be made with several EMS gloves balled up or newspaper wrapped in duct tape. It must not be bigger than the interior diameter of the hoseline you are using so it can travel up to the nozzle.

Drill procedure:

1. Have one firefighter assume the nozzle position with his or her facepiece blacked out.

2. Allocate one 1½-inch discharge on the engine company for the hoseline. If the engine company has a front bumper line it will work well.

3. Place the obstruction into the end of the hoseline that will be attached to the engine company (fig. 18–9).

Fig. 18–9. A simple obstruction can be made out of EMS gloves.

4. Attach hoseline to the discharge and discharge the appropriate pressure for the hoseline.

5. Operates the nozzle in the fully open position. The obstruction quickly passes through the hoseline and lodges into the tip of the nozzle.

6. Ensure that the nozzle firefighter identifies the obstruction by the sound of the stream and the broken stream.

7. Ensure that the nozzle firefighter notifies the officer of the obstruction and need for second line to be advanced.

8. Rotates fog nozzle to the flush function, if available, to try to flush the obstruction past

the tip. This may be repeated several times if necessary.

9. If unsuccessful, remove the breakaway tip, if applicable.

10. With the tip removed, inspect the tip and shutoff with gloved hands for the obstruction. If felt, remove the obstruction and reattach the tip. Determine if flow is appropriate (fig. 18–10).

11. If unsuccessful, open the bale fully and maintain firefight until second line advances.

Fig. 18–10. Remove the tip to clear the obstruction.

This simple drill requires limited equipment that can be found in any firehouse. It provides the skills that could save lives and property. Practice this often and we can ensure that when this occurs on your fireground, you can overcome the obstacle.

PART 4

Truck Company Operations

19 A Bull in a China Shop: Truck Work

Ah, the firehouse. It is a comfortable place to be. You are largely among friends, insulated from the outside world. At times, the firehouse can be the most loving place on earth. Only we who are firefighters truly know what it is that we do when we get on the rig or make the push down the hall. Politicians, neighbors, family, and friends may think they know what it is like, but they don't. They can't. Looking out for each other is what we do.

However, the love can be "tough love" at times. If your firehouse has more than one piece of equipment in it, you might know what I am talking about. If you have more than one rig inside those firehouse walls, many times you will find that there is a stark separation between members in the company, like the differences between Oscar and Felix from the TV show *The Odd Couple*. Sometimes it's hard to feel the love. Occasionally this divide is as obvious as a line painted right down the middle of the apparatus floor, sometimes it is a bit more subtle.

Together, Separately

Career or volunteer, the split can be palpable. What is this divide to which we refer? Are we not all firefighters, at the firehouse to fulfill a common goal? Of course we are, but what we are referring to is the different manner in which we operate to accomplish that goal. We know that engine companies are tasked to bring hose and water. Truck or ladder companies are normally associated with forcible entry (FE), search, and ventilation. The historic actions of engine company and ladder (truck) company members and their functions at fires are indeed different (fig. 19–1).

We know that throughout our firefighting history many things have changed and continue to evolve. As with many public safety organizations in society today, the fire service is tasked to do more with less. Multi- and dual-role apparatus are becoming common in many places. But if you are a firefighter who has the luxury to be assigned to a specific company (engine or ladder), you will obviously spend more time developing specific mastery of the tools and positions for that assignment.

You should know what your function on the fireground will be before you even get to the fire. You should be well prepared to carry out your role in your specific company, whether it is with the engine or the truck. While we should not focus on our specific role so intently that we put blinders on to those operating around us, for the best team effectiveness we need to be masters at what we are required to do.

Fig. 19–1. Engine and truck companies sharing quarters

As humans, we all have varying levels of desire when it comes to belonging to groups. Psychology tells us that we all have certain needs and wants as it relates to acceptance, association, and fitting in. In general, we like to be part of such groups. Many times that is one of the reasons that we join the fire department. Within the department, some firefighters naturally gravitate to the engine company. Others are more inclined to enjoy the work that the ladder company is tasked to complete.

Any time we create separate and distinct groups we run the risk of causing tension between them, even if we are ultimately working toward the same common goal.

I relate a famous quote from the Greek philosopher Aristotle regarding the adaptation that occurs with individuals as they become masters in their particular company. He stated, "We are what we repeatedly do. Excellence is not an act but a habit." In garnering company-specific mastery, there is a trend to adopt the persona of that company. The firehouse can be described by the persona of its members. Some are "truckies" and some are "enginemen." These are just a few of the nicknames attached to those firefighters who are in the process of becoming masters of their specific craft.

There have been epic pitched battles between the engine company and the ladder (truck) company. Some can be aired publicly, others should remain in-house. As seen in many firehouses, we are of the belief that the slogan "What you see here, what you do here, when you leave here, stays here" is good for all firefighters. But in-house or not, the rivalry between the engine and truck becomes fodder for T-shirts, stickers, and slogans. I am sure that you can recite a few as you read this. Some sayings and slogans that pit one side against the other are "Truckies are real firefighter helpers," "Engine firefighters are truckie speed bumps," and "Trucks go to medal day, engine companies to the burn center."

In many instances around the firehouse, the chores are divided up into an engine vs. truck concept, just like on the fireground. If the engine company has the house watch duties, the truck is tasked with cooking the meals (figs. 19–2 and 19–3). One month the truck cleans upstairs, the engine downstairs, and then they switch. The perceived divide can actually add to the esprit de corps of the firehouse. The friendly jovial firehouse back and forth banter "across the floor" is usually healthy and can keep members on their toes. Senior members and company officers may need to step in from time to time to keep things civil, but as we mentioned in a previous chapter, taking pride in what you ride is absolutely a good thing!

"Truckies" have earned a dubious reputation from their engine brethren for acting as cavemen and barbarians at fires, hence the title of this chapter. Many of the tasks ladder companies accomplish at fires involve opening up and breaking through things, so we can see where the parallels are drawn. Those who gravitate to the ladder company, like all firefighters, should not only be fundamentally strong physically, but also have a firm grasp of why they are breaking what they are breaking (windows, walls, or doors). The key point of most operations is having your members understand what they are doing and, more importantly, why they are doing it.

When the bell sounds or when the tones go off, the "engine vs. truck" firehouse banter takes a backseat to the response. From the receipt of the alarm, all members should start to begin their own personal size-up, focusing on what their roles will be at this fire. For the purposes of this chapter, we will focus on the roles, positions, and functions historically assigned to ladder or truck company members.

Fig. 19–2. Engine member doing house watch duty

Fig. 19–3. Ladder company member cleaning up during committee work

At residential fires, there are typical duties for engine companies and truck companies to accomplish. We will use the words *truck* and *ladder* interchangeably throughout this chapter. In fact, the rig that brings you to the residential building fire does not even have to have a hydraulic ladder of any variety (straight stick or ladder tower) on its roof to do what would normally be considered "truck work." Truck company operations are much more than the just the vehicle that you arrive on.

Ladder Company Functions

While the ladder company's apparatus are often the most expensive pieces in our inventory, any trained member on the fireground can accomplish the duties that ladder company members perform at fires. In fact, if a ladder company is not present early in the incident, the incident commander should assign an able-bodied engine or rescue company to fill in and accomplish the roles that the truck would normally complete, as most of these duties need to be occurring simultaneously with the engine company's fire attack.

While members of engine and ladder companies have historically had separate and distinct duties, they are absolutely interdependent upon one another. Truck companies that open the ceiling and walls in a room with fire behind them will be run out of that room without the protection of a hoseline. Conversely, the engine company cannot open all the walls in a room with the nozzle and hose alone.

Ladder companies operating independently from the engine company advance will increase the exposure to firefighters and civilians alike. If we do not have proper coordination and communication, we will likely cause greater damage throughout the home. Engine and ladder companies must work together in a delicate balance of harmonic movement for ultimate success and completion of the mission.

The successful outcome of lives and property saved is the result of our combined efforts at the fire. Most of the duties which are assigned to the truck company should be occurring simultaneously with fire attack. Think about the duties that need to be performed at every fire. We need to address the "LIP" principle in fighting residential fires: life safety, incident stabilization, and property conservation.

Think for a moment how the typical ladder company actions best fit into these principles. What follows is a list of typical actions (in varying degrees, dependent upon conditions found). It could be perceived as an updated view on Chief Lloyd Layman's 1955 RECEO-VS concept. Let's address typical ladder company actions individually and discuss the associated operations when faced with the residential building fire. They form an acronym that may be recounted as FR L VOSU. These are:

- Forcible entry
- Rescue
- Locate seat of fire
  - Confine
  - Control
  - Extinguish
- Ventilate
  - Life
  - Fire
  - Ladders
- Overhaul
- Salvage
- Utilities

Forcible Entry

The only thing that is consistent in residential forcible entry is that it is consistently changing. We note two things happening on this front. With the downturn in the global economy and residential break-in thefts on the rise, homeowners are increasing the security factors for their dwellings to keep their possessions safe. It is certainly rare to find an unlocked or unsecured private residence today. Secondly, the availability of home security systems, which can now be bought on the Internet and in big box hardware stores in most communities, ensures easy accessibility to security hardware.

While urban firefighters may have seen the increased level of security devices on residential dwellings in the inner cities for some time, forcible entry challenges are moving at great speed toward the normally tranquil, rural suburbs.

One of many underlying themes that run through several chapters in this book is that knowing *your* buildings gives you an enormous advantage. Study your buildings to know what is in your communities. Understand and plan to overcome the greatest frequency of FE challenges you will face. If you want to see what is new in front door construction or the latest set and style of locking mechanisms, all you need to do is take a stroll through the local big box hardware store (fig. 19–4). Take the company over and look down the aisles of entrance doors. Note the style, design, and construction methods used. Peruse the hardware and lock section. By doing so, you will greatly increase your odds to stay on top of the newest security devices coming to a residential home near you. This will keep you thinking ahead of the next challenge that may be on its way to your neighborhood.

Fig. 19–4. Big box stores can offer a peek at what you may see coming with regard to home security.

Let's address the style and type of doors that we will likely encounter in private dwellings. First of all, the most prevalent residential buildings doors are normally all inward opening, whereas most commercial occupancies swing out. The directional swing (either left or right) of the door can vary depending on building layout and homeowner needs (fig. 19–5).

Private dwelling entrance doors are most commonly composed of either wood or fiberglass and are normally seated in wooden jambs. The primary locking mechanisms found in residential doors are knob locks, tubular deadbolts, slide bolts, and chain devices (fig. 19–6). Some newer products have hit the market such as the "door club" (www.theclub.com). This residential locking device is located in the floor of the home, just at the lower area of the door. Devices like these are not particularly difficult to defeat, but it and others like it are new. As mentioned earlier, we must stay abreast of the new forcible entry challenges as they come to market. Take time to look for them, share them with others firefighters, and devise plans to defeat them before the alarm sounds.

When it comes to forcible entry, we must have options. We must have a plan A, B, C, D, and so on. Even residential forcible entry does not have a "one size fits all" method to success. In order to have multiple plans, we must be adaptable with our tool selection. We also must recognize when our methods aren't working and when it is time to move to the next plan.

Our forcible entry tool selection and the methods we use to overcome FE challenges must be compatible with the mechanisms that we face on a case-by-case basis. Just as no two fires are exactly the same, there is a great deal of variation between doors and locks in the residential building.

Fig. 19–5. A standard residential front door

on a wooden door may work, but it will likely cause the wooden doorstop to become detached from the rest of the jamb. If this happens, we still need to get the door open. It's not panic time; it's time to move to plan B.

Fig. 19–6. Note depth on residential locking devices.

FE has been categorized into two subsets: conventional and hydraulic. Each subset has advantages and disadvantages based on the incident conditions. Your job, in order to become a master at your FE craft, is to decide what conditions encountered will best match up with the tools that your company has at its disposal.

When deciding on the tools of choice for residential forcible entry, there is a need to have options. We cannot bring every tool on the rig to every door we encounter, so we must choose wisely. We must know the pros and cons of each such tool in our FE inventory. Firefighters should ensure that the right tool is being applied in the right situation.

Certainly no one will deny the fact that hydraulic forcible entry tools are superb at forcing inward opening doors, but they also are not infallible and have limitations. Take, for instance, the Hydra-Ram, also known as the bunny tool (fig. 19-7). This hydraulic FE tool was designed to open metal doors in metal jambs. The moving teeth on the jaw of the tool use the stability and structure of the metal jamb to push the door inward. The force applied to the door relies on the stability of the base, in our case, the doorjamb. Attempting to use a Hydra-Ram

Fig. 19–7. The Hydra-Ram hydraulic forcible entry tool. (Courtesy of Traditions Training, LLC, www.traditionstraining.com.)

While we do not want to limit ourselves with FE options, our experience leads us to believe that a set of "irons"—the marriage of the 8-pound axe and Halligan bar—are the best tools suited to force the great majority of residential doors (fig. 19-8).

Fig. 19–8. The "irons": an eight-pound axe and Halligan tool

Please don't overlook the obvious. Be sure to try the doorknob on every door you come across first! Complete a quick size-up of the door. Take a quick look to see what materials comprise the door and the doorjamb. What type and style of locks do you see? Can you tell which ones are engaged? Also, another quick point of attack may be on either side of the entrance door. Many residential front doors have decorative sidelights on either side. A quick FE solution may be to just to break out a portion of the sidelight near the lock, then reach in and unlock the doorknob and deadbolt through that sidelight (fig. 19–9).

If the door is locked and we are tasked to force it, the forcible entry techniques we will discuss involve two firefighters and a set of FE irons. Forcible entry is as much about technique as it is about brute force. Skills must be practiced and honed to become flawless. Investigate the details of your tools and study the reasons why they work. For instance, look at the Halligan bar, with its beveled and curved fork end. It is made this way by design, as is the curvature to the adz and pick end (fig. 19–10). This tool pairing, with proper technique and force applied to the locking mechanisms, will get us through most residential doors.

Fig. 19–9. Residential door sidelight panels

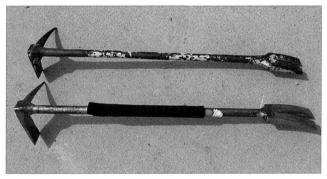

Fig. 19–10. The modern pro-bar: a one piece 30-inch Halligan bar

Conventional forcible entry

After sizing up our door's construction features noting the locking mechanisms and having selected our tools, it's time to go to work. We have narrowed the conventional forcible entry method into a four-step process: *shock*, *gap*, *set*, and *force*.

Now just because we have laid out a four-step process does not mean that the door will only break open at the last step. Entry may occur at any one of these steps along the way. Because of this, we must be prepared to control the door once it opens. Utilizing a rope hose tool, hose strap, or tubular webbing as a rope to wrap the doorknob can assist us in controlling the door.

There have been numerous occasions where a crew has lost control of the entrance door and were literally blown off the front steps by a rapid fire advancement that occurred in the dwelling. If it is at all possible, secure your ability to close the door prior to opening it. As you put force on the door, it may spring open very suddenly and overrun your position, especially if fire is right on the other side.

While it is difficult at best to describe forcible entry techniques on paper, the thought here is to give you certain typical benchmarks that should yield success. The best path to becoming a master at forcible entry is by really doing it, repeatedly, having your hands on the tools and the doors as often as possible, both in training and at real incidents. Study the processes and techniques described in the next few chapters and drill with them at your next opportunity. If you can't get in, you really can't do much of what you are called to do at the residential fire.

Don't forget that forcible entry is a form of ventilation. Don't think for a second that the fire will not have a reaction to either the influx of fresh air or the creation of an exhaust port that you have provided by opening the front door. Forcible entry is ventilation; therefore controlling the door is paramount. We were once taught that when we opened a front door, we were to chock it wide open immediately. We feel that due to information gathered from several recent fire travel studies, we may need to revisit this tactic.

We are not saying that you need to wait for the hose-line to enter the building; just do not chock it open as we are used to doing. Take a second to gently close the door behind you after you get inside (ensuring that it does not lock behind you by throwing the latch or using a doorknob strap) (fig. 19–11). This will limit the potential ventilation effects that an open doorway would provide until the hoseline is in position and ready to move in. We talked earlier about the need for coordination in our attack. This is one of the first steps we can take to coordinate our efforts.

Let's go through the steps for conventional forcible entry (utilizing the irons). The four-step process we discuss is: shock, gap, set, force.

Shock

First, see how and where the door's locks are engaged. Use the fork end of the Halligan bar to strike the door high (top third), middle, and low (bottom third) on the locking side near any locks. Doing this gives you a feel for the stability of the door and door frame. You may also be able to feel and gauge resistance from any locking devices. You may note locks that are not actually engaged. When shocking the door, avoid placing your free hand on top of the adz end of the tool. Keep both hands on the shaft of the Halligan, utilizing a cross-handed grip (fig. 19–12). If you cover the adz end with your hand, the reverberating shock after making contact with the door will continue back through the tool and into your wrist, arm, elbow, and shoulder, bringing the potential to cause injury.

Fig. 19–11. Webbing strap used to help control door during forcible entry

Fig. 19–12. Shock the door using a cross-handed grip

Gap

Use the Halligan's adz end to gap the space between the door and the jamb. Occasionally enough of this gap may have been created during the shock step and may not be completely necessary. You are attempting to create enough space for the fork end of the Halligan to be driven between the door and the doorjamb (fig. 19-13). Depending on the position of the door (left or right swing), insert the adz into position between the doorstop and the door and push up and or down to create the necessary gap. Don't ever give up your gap! Inserting a wooden door chock or the blade of the axe into the space before the Halligan tool is repositioned can hold this gap.

If the doorstop comes off in the process, do not worry. With the doorstop removed we can see exactly where we need to place the forks of the Halligan. We can now have a second firefighter drive the fork into position through the gap with an axe.

Fig. 19–13. Adz end of Halligan creates gap

Set

Setting the Halligan bar involves striking the adz end and driving the fork of the tool through the established gap in the previous step (fig. 19-14). Take a look at your Halligan. How thick are your forks? All Halligans are not created equal. Differing brands have varying thicknesses to their forks. A thicker fork will require a greater gap and/or greater force to set it in place (figs. 19-15 and 19-16). When deciding on where on the door to set the forks, they should be placed slightly above or below the position of the locking mechanism. If there are multiple locks, place the forks in between the engaged locks.

The Halligan bar's forks are beveled by design, adding increased leverage when forcing the door. To get the greatest leverage possible, keep the bevel side of the forks to the door (bevel of the tool to the door itself). If the gap between the door and doorjamb is particularly tight, initially place the bevel of the forks to the jamb. If attempts to open the door do not yield success, hold the gap with a wooden chock or the head of the axe as you reposition the Halligan with the bevel to the door.

Fig. 19–14. Set the fork of the Halligan between the door and the jamb.

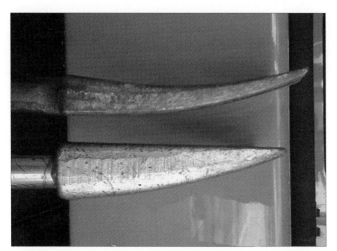

Fig. 19–15. Note thicker fork end with less curvature. (Courtesy of Traditions Training, LLC, www.traditionstraining.com.)

Fig. 19–16. Note the details in this Halligan, thinner forks with a natural bevel. (Courtesy of Traditions Training, LLC, www.traditionstraining.com.)

to "drive" or continually hit the Halligan until told to stop once the tool is in position to force. Another tip to assist your firefighters in knowing just how far to drive the forks into the door can be done easily with a small grinding tool. Score a line along the side of the forks (fig. 19-18). This will indicate an approximate depth to drive the Halligan past the door behind the jamb.

Fig. 19–17. Forcible entry team communication is paramount. (Courtesy of Traditions Training, LLC, www.traditionstraining.com.)

The striking member in the *set* step must listen to directions from the Halligan holder before initiating contact with the axe (fig. 19-17). The striking member, and the striking member *only*, should call the hits to the Halligan. The Halligan holder should keep his or her hands mid-shaft and may need to push or pull the tool slightly to guide the forks into position around the backside of the doorjamb. Utilize standardized commands in your unit to have consistency in striking the tool. During the hit, the Halligan tool holder will call for hits one at a time while watching the forks move through the gap. Once the forks are past the leading edge of the jamb, the striking firefighter can be instructed

Fig. 19–18. Halligan with depth gauge scored in forks

Force

Once the forks are set behind the jamb, we need to use the angle created with the bevel of the Halligan tool against the door. The pressure created on the locking mechanisms should break the hold the locks have in the jamb. Firefighters should position themselves to the outside of the Halligan and use their legs to drive their bodies into the tool (fig. 19–19). As with each step, remember to hold the increased gap you have created if the door does not open with the initial force applied. It should be done with a sharp action. This new gap may now allow you to put the adz of the Halligan into that gap behind the jamb and force up or down. You may also place the head of the axe behind the forks to increase your angle for more power.

Fig. 19–19. A firefighter positions himself to force the door.

When it comes to forcible entry, give it 30 seconds of focused fury. If after 30 seconds the course of action is not creating the FE success you were anticipating, it may be time to adjust your plan of attack. You must have a plan B, C, and so on. Sometimes a forcible entry problem that may appear to be very simple becomes extremely complex when stressors are applied, such as a few engine members ready with the line all shouting, "Do it that way!" Company officers controlling the forcible entry team must be certain to let the members know when it's time to move to the next plan.

Company officers must not be completely focused on the locking devices and the door itself. Those are the forcible entry firefighters' tasks. They must keep their heads on a swivel. While they must offer calm, focused advice to the FE team, they must also be watching smoke and fire conditions and watching the status of the hoseline.

Let me digress for a moment in reference to bosses. We have all had excitable bosses. They usually do not add much to the situation, but instead create more chaos. They raise everyone's emotions, and yet help no one. A firefighting title you never want is "the screamer." I am fairly certain that we have all experienced a screamer at one time or another in our fire service careers. In FE, a screamer is one of the worst people to have around.

Here is a personal screamer story. We were working through a tough door and our screamer was frantically yelling at us to hurry up! Finally, my senior guy who had the irons told me to stop hitting. He turned and said, "The only way we are getting into this place is if I open the door, and I can't do it with you (insert expletive here) screaming over my shoulder. Back off!"

The company officer must offer calm, concise advice and not just scream, "Do it faster!"

Window Security

While the forcible entry techniques discussed above can be applied to any locked door on the front, side, or rear of a private dwelling, the building has other avenues for us to utilize to enter the structure. Windows can be advantageous avenues for us to pursue entry and should also be addressed as potential exits. If using the window, we also want to be certain to fully clear it. This includes removal of the sash and any decorative blinds, effectively changing the window into a door (fig. 19–20).

If we have to exit through the window, we want nothing to hinder or delay our escape. Our initial actions when forcing window security devices begin in the same way as our FE methods begin on doors. It starts with our tool selection and size-up of the device. Again, firefighters need to know the tool options and the limitations of all their tools. Many times these window devices can be removed with simple, well-applied hand tools such as the irons and metal firefighting hooks. While we cannot rule out the use of the rotary power saw with a metal cutting blade, or even hydraulic FE tools, we need to know all options that are at our disposal.

However, before we commit to enter any structure through window openings or announce them as potential egress routes, we must be sure to remove any security devices that may be present. Owner- or occupant-improved window security measures can include window guards, gates, or bars.

Fig. 19–20. Never leave windows uncleared of glass or sashes. (Courtesy of Nate Camfiord.)

Guards

Window guards are designed to keep occupants in, not intruders out. Primarily found on upper floors in private dwellings, their primary purpose is to keep small children from falling from the windows to the street below. They should not be viewed as a significant deterrent for firefighters to make entry. They are composed of light gauge steel or aluminum horizontal rods that slide past each other to fill the needed window space and are normally found only in the lower half of the window (fig. 19-21). They are normally found on the exterior side of the window glass, exposed to the elements.

Window guards are usually secured into place with whatever cheap metal or wood screws the homeowner had laying around. They are attached to the window frame at a few points on either side with said screws. While child guards may initially look intimidating, they are easily removed as long as you know how they are put in place.

Fig. 19–21. Child guards are designed to keep kids in.

If these guards are located on the outside of the window, attempt to keep the window glass in place until you remove the guards, if possible. Most child guards can be defeated with simple hand tools. The simplest method to remove them from the window is to strike the end of the sliding section (depending on window width, this position will vary) with the Halligan tool using force in the opposite direction from the slide movement needed to fill the window (fig. 19-22). If this does not work, choose one particular side and concentrate your work there with the fork or the adz end. Once one side of the guard is cleared from the window, we can bend the guard out to 90 degrees and remove the other side using the torque and leverage created. These guards are fairly malleable, so they may bend and deform when struck with tools.

Fig. 19–22. Firefighters forcing open a child guard with Halligan

Gates

Window gates are designed primarily to keep people out, but are able to be opened from the inside for escape by the occupant should the need arise (fig. 19–23). Gates are normally one of two styles. They are either the accordion sliding scissor style or the type that have a portion of the gate itself, like a small door, which opens into the room side of the gate (figs. 19–24 and 19–25). They normally are the size of the complete window, top to bottom and side to side. Locking mechanisms vary for these style gates. Some operate with a simple slide bolt, some with a turn latch, others with a key. Sometimes they may even have a padlock on a hasp. The locking mechanism may or may not be additionally protected with a metal shroud or cover plate device.

These window gates may not be initially noted in your outside survey since they are normally located behind the window glass on the occupant side. They are installed into the internal frame of the window. This is certainly a factor to consider as it may complicate matters. You will have to remove the glass of the window before you can attack the gate. You now may have rough glass edges and a smoke condition to overcome in addition to the gate itself. Gates are comprised of stronger gauge steel. They are more difficult to force from the outside than the previously mentioned child guards. But again, as with the guard style, the attachment points are usually the greatest weakness of the gate.

The accordion scissor style gates run in a track that has a locking mechanism on one side (generally with a padlock of some type) and is affixed to the window frame on the other side with screws. It is advisable to initially focus your forcible efforts on the non-locking side of this gate. Again, the weakness is not normally in the locking mechanism but in how it is fastened to the window jamb. Depending on the amount of space in which you have to operate, the fork or adz end on the Halligan worked in behind the mounting plate on the non-locking side will either shear the screws or pull them from their wooden anchor points. With the gate no longer attached, one of two things can be accomplished. You may be able to slide the gate towards the locking side, creating the opening you need, or you may be able to hinge the scissor gate with an in-out motion that can release the locking mechanism on the other side or break free the attachment points on that end.

Fig. 19–23. Window gates on a residential home

Bars

The highest level of window security is the window bar (fig. 19–26). While some may appear decorative in style, these are affixed strictly as security devices. There are no means to open these bars and they are most often found on the lower floors. They are attached to the exterior of the building and located in front of the window glass. Some residential buildings are set up as literal cages with bars on multiple floors covering nearly every window (fig. 19–27). This has been a common sight in the urban setting, but we are increasingly seeing it in the suburbs.

Fig. 19–24. Scissor-style gate

Fig. 19–26. Window bars

Fig. 19–25. Door-style gate

Fig. 19–27. Residential home covered with bars

If you come across multiple bars on a residential building, this must be transmitted early to all who are responding, and perhaps included in your on-scene report. Request additional units immediately. This is a dangerous dwelling. While the initial outside firefighting teams can work to remove at least one set of bars from each floor right away (fire floor first, then floor above, then top floor), you may consider assigning a complete company to specifically address and remove window bars if resources permit. If you have options for responding companies, tower ladders' buckets make more stable work platforms for removing the bars from all floors (rather than portable or rear mount ladders).

Bars are normally constructed of iron or heavy gauge steel and affixed in two different manners, depending on the exterior of the home. If the exterior walls are masonry and brick, the attachment points may be set into the mortar joints on the exterior of the building (fig. 19-28). This attachment variety may prove a bit more difficult than the more common wood frame structure. Bars attached to wood frame constructed buildings are normally secured into the framing that surrounds the window itself. They are set with lag-type bolts through the exterior sheathing. These lag bolts may or may not hit the studs in the wall. If not, they may be easier for us to remove. Remember that most times these attachment points have been exposed to the elements since they were installed. Do not be scared or intimidated. You will remove them. We have had experiences where these bars, while appearing daunting at first glance, literally fell off in our hands.

Fig. 19–28. Connection points for bars can look intimidating, but in reality come off rather easily. Look at the deterioration in the brickwork surrounding these attachment points.

In either style of construction, with some variation, there is usually a minimum of four attachment points. The great majority we have seen have this type of setup with two points on either side of the window, one low and one high. These attachment points will be the focus of our attack. When removing these bars, it is best to attempt to leave the window glass in place until you have effectively removed the device so as not to complicate the removal with the presence of smoke obscuring your view.

Size up the bars as you size up a door for forcible entry. While we mentioned four attachment points, we normally need only concern ourselves with releasing two of them. Choose the side of the bars that looks weaker. Perhaps one side is missing a lag bolt, or the bolt is visibly loose. Often you can use hand tools to create enough leverage to break these attachment points. You can use the Halligan adz end or forks to create leverage, then use the hook end to pop the lag from the attachment point.

Tools such as the Lincoln bar with its extended leverage-applying handle are very useful here (a Lincoln bar is a 6-foot-long Halligan-style hand tool). Hydraulic tools such as the Hydra-Ram may be used to "push" the bars off the building. Metal cutting forcible entry saws, torches, and rebar cutters also will do the trick. Bashing the connection brickwork or caving in plywood adjacent to the anchor point can also soften or release the bars from their attachment points. The key is to have options and know when it's time to move on to the next plan.

Once we remove two connections points on one side of the gate, either two on the same side or the top or bottom two, we can use the bar's weight and leverage it against itself. Take the gate and pull it back and forth against the remaining attachment points to remove the entire gate from the window.

Remember, anytime we force an exterior building opening (door, window, scuttle, etc.), we are also creating a ventilation opening. Our creation of an entry or exit point is going to have an impact on the fire conditions in the structure. Even forcing open the front door is ventilation. It can and will have an effect on the fire. For every action we perform, the fire will have a reaction. We must plan for, practice, and anticipate (PPA) the fire's reaction to these events. We must know the limitations and design of all our forcible entry tools. We must know what we will be facing most often. Be a student of the area to which you respond and be ready to go to work. Adopting this mentality will help you to think and to overcome obstacles when forcing entry.

Rescue

The topic of rescue is deserving of its own chapter so we have dedicated chapter 25 to fully cover the residential search for civilian rescue and locating a downed firefighter. Obviously saving lives is the main function of *all* units on the fireground. Truck companies are often tasked with conducting primary and secondary searches of residential dwellings. Not only must members practice good search tactics, they must also train on civilian and firefighter removal techniques. Fireground search and rescue is covered in depth in chapter 25.

Locate the Seat of the Fire

Once forcible entry is complete and we have gained access to the building, our next task is to find the seat of the fire. Sometimes this isn't as easy as it may seem. While fire blazing from two or three windows may be impressive to a civilian, what you see is a ventilated fire that you can now easily locate (fig. 19–29). The building with that grayish smoke coming from everywhere indicating smoldering or decay stage fires is where we can run into difficulties in locating the source. We discuss the use of our five (or sometimes six) fire senses and the use of the thermal imaging camera to aid us as we initially move through the structure to find the fire at length in chapter 21.

Fig. 19–29. Fire showing helps us get to it faster. (Courtesy of Kentland Volunteer Fire Department, www.kentland33.com.)

As we move to the seat of the fire we should attempt to aid in the creation of a path for the engine company to facilitate the hoseline advance. Once you have found the fire, be sure to communicate that information to the engine officer and the incident commander. Location of the fire origin is paramount! Transmitting this information allows incident commanders and company officers to evaluate strategies and the culminating tactics that need to be carried out.

Fires in specific areas of residential buildings, such as basements and attached garages, may require specific detailed strategies and tactics for operations. Having pre-planned SOPs and pretrained for these specific fires by conducting practical, hands-on evolutions will increase maximum operational efficiency and lead to rapid fire extinguishment.

Being a student of past fire events and knowing the dwellings in your first due, you should also identify the prevalent origins of fires in private dwellings. We know where the greatest frequency of fires in the residential building are prone to occur. *The USFA Residential Building Fires 2008–2010 Topical Report Series Vol.13, Issue 2* issued April 26, 2012, notes that 45% of fires begin in the kitchen. Based on our knowledge of home styles in our areas, we should know where the likely place is to find the kitchen. Likewise, bedroom locations should be principal in our minds (fig. 19–30).

Arming our crews with information that is readily found in residential fire history should make our job of locating the fire easier. Pair this historical data with our residential building knowledge and the ability to process live fireground information and we will quickly find the source of fires. We will find any civilian victims. We will be operating in the most expeditious manner possible.

Confine, Control, Extinguish

Even though truck companies do not regularly stretch and operate hoselines to extinguish fire, they can absolutely confine and control aspects of it. We know from being students of residential building construction that the great majority of residential buildings are subdivided into smaller rooms. While the modern trend in new construction and home renovation is the open floor plan concept, ladder companies can still utilize tools to limit fire spread. Fires in their incipient stage can be controlled and many times largely extinguished with minimal application of water.

Fires need oxygen, fuel, heat, and the self-sustaining chemical chain reaction to grow and spread. How can firefighters limit fire growth without an extinguishing agent? Obviously we cannot remove the physical fuel from the fire area, nor can we reduce heat without extinguishment. Most fires today are not fuel or heat limited, they are air limited.

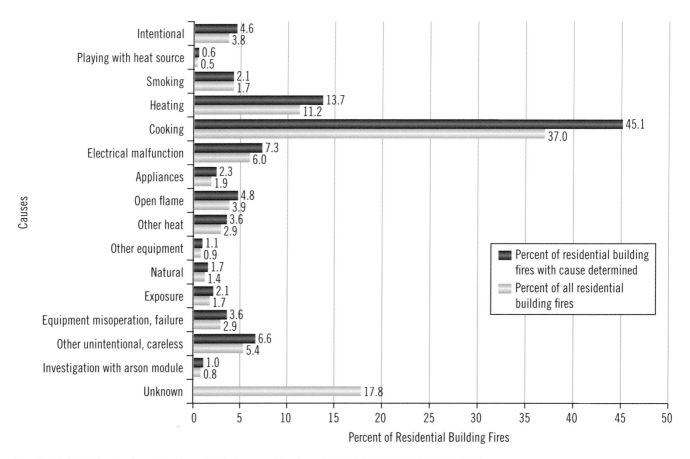

Fig. 19–30. USFA chart on fire origins in residential homes. (Courtesy of United States Fire Administration.)

We can further stymie fire growth by confining and controlling the amount of air we introduce to it. The ladder company can confine fire growth with something as simple as closing an interior door (fig. 19-31). If no door is present, a remote door may be removed from its hinges to cover an opening to an area where fire is present. Closing the door will reduce the amount of fresh oxygen and starve the fire. As discussed in many chapters in the book, forcible entry must be considered a form of ventilation. Ladder companies may reconsider the seemingly routine manner we were taught in chocking the front door wide open once we get it open, unless the engine company has the first hoseline charged, in place, and ready to operate.

Many departments have a member of the ladder company team carry and utilize a 2½ gallon pressurized water fire extinguisher in their complement of hand tools (fig. 19-32). This water filled "can" could be utilized to confine a fire to an area of origin and to extinguish or knock down smaller incipient fires. You may be able to hold a fire in a couch *to* the couch, a fire in a room to that room, and so on. In conjunction with closing a door to the fire area, the water can affords the ladder company some protection while conducting the primary search. The firefighter with the can must let the company officer know when the can is nearly out of water or if the fire is extending beyond the team's extinguishing capacities.

Fig. 19–32. 2½ gallon water extinguisher with strap attached.

We are not suggesting that a 2½ gallon water can will completely suppress a room full of fire, only that it will assist in confining and controlling it until the hoseline arrives. It is a tool with limitations. Practice with your members in controlling the stream application from the water can to conserve the water that you have (fig. 19-33). Placing a gloved finger over the tip disperses the stream and may be more effective in certain situations. Remember to never leave a water can in the immediate fire area since when heated, the vapors inside may cause the container to rupture under extreme pressure.

Fig. 19–31. An open door or window is a potential flow path.

This explosion can cause shrapnel to be thrown with great force.

Fig. 19–33. Members practice with the 2½ gallon water extinguisher.

We can also confine and control the extent of fire with ventilation tactics. While we discuss the art of coordinated ventilation in chapter 23, it deserves some mention here. Any opening that we create in the dwelling will have an effect on that fire's behavior. Sometimes our ventilating actions will aid in successful extinguishment. When uncoordinated and haphazardly conducted, ventilation efforts will extend fire, causing greater fire spread and damages to the home and, more importantly, place operating members in positions of unnecessary and unexpected danger.

Ventilation

Ventilation, in the simplest of terms, is causing or creating openings to allow smoke, fire gasses, and heat to escape from the fire dwelling. Ventilation on the fireground should occur for one of two tasks. Ladder companies are known to ventilate for life and ventilate for fire.

Venting for life is accomplished with the understanding that we may inherently draw fire (as we have now increased the amount of fresh air in, or caused an exhaust port out) and smoke to our position in the home. We take this calculated risk in order to improve conditions in the dwelling. It assists us with visibility during our primary search. It also hopefully gives any occupants increased chance for survival by removing the contaminated environment. This form of ventilation may occur, if warranted, prior to the establishment of the engine company hoseline. Members searching inside, or members outside conducting vent, enter, isolate, and search (VEIS) techniques (which is discussed at length in chapter 24), may vent for life if conditions require them (fig. 19–34).

Fig. 19–34. Members vent for life, preparing to vent, enter, and search. (Courtesy of Kentland Volunteer Fire Department, www.kentland33.com.)

Venting for fire is the process by which members open ventilation routes (largely windows) when fire is behind them (fig. 19–35). This action should be paired and conducted in harmony with the advance of the hoseline. If completed prematurely, fire will likely extend upward and outward, obtaining new fuel. It is largely joked that ladder company firefighters are known to love to the sound of breaking glass. However, it takes a knowledgeable and patient firefighter to properly time the execution of glass breaking. Firefighters must listen for cues and commands from the interior rather than breaking glass just for the sake of breaking glass.

Positive pressure ventilation (PPV) is a topic worthy of some discussion (fig. 19–36). Ladder companies that carry PPV fans must be well versed in their use. There is a tendency that if someone sets a PPV fan at the front door, it should be started and run. This is not always the case. PPV fans are excellent tools for removing smoke from a dwelling, but should not be used prior to the fire's full extinguishment. While companies are still checking for pockets of hidden fire is not the time to fire up the fan! The influx of air forced into the dwelling can rapidly cause a small hidden fire to exponentially and violently appear to the forces inside. Like many fire departments' actions these days, the results of improper PPV fan usage are easy to find on the Internet and should be viewed as a training tool for all members. Just because the ladder company member places the fan in the doorway does not mean that it should be started! There has been such great recent discussion about wind effects on fire of late. If we fire up a PPV fan at the wrong time, we can be creating the same effects as a wind impacted fire.

Fig. 19–36. Positive pressure ventilation fan

Fig. 19–35. Members vent for fire opposite the advancing hoseline. (Courtesy of Kentland Volunteer Fire Department, www.kentland33.com.)

Ladders

Operating in a ladder company, it is rather obvious that apparatus mounted hydraulic ladders and smaller portable ladders are part of our bread and butter operations and a mainstay in our training regimen (fig. 19–37). Ladders are used to help us in searching. They play a role in getting us to positions to ventilate and become platforms for access and egress to upper floors. In chapter 22 we address the function and form for all fire service ladders, and explain how they are best utilized in the residential building fire.

Fig. 19–37. Members routinely operate on ladders at residential fires. (Courtesy of Nate Camfiord.)

Fig. 19–38. Members opening up in the fire area.

Overhaul and Salvage

It is not an accident that we list overhaul before salvage. Overhaul is the systematic opening up of areas that have been exposed to fire. Overhaul must be completed before we can say that we have fully extinguished any fire (fig. 19–38). Thermal imaging cameras can help us locate "hot spots" to open, but we must be certain that there are no hidden pockets of smoldering debris. We must open up whatever we need to open up. Once units take-up from the fireground, we need to be absolutely certain that we will not be called back. Rekindles are embarrassing and just plain unprofessional.

Insurance companies will not give the homeowner nor any fire department credit for half broken windows or partial boards of broken sheetrock. We must open up affected areas of the building until we are certain we have covered it all. Trim the glass from all windows. When it comes to opening walls and ceilings, our rule of thumb is that you should have two clean wall studded bays (with no fire damage present) on either side of where you have fire damage. When opening up around doors and windows, remember how they are assembled as this will help you get to the guts behind the wall. Tips to opening up walls and ceilings are plentiful. Simple hand tools such as the Halligan bar and 6-foot hooks will make short work of most residential coverings.

While we must open up what is necessary, we should be doing so with a charged line standing fast. There is no bigger danger than packing away the line before the fire is fully overhauled. Invariably, once the line is all buttoned back up on the rig, a line will be needed in the fire building. Once we have opened up sufficiently, wet all areas down with the line to mop up. As with a secondary search, leave no doubt that the fire is completely extinguished. While during overhaul we can certainly take steps to start salvage, we can do things such as moving furniture pieces, put pictures in dresser drawers, etc. Take salvage into consideration, but do not let it delay the overhaul process when doing so. Salvage can, however, be occurring simultaneously elsewhere in the house. Additional companies, if sufficient manpower exists, may begin salvage operations in remote areas of the fire building not affected by fire.

During overhaul, when do you remove your facepiece (fig. 19–39)? Do you have a department policy when to do so? Should you? Refer to chapter 5 for discussion and reflection on recent studies on SCBA facepiece removal during the overhaul phase of the fire.

While our application of water leading to final extinguishment ends the burning process, the human emotional trauma for the homeowner has just begun. Why do we sometimes hear of homeowners going back inside a burning dwelling when every human and animal is already out? As insane as it may sound to us, it's for their possessions! People have died going back into building fires attempting to rescue their own possessions.

The goal of fire department salvage is to attempt to soothe the burn, an attempt to minimize the impact for our homeowners. It is our way to prevent further damage to the valuables that remain, as much as feasibly possible. Remember, our oath is to protect life and property. Once the fire is out, which is the best way to conserve civilian property, and the lives have been accounted for, we should aim to protect what is left. Often, it is the little things we do to help after a fire that go a long way.

Take a few extra minutes to push items in the room together and throw a salvage tarp or sheet of plastic over them (fig. 19–40). Within reason, and with the authority of the fire investigators, assist the homeowner in retrieving items requested. Ask anyone you know who has experienced a fire from the homeowner's side what is feels like. You may just have a touch more compassion and understanding the next time "Harry Homeowner" has a request. Think about how you would want your family items handled if you were at work and got a phone call from your loved one that you had a fire in your home.

Fig. 19–39. A firefighter takes a break during overhaul. (Courtesy of Nate Camford.)

As firefighters we often grow callous to the devastating effects that are the end result of building fires. For the common homeowner, fires are unthinkable. They are unthinkable, that is, until they happen. They are life-changing events. Photos, family heirlooms, and prized possessions are damaged, broken, destroyed, and gone forever. Fire's path is destructive, and our sworn duty is to protect our citizens from its wrath.

Fig. 19–40. Salvage operations. (Courtesy of Nate Camford.)

Utilities

Utility control is not in the forefront of the initial attack on the fire and is often an unappreciated afterthought at residential building fires. This is a task normally assigned to a later arriving company. To the incident commander, it may be just another of the many "check boxes" that they must account for at every fire.

Sometimes utility control is a more prevalent matter in the residential fire operation (fig. 19–41). Leaking natural gas from broken or cracked lines that feed fire conditions will absolutely impact the firefight. Arcing and sparking home service feeds can pose electrocution hazards for members. While sometimes we may have to wait for the arrival of the utility company to fully mitigate the problem, there are some tasks we can accomplish as firefighters to control many situations. Typical utility control refers to minimizing the dangers to operating forces and the residence caused by gas, electric, and water. We can succeed at minimizing the danger by controlling the utilities as early in the incident as possible, although true and total control will come only with the response of the local utility authorities.

Fig. 19–41. Gas meters

Natural gas

Several past incidents stand out with regard to utility control. The first event we discuss is fairly common. Countless times, especially while operating at fires in kitchens and those found in utility areas—some of the most common areas for fires to start in the private dwelling—we find natural gas appliances. Such residential uses of natural gas in the home can be found in kitchen stoves, ovens, dryers, fireplaces, and home heating units.

Many natural gas-fed fires are often not initially recognized. As a young firefighter I can remember wondering why a particular kitchen fire kept flaring back up after we knocked it down. A small, gas-fed fire had originated on the stovetop and the end result left the rest of the kitchen incinerated. What we didn't originally recognize was that the fire had melted the natural gas flex tubing that came off the cast iron piping behind the stove. After the officer noticed the second reignition, he ordered members to look for the gas shutoff behind the area of the stove, which they did.

They quickly extinguished the flames and then continued to use the line to mop up hot spots in the house. Occasionally, once we knock the fire down, we don't realize that natural gas is continuing to escape. This is largely due to the fact that we are wearing our self-contained breathing apparatus (SCBA).

All that is needed is the right concentration of vapor and heat to reignite the gas. If this does occur, it is best (under the protection of the hoseline) to keep the fire in check but not fully extinguish it. The best way to fully extinguish the fire is to remove the fuel (in this case, shut off the gas). Gas valves are often located behind the appliances and can be initially isolated with a quarter-turn valve in this location. If this shut off is damaged, not present, or otherwise unable to be closed, the next location to shut off the gas is at the location of the meter.

Before leaving the scene, natural gas utilities to the entire structure should be turned off, locked, and tagged in the off position. The location of this shutoff should be at the meter. The gas meter may be inside the structure (sometimes in the basement) or on the building exterior. Most gas shutoffs are quarter turn ball valves. To shut down the gas, place the turn valve perpendicular to the run of the piping. In most instances, hand tools (such as channel locks or vice grips) will be adequate to manipulate the shut off valve. You may find the fork end of the Halligan placed over the exposed stem will work, or a larger pipe wrench may be needed (fig. 19–42). Regardless of how and where you shut off the gas, the local utility company should be notified to secure the initial actions.

Electricity

The great comedian George Carlin once said, "Electricity is really just organized lightning." As firefighters, we should treat electricity in the same manner. Electricity powers most items in the great majority of our residential structures. Electricity is one of the top five causes of residential building fires according to the USFA 2008–2010 report. Not only is electricity a cause of fires, but it also can pose a hazard to firefighters operating in and around the fireground. Both on the exterior and the interior of the structure, electricity is always nearby while we are on the job. We must be consciously aware that electricity is traveling behind most walls and ceilings where we are operating. Sometimes it is in places where we least expect it.

Fig. 19–42. Member shutting off gas valve with a Halligan bar.

Such an unexpected incident happened to a ladder company officer in my truck when I was a firefighter. Our captain was conducting a primary search in a fire apartment when he was zapped with an electrical charge while moving through a bedroom. While searching the room, he ran into a bed. All sounds normal so far. A bed is certainly not an unusual thing to find in a bedroom, right? But what happened next was unexpected and rather painful for my boss. Unbeknownst to him, the mattress was on top of some kind of electrically operated hospital bed frame, which was plugged in. Somehow, during the fire the bed frame and rails became energized as a result of the compromised electrical circuitry in the apartment. Once this became known, the incident commander made the announcement over the radio for members to exercise caution and quickly assigned a company to remove power at the breakers to the entire apartment.

Electricity is certainly a utility that must be accounted for and controlled at every residential fire. Most times this is best accomplished at the main circuit breaker panel for the home (fig. 19–43). Fire department personnel should not attempt to pull the electrical meter on the outside of the home. Only the utility company personnel should attempt this. Meters measure the amount of current that the home uses. In some instances, removing the meter itself does not remove the power, only the ability of the power company to record usage.

Fig. 19–43. Residential electrical breaker panel

Let the power company handle any concerns at the meter. As stated above, our focus should be directed at the breaker panel. This should provide us an acceptable margin of safety to operate in the fire building. If at all possible, kill the main breaker for the home. If there is no main power disconnect at the panel box, note any tripped breakers before killing the power to all the breakers. Fire investigators will need to know what breakers were tripped prior to or during the fire, so note them prior to your company shutting off all the individual breakers.

As we discuss in chapter 22, all laddering operations should start with a survey for the presence of overhead obstructions, including electrical wires. Several firefighters have been seriously injured or killed when portable or apparatus mounted ladders made contact with, or were even in proximity to live power lines. Due diligence must be exercised when moving and operating on ladders in and around electrical lines.

Water

While we bring our own water to extinguish the fire, the water utilities in the home are normally a lesser concern while operating. Residential water supply, especially as it relates to managing gas and electric in the dwelling, is not a primary concern. This is not to say that it may not ever need to be addressed.

The intensity of the fire may rupture water supply lines throughout the building. Most water supplied to residential homes comes from one of two sources. Either water is pumped from a well located on the property or fed from the residential line connected to the main in the street or curb. In both cases, a shutoff is usually located inside the dwelling in a utility closet or in the basement. These are either wheel type or quarter turn valves. If the home is fed from the street, a water meter is often affixed to the exterior of the structure for easy reading by the water utility meter reader (fig. 19–44). Most of the time the shutoff is directly opposite this meter.

The application of water from our operations may add an accumulation of water, especially on upper floors. In the private dwelling this may lead to concern. Water can add substantial weight to the structure. Using a single 1¾-inch hoseline with a nozzle flowing over 100 gallons per minute (GPM) for approximately 10 minutes means we will effectively flow nearly 1,000 gallons of water into the structure. At 8.33 pounds per gallon, we will be adding approximately 8,330 pounds to the dwelling. This added load, in addition to the weight of the operating members, may be cause for concern.

Fig. 19–44. Many times the water shut off is directly opposite the meter reader.

Removal of water from upper floors can be accomplished by a number of methods. Standing water can be removed by the removal of toilets in bathrooms. Take the toilet off its base and allow water to drain into the waste line. Holes can also be made in exterior walls and/or floors to remove water. Obviously this must be carefully communicated to operating units because, in addition to the cascading water, we could be creating openings that may become trip-, slip-, and fall-type hazards.

Common Goals

Whereas ladder company and engine company functions are fundamentally different on the fireground, we are working towards the same goal. Saving lives and property is our sole focus; we take different paths, but we rely on each other to reach the final goal. We know that what has been labeled historically as truck work needs to be completed in concert with the first hoseline's operation. If no physical ladder company is present, there should be no hesitation in waiting to assign these tasks until they arrive. Any well trained engine or rescue company can accomplish the "typical" ladder company assignments that we discuss in this chapter.

Nearly every piece of fire apparatus that arrives on the scene of a residential building fire has a compliment of forcible entry tools, ground ladders, and members who know how to search, vent, and control utilities. All firefighters must be familiar with the actions that each other are taking on the fireground. We must continue to explain the *why* behind what we do.

The Backstep

Ladder Company Operations

Successful operations at residential building fires are the result of teamwork. Engine companies are successful with help from ladder companies and vice versa. Together, we must all work under the goals of the "LIP" principle. They are:

L: life safety
I: incident stabilization
P: property conservation

How do ladder companies operate within the confines of the above principles? While life safety can be viewed as a rather broad-based goal, it starts in truck company operations with our teams of two in the IDLH for *our* safety. We have to recognize that our lives fall into the principle as well. But, alas, our goal is to save lives and property, and the principle is carried out primarily with civilians in mind through the ladder company's ability to conduct search and rescue operations within the home.

Good search techniques in the residential dwelling evolve from experience and understanding. Most times, the fire academy concrete burn building does not give us a good foundation for realistic residential search. Your searches must be refined and practiced in anticipated forms to match conditions that you will likely find in the houses you are searching! As you discuss your ladder company residential tactics, take the time to talk about the effect your actions will have on life safety within in the structure.

Ladder company members, through multiple avenues of tactical operations at residential building fires, lead toward incident stabilization. Forcible entry, providing avenues for egress, and facilitating engine hoseline advancement leads down that road. Laddering, ventilation, confinement, control, and extinguishment also fall in the path to stabilizing the incident. Think of what other operations you and your ladder company members perform that lead to successfully stabilizing the incident.

Property conservation from ladder company operations include ops such as utility control, salvage, and overhaul. While not as glorious as rescue, once we are certain human life is accounted (life safety) and have the fire in check (incident stabilization), we shift gears to conserve and protect property from further damage. As you look through your own home, look at the potential avenues for minimizing damages. Where do the utilities enter your home? Are your breakers and water shutoff valves labeled? You never know when you may be at the firehouse and a utility emergency happens in your own home.

Ladder companies address the "LIP" principles through the actions and tactics that they carry out. Well-trained, disciplined, and aggressive ladder companies save civilian lives. They understand that they are not merely "bulls in china shops." They have a clear understanding of the "why" behind what it is they are breaking or opening up. They communicate well, and through the use of a two-team concept (which is further developed in chapter 20), they attack the fire dwelling with purpose and clear and precise focus. Keep the "LIP" principles in your mind when carrying out ladder company tasks to save lives and property.

20 Preplanned Riding Assignments: The Truck Company

As the first-due units arrive for a fire in the residential dwelling, the radio crackles to life, "Engine 78 on location 123 Main Street. Looks like we got fire showing from the first floor Baker side of a 20×30-foot, three-story peaked roof single-family frame dwelling. We have our own hydrant at 125 Main and will be advancing a 1¾ with a crew of three to the front door for interior attack."

The lumbering ladder company arrives, its company officer getting out of the cab as the engine officer hits him up on the radio (fig. 20–1). He is on his way back from taking a quick lap saying, "Engine 78 to Ladder 64."

Fig. 20–1. Ladder company apparatus arriving at a residential building fire

64's officer replies, "Ladder 64 to Engine 78, go ahead with your message."

The engine boss states in a hurried voice, "Ladder 78, be advised we have one person showing at one window on the third floor, looks to be occupied attic space on the Charlie side, and there are reports of people trapped on the first floor in the rear." He takes a quick breath and continues, "The fire is showing the first floor Bravo side, looks like it might be coming from the kitchen area."

Ladder 64's officer replies, "Ladder 64 message received, Ladder 64 to command, did you copy Engine 78's message?" He looks over his shoulder, "Let's go to work, guys."

So far everything seems to be going smoothly. We have heard a well thought out, clear, concise on-scene report. We had a first arriving company officer take a quick building lap and give a situation report of what was found to incoming units. The arriving engine company knows what they will be doing: stretching and operating the hoseline. But in the ladder company, who is going to be doing what? Ladder 64 has no predetermined operational plan in place. Who is responsible for who will be going where in the building at this fire? What tools are they going to bring? Where are the members going to be operating and with whom? Is this where freelancing originates? Let's see . . .

The officer meets his team in the front yard. They had scurried off the rig, grabbed some tools, and all are heading to the front door. He tries to huddle them up, corralling them on the front steps. There are three other firefighting members of Ladder 64 (with the officer making four in total).

He starts handing out the team assignments, "Luke, you and I will head inside. Oh, wait a minute, I would have preferred if you had brought the irons instead of the 3-foot closet hook, as we have to get inside this locked front door. I have a 3-foot closet hook! Go back to the rig and grab an axe and Halligan. We are going to search for that person reported trapped on the first floor initially, but also for the fire. Luke, grab me an extinguisher while you are at the rig, too, just in case we need it. Jason, Mark, drop those cord reel and lights right here. Put down the PPV fan and give me that TI. I want you two guys to work on getting the person at the third floor rear with portable ladders. Bring a Halligan bar and a metal hook for each of you. VEIS the room with the person in it and report to me before and again after you are finished. For now, when you leave the room, also leave the door to the room closed. Understood everyone?" The faces blankly stare back at the officer.

In a previous chapter we discussed the typical responsibilities that have been historically assigned to members operating in the ladder company. From that dialogue we know that there are a myriad of operations that the ladder company is tasked to complete at the residential building fire. We must remember that these important tasks should not have to wait for the physical ladder truck to arrive. These tasks are so paramount they must be orchestrated early and in the initial stages of every residential fire to which we respond. Those fireground responsibilities and functions should not wait the "X" additional minutes for the truck to arrive. Competent engine and rescue companies can certainly get truck duties underway if there is going to be any significant delay. Many times at private dwelling fires the hydraulic aerial laddering device on top of the truck never leaves its bed.

We know that for the safety of our members and civilians' lives, along with the greatest successful fire extinguishment outcome, truck functions must occur in conjunction with the engine company's line movements. During the attack phase of every residential fire, the initial ladder company tasks are nearly the same. They are almost identical for each fire we fight regardless of the home style, whether it is ranch, colonial, or split-level. We have to get inside, we have to search, we have to coordinate our ventilation efforts and place ladders for access and egress.

How can we best accomplish the great multitude of these fireground tasks quickly and safely in conjunction with the engine advance? In order to quickly, safely, and efficiently complete the duties of the truck company at the private dwelling fire, the best option is for the company to break into smaller groups and have functional positional assignments. Small functional teams that are educated, practiced, and properly trained in utilizing solid tactical techniques can get the fireground duties at different points in the fire building completed at the same time. We can even delineate the tools that should normally be assigned to these smaller group positions to streamline and reduce wasted time. These tool assignment and positions need to be based around your first due.

As we can see, ladder company fireground tasks are occurring both inside and outside the structure. This is the obvious place, and the obvious names (inside and outside) for each of the teams that comprise the company. When we create this inside and outside team we can utilize a two-pronged approach. In this way, members can complete these necessary tasks simultaneously. Obviously, in smaller groups we need to make certain all members are up to performing their ladder company tactical skills effectively since there will be fewer people to give input on each task. Also, and perhaps even as important, team members must have good communication skills.

It is important to note that concise communication must exist amongst the smaller teams and between the teams and the advancing engine company. Coordination through good communication is paramount for success. Portable radios should be given priority to each position in the truck company if at all possible. At a minimum, one portable radio must be with at least one member that comprises a subset of each team.

While the engine company's true focus should be to get water to the seat of the fire, in reality the truck is doing everything else. Engine companies must work together to get the line ready to attack the fire. The smaller truck company teams, while completing our other duties, are in part supporting the engine company's fire attack. Our two-pronged attack best utilizes the capabilities of the truck company to facilitate extinguishment and save lives.

In general terms, members assigned and operating in the ladder company need to be a bit more flexible in their approach to the residential building fire. They need to be able to act more independently when making decisions on the fireground. Truck members will not be working all together like the engine, holding on to what I call the "umbilical cord of life," the hoseline. When assigning duties to members at the station, take into consideration the members' fire experience and capabilities. You may want to place your more seasoned, senior, or "independent thinking" firefighters to positions on the truck.

When carrying out tasks in the ladder company, firefighters must be able to make educated decisions within their smaller functional groups, as they more often than not won't have an officer standing over their shoulder at every moment. They must have good communication skills and the ability to be flexible in determining the best way to attack their assigned tasks. They should report back to the company officer (or incident commander if appropriate) when fireground tasks are completed or if tasks are unable to be completed, and relay any pertinent information as it presents itself on the fireground.

I know many of you are thinking, "We are not a big city fire department with six members on each ladder company. How can we accomplish such lofty two-team goals?" If your department advocates strong, company-level training and isn't afraid to write down a few tactical department policies, there is no reason why you can't do it, regardless of your staffing. The first step is to write out a few things out that will relate to your area's buildings, your staffing, and your goals for the team. Write out your goals and then marry those goals with your staffing, then you can start making plays.

Make a Playbook

In this chapter's beginning, we shared a story about Ladder 64 arriving first due to a fire in a private dwelling. How prepared were the members of Ladder 64 to rescue those two civilians? In this chapter we show you how to best maximize the effectiveness of the ladder company. We will show you how to best utilize your members assigned to the company through position-based riding assignments and basic tactical ladder company deployment in the private dwelling (fig. 20–2).

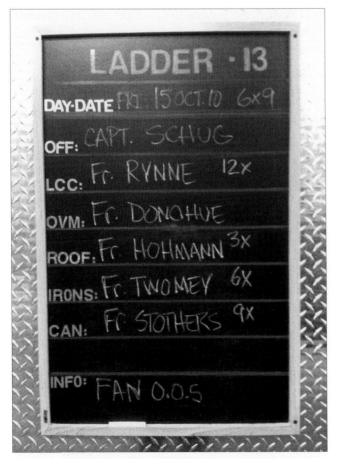

Fig. 20–2. Ladder company preplanned riding assignments

We give you names for such positions, but please do not get hung up on the names. They are only given as a template for you to adapt in your fire department. You must make the positional names work for you, not just do it our way. Colors, numbers, or physical seats on the rig could be used instead of names. The point is to let everyone know who is doing what and where they will be operating.

The bottom line, as you can hopefully now recognize from the opening story in the beginning of this chapter in reference to Ladder Company 64, is that the front yard of the fire building is not the time to be making determinations as to who will be going where and what tools they will be bringing. It is absolutely, unequivocally the least desirable method for ladder company operational effectiveness of the team at residential fires.

Reread that brief story. Is this what you are still doing for your ladder company's firefighting team? Are we best prepared using this method? Is it fair to those who are trapped inside? If your loved one was in that house, is that what you would want to hear happening on the front steps? Seconds count! This is our job here!

Who is responsible for choosing what tools the individuals assigned to your truck will carry for residential fires? Is it done by the personal choice of the member? Is it what the officer shouts out when you arrive? Or is this not addressed at all in your company? Is it okay that "Firefighter Buffy McBuffsalot" brings his 4-foot long, eight-in-one, solve-all gizmo tool that he bought at the last fire buff convention he attended because he thinks it looks cool? Will your team members bring something lightweight because they are out of shape or overburden themselves with so many tools it slows them down?

There is a method to the madness. There is a clear rationale for making tool assignments mirror the positions. Imagine if everyone on your truck company showed up to your next fire crowded around a front door, all armed with 3-foot closet hooks as their tool of choice. Now this may prove to be great if all the fire happens to be isolated in all the home's closets (please note the underlying sarcasm), but obviously that will not likely ever be the case. The point is, like Ladder 64 in the intro, have you ever tried to force a moderately secured front door by banging two 3-foot closet hooks together?

Who now has to go back to the rig to get the set of forcible entry irons? Is this time we can afford to lose? Is this time that the citizens we are sworn to protect can afford to lose? Why should we have to make up these assignments and tool choices again and again at every fire we attend? Must we reinvent the wheel, over and over again, every time out the door?

Having predetermined position assignments and a general tool list that accompanies each position will allow any ladder company to increase efficiency in accomplishing all residential fireground tasks. We will demonstrate how these assignments can be easily adapted to meet flexible staffing levels. We relate stories and show you how this works from various fire departments that have made a successful transition to this concept. We show you how positional truck company assignments at the private dwelling will yield your members the greatest measure of safety, efficiency, and operational ease.

Effectively, we attempt to give you a framework to get you on your way to setting up your own ladder company playbook. We show you the typical compliment of tools for each position and where each member should be operating for the great majority of residential building fires you will encounter.

All effective teams operate with playbooks. Having a playbook allows the team to prepare, practice, and anticipate (PPA, as discussed in chapter 6). Having a predetermined playbook gives everyone insight into each team member's position and the ability to know each person's responsibility for every given play. At fires, it provides direction for firefighters. It provides a framework for each member and ownership of his or her assigned position. Positional assignments also allow company and chief officers to initially address a certain level of member accountability and can limit freelancing. If a member's position is supposed to be on the fire floor searching, then that's where the member should be.

A Team That Fails to Plan Is a Team That Plans to Fail

All teams that want to succeed plan for success. Take any National Football League (NFL) team for example. They have pre-set playbooks, take on-field notes, and have sideline play-by-play reviews. Let's take a look at the offense for one minute, specifically the tight end position.

If and when a running play is called, most of the time the tight end will head down the field to fulfill the role of a blocker. He does this not because he thinks it's the best thing to do, but because that is what is expected of him in the playbook. He has practiced the play and he knows what to do. Conversely, on passing plays, he does not block. He runs short seam or slant routes looking to get open to catch a pass. Again, he knows where to go on the field, and the passing route is based on his ability to read and comprehend the playbook. Great tight ends know their team's playbook inside and out. They know their role. They know where to be and how to get in position for each play that is called. They have mastery of their position.

That team's tight end knows what to expect for the next particular play because he has an on-field boss, the quarterback (QB). The tight end does not run whatever play he feels like; he listens to the quarterback for direction. The on-field boss takes control of the offense and calls a pre-determined play from the playbook in the huddle. The QB then sizes up what he sees in the defense and decides if the play that he has called matches up with potential success based on where the defenders are lined up. The offensive players, like the tight end, listen intently to hear the cadence and the play that is called out, either verifying or changing the play from the playbook.

Sometimes the play needs to be adjusted and it can happen with just seconds to spare before the ball is snapped. If there is a need to change the play at the line, perhaps the defense has changed its configuration, then the QB calls an audible. He is going to change the play based on what he sees before the ball is ready to be snapped. All members hear the change and adapt the formation for the new play that the "boss" called.

What would be the result if team players didn't know the playbook or didn't hear the audible play change at the line? A busted play, a play that was ultimately unsuccessful, would result. Who is to blame if the tight end went out for a pass on a running play when he should be blocking? Who is responsible for the miscue? What was the problem? Was it a communication breakdown in the audible or was it a misstep on the part of the tight end that just didn't have mastery of the call from the playbook? If I were the team coach, I would want to know why we were not successful in achieving our goal, a touchdown. If the team has playbooks, the coach should be able to hold that player accountable for his action or inaction based on the designed play (fig. 20–3).

Fig. 20–3. Fire department operational guidelines are playbooks.

Firefighting is a team activity, much like professional football. But that is where the similarities end. There is no other team that has as much on the line as we do: human life. There is just too much at stake and we must strive to be at the top of our game as professionals. Whether you are garnering a paycheck as a dues-paying union member or volunteering to protect your neighbors' lives and property, we must all strive to be professionals at what we do.

In fact, being a firefighter is more dangerous than any NFL hit, tackle, or scrum. How many NFL players are killed on the job every year? These men practice all week long, whether on the field or in the film room, every day for hours on end. They practice every week, full well knowing that their battle will come on Sunday. These players practice, gearing up for those 16 games a year. They also know their scheduled opponent ahead of time (the opposing team). We have no such luxury of scheduling either our fires or where in the home these fires will occur. We may go to two in one day, one this week, or one in a year. Whenever fire happens, we must be ready. For firefighters, every day is potentially game day, and we must be ready to play.

We must be combat ready at every turn, ready to fight our adversary at every run out the door, whether it is 10, 100, or even 1,000 times a year. We took that oath to protect lives and property and that, we know, is far more serious than any oath or contract that an NFL player makes with his respective team. While it's obvious that we do not make the same wages as professional athletes, there are parallels in the professional team concept. Money is not our focus, saving lives *is*.

Firefighters are some of the smartest people I know. If they all just wanted to be rich monetarily, they certainly wouldn't be here.

Building the Two-Team Truck

Truck companies do not operate at fires like our brethren in the engine. Engine members must work in unison to stretch the attack hoseline. Ladder companies can work in smaller functioning teams to carry out multiple tasks simultaneously. Company officers can and should establish predetermined functional teams, even delineating tool assignments for each group to best accomplish their assigned tasks. The areas in which they will operate at the fire can characterize these groups. Inside positions will obviously be inside the fire building. Conversely, outside members will accomplish tasks around the perimeter of the structure.

Due to the nature of smaller functional groups, operating independently, yet under the guidance of the ladder company officer, members assigned to the truck should be the more experienced members of the department. At a minimum, each team must be equipped with a portable radio.

An attempt to thwart freelancing and a provision for initial accountability is built into the two-team truck concept. How will a two-team truck work reduce freelancing on the fireground? If everyone has a predetermined area in which to operate and tasks to complete, then that is where they should be operating. If there is non-compliance with these tasks, they can be addressed post-incident. If there is no plan, how can you hold anyone accountable for where they were and what they did? Incident commanders and company officers know initially, based on the plan, where units should be operating. Again, we know that there may be variations and that's why we set up guidelines. But if an inside or outside team makes adjustments in the plan (calls an audible), they should let their commanders know.

Let's create a plausible scenario. First-due ladder company arrives and goes to work in the two team truck format as indicated by the existing SOG. As the chief arrives, an isolated collapse occurs in the rear on the fire floor. The chief has no contact with the first-due ladder company inside team. The IC calls the RIT/FAST team to intervene. Where do they need to start their focused search for the truck? Obviously, where the truck was last working, but where is that? Well, the SOG states that their initial responsibility is the fire floor. Here is where the focused search for that inside team should begin. Perhaps the outside team heard them operating in the building and can further isolate a quadrant to help direct the RIT/FAST team. If there are no initial operating guidelines, they could conceivably be anywhere in the home. While the two-team truck concept isn't going to address full operational accountability, it will give the IC a place to start.

The following chart lists the typical two-team positional riding assignments for firefighters in ladder companies, from six down to three personnel (fig. 20–4). If you have more than six firefighters, you can certainly add members to the sections of either the inside or outside team, as deemed necessary for the dwellings in your area. If you have two personnel responding on a ladder company, they will only be able to make one team. Incident commanders hearing a ladder company respond with two members should request additional companies or pair this group up with another team on their arrival to complete the two-team truck.

	Inside Team	Outside Team
3 Firefighters	Officer, FE	Chauffeur
4 Firefighters	Officer, FE	Chauffeur, OV
5 Firefighters	Officer, FE, Can	Chauffeur, OV
6 Firefighters	Officer, FE, Can	Chauffeur, OV, Roof

Fig. 20–4. Ladder company riding chart. (Courtesy of Michael Stothers.)

We certainly know that we can only fill spots in the matrix with the number of personnel we have. Following are several staffing variations for the ladder company and a discussion of the differing roles each firefighter will carry out in the less staffed truck company. If there is less staffing, the incident commander and the citizens you serve must know that it will take longer to complete all fireground tasks, which includes searching. It is that simple. No longer should we be the ones to take the reduction of staffing on the chin. We must be advocates for our communities and ourselves by making the public aware of the drastic effects that the reduction in staffing has on our fireground effectiveness.

For purposes of this book and to address the common tools and tactical assignments spelled out in the aforementioned ladder company duties chart, we discuss the staffing of the ladder company with six personnel. It is to be assumed that members will be wearing full PPE, utilizing SCBA, and carrying a large hand light on a strap and a portable radio.

SIX-MEMBER TRUCK
Inside Team:
Company officer (OFF)
Forcible entry firefighter (FE)
Hook/can firefighter (CAN)

Outside Team:
Ladder company chauffeur (LCC)
Roof firefighter (ROOF)
Outside vent firefighter (OV)

FIVE-MEMBER TRUCK
Inside Team:
(OFF)
(FE)
(CAN)

Outside Team:
(LCC)
(OV)

FOUR-MEMBER TRUCK
Inside Team:
(OFF/FE*)
(CAN)

Outside Team:
(LCC), (OV)

THREE-MEMBER TRUCK
Inside Team:
(OFF/FE*)
(CAN)

Outside Team:
(LCC)

*officer to bring FE tools

We can only fill seats with the number of members that we have responding. If your department has the luxury of fixed staffing, those numbers will give you the baseline. You can then develop your plan based on them. If the company has set persons in the company per day or per tour, having a posted riding assignment board makes recording the assignments easy to hand out. It serves as a reminder for all who see it about the responsibility of each person on the apparatus. Even in the volunteer setting we have seen this concept applied. It also works well in fire companies that utilize a duty crew system.

The Inside Team: Officer (OFF), Forcible Entry Firefighter (FE), Hook/Can Firefighter (CAN)

The duties of the first arriving ladder company's inside team are difficult at best. The inside team will be faced with many task level functions within the fire building. Let's discuss their role, position, and tool selection for each possible position that we may have on the inside team. Once we have discussed the tools for the individual members of the team, we describe where they should be operating in the dwelling. We also cover the assigned duties for both the first and second to arrive ladder companies.

The company officer

Someone must always be in charge. While we know in different fire departments you may not need to hold an official rank or an officer title to be in charge, someone has to be the definitive team leader (fig. 20–5). We do not delve deeply into the interworking of the habits of good and bad leaders here, but from the moment you get off that rig, you are leading your members into the most dangerous place on earth. This fire is *the* fire. It is the fire of the crew's career, until it isn't. You are counting on them and they on you. Lead them, having confidence in their training and in your ability as an officer.

As such, you must have the ability to step back from the task-oriented firefighter and look at the bigger picture to observe what is unfolding around you. You must make sure that the team you are guiding is putting the right tools into operation in the right places and ensure they are operating in the safest manner possible. Keep your "look, listen, and feel" senses alert to changing conditions, and optimize your fireground situational awareness.

Fig. 20–5. The ladder company officer

Think about the previous references in this chapter with regard to planning the NFL offense and about the duties of the truck boss at residential fires. One can certainly interchange ladder company officer in place of the NFL quarterback, as this person is our on-field boss. Company officers are the ones who largely call the fireground plays.

While all members are conducting their individual size-ups of the structure, the officer is responsible for audibly airing what he or she sees to incoming companies through an on-scene report (see chapter 9 for more information regarding on-scene reports) and running the action, calling an "audible" if conditions warrant. If nothing is strikingly different initially, then members will run the play from the established playbook.

While the company officer is an integral part of the inside team, he or she must also check in to functionally supervise the outside team and receive reports from the outside team via the portable radio as needed throughout the incident.

Regardless of your staffing, the company officer is never really just a supervisor. He or she takes the lead role directing the searches, communicating with command, and so forth. However, in many departments, staffing numbers (or lack thereof) forces the officer to perform as an active, hands-on member of the inside team. Take a look at the tool selection for the company officer for fires in private dwellings listed below.

Ladder company officer's tools:

- Halligan bar (perhaps Sunella lock-puller or smaller variety if warranted)
- Thermal imaging camera

The forcible entry firefighter

The forcible entry firefighter, a key member of the inside team, is required to utilize a delicate balance of skill and strength (fig. 20–6). Many of the best forcible entry firefighters I've known are certainly not those who spend their entire tour in the weight room bulking up. They read and study techniques as much as they work out. Forcible entry is as much about placement, technique, and skill as it is about brute physical force. As a friend and an excellent forcible entry instructor, Chief Nicholas Martin of the Columbia, SC, Fire Department always touted, "The devil is in the details," when knowing and describing the tools of choice for forcible entry. As students of our profession, we must investigate and understand the "why" behind those little details found in and on our forcible entry tools and how they can be applied correctly to increase our skill.

While the officer may give you hints or tips while forcing the front door, ultimately you are the one with the tools in your hands. It is you, not the officer, who will be calling for hits to drive the Halligan bar past the jamb. You must control your domain, then recognize and continue the attack when things are working. You must equally recognize when it's time to change the plan. The bottom line is that we have to get the job done. We will not be able to advance hoselines or conduct searches until we get into the house.

Every firefighter in the truck always loves to be assigned the FE position. They are the one to get in that door and they don't want to let the team down. Many eyes will be on you as the FE firefighter. Be a professional at your job with your tools. Know the limitations and proper applications for each one in your inventory. As discussed in several previous chapters, you must stay aware of the changes in home construction and security upgrades to private dwellings in order to properly arm yourself with the correct tools for each situation.

Fig. 20–6. The forcible entry firefighter

For the great majority of residential building fires, the tool assortment for the forcible entry firefighter should include at the minimum:

- 8-pound flathead axe
- Pro bar style Halligan bar (married with axe as "irons")
- Hydraulic forcible entry tool (Bunny/Hydra-Ram)

Once inside, with the initial forcible entry now behind us, the FE firefighter's focus becomes the search. Unless an obvious rescue is apparent or known prior to entry, the search is an initial one to find the location of the fire, then a search for persons. Search techniques for the residential building fire are further developed in chapter 25.

The can/hook firefighter

The can/hook firefighter normally brings just that, a 2½-gallon water can extinguisher and a 6-foot hook (fig. 20–7). Why do we select these two tools for this firefighter to bring? Let's talk about each tool individually—first the can, then the hook.

Fig. 20–7. The can/hook firefighter—note strap attached to water can.

Many times small or incipient type fires in residential structures can be quickly knocked down and nearly fully extinguished by the can/hook firefighter with the 2½ gallon water extinguisher (a.k.a. the can). If you haven't drilled with the utilization of the can, you may be surprised at how much fire it will actually help you control. We recommend the addition of a strap to this tool to assist in transporting it to and from fires. Imagine, as the engine company hoseline is being stretched, you may actually be able to put the fire out!

That said, we know that just because your ladder company inside team is bringing a can in, it certainly isn't going to replace the need for the engine company hoseline. It should, however, bring the advancing search team just a slight bit of comfort knowing that there is someone at the edge of the fire area with a bit of water and the goal to keep the fire in check.

The can/hook firefighter has to use good judgment and tactical skill with the can to best utilize its powerful punch. If, in the estimation of the can/hook firefighter, the team can't knock nearly all the fire down and extinguish it, they really shouldn't try. If you have an entire bedroom on fire (mattress, box-spring, bed, dressers, etc.), you will likely not have enough water to overcome the BTUs presented to you. This is not an incipient fire (such as a pillow or light bedding on fire on top of the mattress), this is a room of fire.

Our goal as the can/hook firefighter may not always be to fully deploy the contents of the water can into a flaming, fully engulfed bedroom. One might attempt to contain the fire to only this room utilizing the room door and the water in the can by sweeping the top of the door and doorframe with the can's water stream. You may need to pull a door from another nearby room to assist you if the room of origin begins to burn through. Your goal is to protect the rest of your inside team as they start their search and be a beacon for the advancing engine company.

Once the engine company has arrived and begun the initial knockdown of the fire area, the hook is a great tool to have at the ready. Once the main body of fire is extinguished, engine company members will be looking to you and your hook to open up walls and ceilings in the fire area. We prefer the 6-foot wooden hook instead of the metal variety here since you may be initially opening up walls/ceilings that still have power running through them.

We also choose the 6-foot hook's length by design. Closet hook lengths are good for closets, but using them on the rest of the ceiling in the room will quickly tire you out. Anytime you have to hold tools up and use your hands above your head, you will tire faster. Using the reach of the 6-foot hook allows the firefighter to hold the hook into the body and pull in short, choppy strokes. Longer, 8-foot hooks become cumbersome to try to maneuver through most private dwellings. They may be needed after knockdown if vaulted ceilings are present, but they can be called for later.

Remember, in searching for hidden pockets of fire in the private dwelling, such as behind walls and ceilings, be cautious if you find fire as you start to open up (fig. 20–8). If you continue to open up without having the protection of a hoseline, you have the potential to quickly spread fire into the room. This may cause unnecessary fire damage, create a dangerous environment for others operating, *and* cause you to be run out of the room. When you open up and fire is discovered, call for a hoseline to be in place before you get ahead of yourself.

Fig. 20–8. The can/hook firefighter opening up. (Courtesy of Nate Camfiord.)

Once we pull a wall or ceiling open, chances are we are not going to be able to put it back. Sometimes entire sheets of drywall or an entire plaster ceiling in a room can come down with one good pull.

The inside team positioning: first due

The first-due inside team at fires in residential buildings will initiate their impact on the fire with command over operations on the fire floor of the dwelling. Whether the fire is in the basement (see chapter 11 for certain exceptions) or on a lower or upper floor, the first-due ladder company is responsible to initially focus on this fire floor. Once operations are complete on the fire floor, the company may have to continue to work the rest of the building (searching, opening up, etc.) if other units are not on the scene.

If not already established by a previous company, they must initiate entry into the dwelling (fig. 20–9). You must keep in mind that opening the front door, whether you force it open or not, creates a form of ventilation that will have an effect on the fire. You must be prepared for a myriad of reactions. In many instances, this door may not need to be chocked open until the hoseline is ready to move in.

Fig. 20–9. Inside team prepared to force entry. (Courtesy of Traditions Training, LLC, www.traditionstraining.com.)

The inside teams search of the fire floor is two-fold. The team is searching for both fire and life. Which is more important to find? Well, the answer is . . . both. While we are obviously looking to save lives, fire extinguishment saves lives as well. They are interdependent. With fire extinguished, there are fewer byproducts of combustion and, therefore, certainly increased civilian survival chances. A quick search conducted, a victim found and removed, obviously also increase civilian survivability.

The initial search of the inside team is to find the fire area and attempt to extinguish/control it with the water extinguisher. If we happen to find a civilian victim along the way, excellent. Obviously we promptly remove them to fresh air. But we still need to find the seat of the fire since it continues to grow while we are in the process of removing the occupant.

While smoke and byproducts of combustion will eventually fill the entire home, the greatest initial danger is within the immediate fire area and the areas adjacent to it. Once we determine the seat of the fire, that is where our primary search should truly become focused, then moving outward to less exposed areas. The inside team must relay the findings of the location of the fire area back to the engine company and the incident commander. They should, as much as is possible, facilitate hoseline advancement by making a clear path for the engine's advance.

The officer of the inside team shall also have the strict authority to control the horizontal ventilation that occurs on that fire floor. Any and all horizontal ventilation should be confirmed and communicated through the officer prior to occurring. While the inside team may horizontally ventilate as necessary as they move through the building, they must keep in mind that they may change conditions by doing so.

The initial actions of the inside team can simplified to these tasks on the fire floor:

- Force entry to fire floor
- Search for fire victims/fire origin
- Control and confine fire/coordinate fire floor ventilation
- Expose hidden fire

If these initial tasks are completed on the fire floor and no additional companies have arrived to bolster the attack, the first-due ladder company is called to search all floors of the dwelling. Start with the floors above, but do not neglect the floors below the fire. Second arriving ladder company officers should check in with the first arriving ladder company officer to ascertain the status of the searches in the building.

Second-due ladder company: inside team

The second arriving ladder company's inside team needs to listen as they are coming down the road. Listen to the dispatch and fireground department handie-talkie radio traffic. They need to hear and process the fireground updates from command to the communication center. These transmissions can help you visually paint a picture as to what to expect when you arrive on the scene.

Generally, with the exception of basement and attic fires, your job as the second-due company will be to search the floors above the fire for victims and fire extension. Be sure to check in on your arrival with the first-due ladder company and the IC to determine where you need to start your searches. Assuming can result in duplication of efforts. There is no need to assume anything. Find out definitively where you need to go before entering the building. If you are headed above, be sure that you communicate that fact to the engine company on the fire floor so that they know that you will be operating above.

We recommend the same tool assortment for each member in the second-due company, but with an attic or top floor job, have the can firefighter bring an additional 6-foot hook for opening up ceilings. This hook can be used by anyone, not necessarily for a two-handed, two-hooked can firefighter. Even engine company members can assist in opening up areas on the fire floor with this extra hook as needed. If you are not needed on the top floor, remain below. Ensure that the fire did not originate on a lower floor, with special consideration to inclusion of a basement check.

While we are on the subject of top floor fires in private dwellings, it cannot be overstated that we as firefighters consistently overload the top floor. The first-due truck and first-due engine normally have enough work to do in what is generally a small space. Add a rescue or squad company, and then second-due truck, and we can quickly clog the stairs and overload the top floor. If fire were to dramatically push down on the operating forces, surely someone would be trapped and cut off from egress. If not trapped, the firefighter will be severely burned while waiting to move that clog of members on the steps. All firefighters, but especially second arriving company officers, need to exhibit control of their teams and check in before showing up to an already crowded top floor.

Another instance where the second arriving ladder company's inside team may not be going above was covered in greater detail in chapter 11. This is in regard to second due ladder company operations at basement fires.

As with most operations, communication is the key to success with basement fires (fig. 20–10). If the first arriving ladder and engine company cannot make the stairs down to the basement (due to fire, smoke, or the structural conditions of the steps) and an outside entrance is present, they may elect to hold the stairs on the first floor. This would then permit the first-due ladder to search the first floor and all remaining floors above. The second-due truck will team up with the second hoseline for entry and search into the basement from that exterior entrance. Preplanning and communication will allow for this to successfully occur.

Fig. 20–10. Basement fires require coordination. (Courtesy of Kentland Volunteer Fire Department, www.kentland33.com.)

Reduced staffing: first- and/or second-due inside team

The absolute minimum number for any interior firefighting team in which crews may encounter and operate in IDHL conditions is two firefighters (unless an immediate life hazard is known). If we have the ability to add a third firefighter to either team (inside or outside), our preference is to add them to the inside. This is where we will have the most impact on saving lives, which along with extinguishment are our primary goals. Upon completion of initial interior search duties, this extra person can later be sent to bolster an outside position if needed.

The reality of a ladder company responding with three or four personnel is commonplace in many areas. If faced with a three- or four-person truck, we can still utilize the inside and outside two-team approach with great success. We will, however, need to modify our inside team to two personnel.

If we have only two personnel on the inside, here is how we carry the same tools and complete the same tasks in the dwelling. Regardless of your staffing numbers, someone needs to be in charge of the company. One firefighter (or fire officer) will still remain the team leader. However, the person will now have to wear an additional hat as the FE firefighter. As such, the team leader is now taking on more responsibility on the fire floor. This firefighter/officer will carry a merged load of tools and equipment of the FE firefighter and the OFF. He or she will need to be actor and director and should carry the FE equipment (irons/bunny tool) as well as the thermal imaging camera.

As the second firefighter on the inside team, the can firefighters' tools remains the same. They still carry the can/hook. As you can see, reducing the team by one firefighter is one less body to assist in the search. Quite obviously this puts additional stressors on the team-leading firefighter (or officer) who is doing double duty as FE firefighter and company officer.

One can see how the team leader could become overly focused on task-oriented firefighting skills, easily losing the ability to step back and review his or her surroundings. The ability to process updates in situational awareness can become clouded. This may lead to poor decisions or inadequate information being passed to other team members.

The Outside Team: Ladder Company Chauffeur (LCC), Roof (ROOF), Outside Vent (OV)

The duties of the first arriving ladder company's outside team are plentiful, but if properly practiced, can be initially begun with even just one firefighter. As the name leads you to believe, the first-due ladder company's outside team is tasked with duties that originate on the exterior of the home. This does not necessarily mean that they will never enter the residential building. In fact, in certain specific situations members assigned to these positions will employ vent, enter, isolate, and search (VEIS) maneuvers. Members assigned to the outside functions must have a strong working knowledge of the buildings in their area. They must understand the likely places where people will be based upon what clues and cues they are presented with on the exterior of the home.

If at all possible, members assigned to the outside team should be the more seasoned members of the company, as they operate at times without the "officer" per se looking over their shoulder. While they function under the direction of the ladder company officer, they have some autonomy as to where to initiate their attack and when to do so. Outside team members should, as needed and on a regular basis, check in with the inside team via the portable radio to update the interior team of their status.

We discuss positional assignments and tool selections again based on the six-person staffing of the ladder company. We then address short staffed company operations and wrap up with first- and second-due outside team actions.

Ladder company chauffeur (LCC)

In many paid departments the ladder company chauffeur is one of the most senior members of the company, and perhaps even in the entire firehouse (fig. 20–11). In fact, in many of the places I have found myself lately, the LCC has more time on the job than most of the bosses. While some may find this unnerving, I always found it to be a great comfort for me as a covering officer. Those senior members knew their area, their buildings, and perhaps more importantly, who else was working. I would often defer to their direction and tutelage when making up the company riding list because they could give me the run down on who was who. In our experience the volunteer setting, the apparatus drivers were the same—the older, more senior members.

Fig. 20–11. The ladder company chauffeur

The LCC's first job is getting the apparatus to the fire. The safe response is paramount and cannot be understated. The chauffeur and the company officer are equally responsible for that safe response. As LCCs, be certain to wear your seatbelt (fig. 20–12). Many firefighters are killed and seriously injured responding to alarms. Assist the company officer in setting the tone for the safe response of all members by buckling up. If the apparatus doesn't make the scene, no one gets to do any work.

Fig. 20–12. Buckle up when operating the apparatus.

At residential fires the ladder company chauffeur's duties are plentiful. If conditions are favorable for ladder company apparatus placement and utilization of the hydraulically operated ladder, the LCCs must remember that a 100-foot rear-mounted ladder is just that, 100 feet. If 105 feet, 106, feet or 110 feet are needed, you are not going to make your objective. While it may seem obvious, engine company chauffeurs who many times arrive first must be cognizant of the fact that they can extend hoselines, but trucks cannot make additional sections of ladders appear. Accordingly, engine companies should practice due diligence to leave the front of the fire building open for the ladder company.

Likewise, truck company chauffeurs, in consultation with the ladder company officer, must set up the rig where they want it to be the first time (fig. 20–13). Slow down and plan your approach as you enter the block. Once the apparatus emergency brake is set and the team is in motion, it is unlikely that you will be able to reposition the apparatus. If you need to move the apparatus slightly after the members have gone into the fire building, get help. Another firefighter from a different unit, a police officer, or even a competent civilian (as a last resort) can assist you in maneuvering the apparatus. This is especially necessary if operating in the reverse gear.

Fig. 20–13. Slow as you approach the building.

Fig. 20–14. Portable ladders

Other incoming responding units must avoid the temptation to crowd the rear of the ladder company apparatus. Many trucks have portable ladders that come out of slide trays directly to the rear of the apparatus. Some ladder companies have a traffic cone on the rear step, which may be placed at a distance determined by the LCC to remind incoming apparatus of this fact (normally slightly farther than the longest ladder in the tray: 20 feet).

If the truck company's outside team is going to be using the apparatus-mounted ladder, place it for greatest overall coverage of the dwelling or as directed by the IC, unless obvious rescues are apparent on arrival. The primary objectives for the outside team are coordinated ventilation and ladder placement. More often than not, you will rely heavily on the utilization of portable ladders to accomplish laddering the building.

The compliment of portable ladders on most truck companies is perfectly suited for the residential dwelling (fig. 20–14). This is good, since many times the bed-mounted ladder stays on the top of the rig at private dwelling fires. This is in part due to the fact that most single-family homes are set back longer distances from the street and have limited access for the apparatus in driveways with trees and landscaping. This certainly does not preclude you from putting your rig in the grassy front or rear yard if conditions warrant (fig. 20–15). Another obstacle, in addition to the building setbacks, can be the presence of the home's electrical connections to the street service overhead. Both of these factors can negate the use of rig mounted aerial devices.

Fig. 20–15. If conditions warrant, put it in the lawn. (Courtesy of Kentland Volunteer Fire Department, www.kentland33.com.)

The LCC in the six-person ladder company's three-person outside team is the team's go-getter. As the more senior member, we prefer to see this firefighter facilitate the outside team. This person is a motivator and unofficial director for the exterior operations, choosing tools as required. In the five or less staffing view (two outside team members), the LCC will be an integral part of the ventilation/egress team.

Tools:

- Six-person staffing: tools as deemed necessary by LCC
- Five or less (two outside team members)
  - 6-foot metal hook
  - Halligan bar
  - Portable ladders

Outside vent firefighter

The outside vent firefighter's main function is to coordinate ventilation and establish secondary means of egress for interior firefighting forces (fig. 20–16). As stated earlier, do not get hung up on the names of these positions. Some think that this firefighter just runs up and down the outside of the buildings smashing out all the windows within reach. We aren't making this up. Go to the Internet and look for yourself. There are many videos showing this being done. Obviously this is terribly incorrect and could not be further from sound firefighting practices.

doors can quickly and without warning cause tremendous problems for interior operating members.

The OVs will live on ladders (fig. 20–17). As such, they must be masterful and comfortable handling and working on them. They will be moving and working from portable ladders and should be able to maximize their work from ladders (practice utilizing leg locks, using the hook from their harness, etc.) while ensuring that they are bringing necessary forcible entry tools to complete the jobs performed from the ladder.

Fig. 20–17. The outside vent firefighter with a portable ladder

If the situation warrants and members are needed to perform VEIS into the residential fire building, the OV can pair up with either the roof firefighter or the LCC to accomplish the task. The complete and quick search of a particular room should rarely be attempted alone (unless a known life hazard exists: see OSHA two-in/two-out for clarification). Anytime we are entering an IDLH environment we must be in a team of two. The VEIS tactical concept is discussed in depth in chapter 24.

The pair will work with the LCC or the roof firefighter to accomplish other exterior functions as needed, such as removing security devices (window guards, gates, bars), coordinated ventilation of windows, and forcing open exterior door openings in concert with the needs of the interior forces. If ordered, they can also assist with the vertical ventilation of the structure and can assist the roof firefighter in opening up roof and attic spaces. Ventilation and tactical decision making for both residential horizontal and vertical ventilation will be covered in chapter 23.

Fig. 20–16. The outside vent firefighter

While we can appreciate the well-worn "bull in the china shop" title given to members of the ladder company (truckies do have a propensity for breaking things), members must exhibit good judgment when breaking windows at fires. We know from our previous discussion of fire travel in chapter 11 (and we further discuss ventilation tactics in chapter 23) that a firefighter who indiscriminately ventilates windows and

Outside vent (OV) tools:

- 6-foot metal hook
- Halligan bar

- Thermal imaging camera
- Portable ladders

Roof firefighter

The roof firefighter is the last position filled in the six-person firefighting ladder company (fig. 20–18). This is not done because of the position's unimportance, but rather its ability to be filled at a later time. Don't think that the roof position will always be going to the roof. It's just a name.

valuable for the outside team. On fires in the peaked roof residential buildings, the roof firefighter should team up with the OV firefighter to perform the VEIS maneuvers as indicated. With this team of two operating independently, the LCC is free to continue to throw additional portable ladders to other positions around the house, outside of the IDLH (fig. 20–19).

Fig. 20–19. The roof firefighter paired with the outside vent firefighter

Fig. 20–18. The roof firefighter

The first-arriving ladder company's outside team has many jobs to do initially. Therefore, if we had this position available, we would rather pair this member up on the outside of the building to address many of the exterior fireground functions with the OV and LCC such as VEIS, ladders, and horizontal ventilation. If this level of staffing is available, this firefighter is extremely

If, however, the IC immediately orders peaked roof work (i.e., the roof cut), so be it. A team of two firefighters, namely the ROOF and the OVM, should accomplish this roof-cutting operation (fig. 20–20). The impact of taking two out of three firefighters from other tasks will seriously impact and delay the other exterior functions and cannot be overemphasized. This will cause the LCC to throw portable ladders alone, and until paired with another firefighter, unable to enter the building.

Fig. 20–20. Taking the saw to the roof

Conversely, on flat roofed private dwellings, the roof firefighter's position on the roof may be more advantageous. Vertical ventilation of smoke from atop the interior stair may be completed without much fanfare with something as simple as the removal of a skylight. For fires on flat roof dwellings, a lifesaving rope should be brought if the fire is on a lower floor. If the fire is on the top floor, the OV and roof firefighter should work to operate the saw over the fire area.

Roof firefighter's tools:

- 6-foot metal Halligan hook
- Halligan tool
- Saw/lifesaving rope (flat roof; dependent on fire floor)
- Portable ladder/aerial device

Outside team positioning: first due

While it may seem obvious again as the name suggests, the outside team works the exterior of the building. They attempt to soften any building defenses and coordinate and time horizontal ventilation with the advance of the interior operating forces. They may, as deemed appropriate and necessary, request permission for and access the building's interior via portable ladders for VEIS maneuvers. They should choose the area on which to initially concentrate their work that is closest to the area of the greatest danger, the fire area. This is where other inside team members are also focusing their work and, ultimately, where fire extinguishment happens. Once extinguishment is completed in this area, work outwards to areas further removed from the fire.

In one-story homes this should mean initial horizontal ventilation opposite the fire when, and only when, the hoseline is ready to move in (fig. 20–21). Doing this in a coordinated manner allows the heat and converted steam and smoke to be directed out the opening you have created. As with any window you take, if you decide to take it out, clean it out completely, sash and all. Once that is complete, you may be tasked to VEIS (in your team of two) a first-floor window in a room that is adjacent to the fire area.

Fig. 20–21. Ladder company members conducting coordinated horizontal ventilation. (Courtesy of Nate Camfiord.)

Most of the time in two- or three-story dwellings, the outside team will be placing ladders to VEIS on the floor above the fire. While we discuss the specifics in another chapter, just know that being above the fire, possibly without the protection of a hoseline, is initially dangerous and not to be taken without due regard and notification. Portable ladders should be initially placed above the fire area.

Members assigned duties on the outside team should be able to manage the situations above, and as time and personnel permit, complete the remaining exterior fireground tasks. Depending on the available staffing, prioritize your completion of these tasks referring back to the "LIP" principles, notifications to and from your ladder company officer and/or the IC, and your good firefighting judgment.

To quickly sum up the duties of the outside team:

- Communicate all pertinent findings/changes to the IC and inside team.

- If needed and/or applicable, position the truck for use of the apparatus-mounted hydraulic ladder.

- Throw portable ladders for access/egress and use for VEIS where applicable.

- Coordinate necessary horizontal and vertical ventilation with inside teams.

- Soften building entry points (remove gates, window bars, force exterior doors).

- Manage control of exterior building utilities (gas, water).

- Place fans/increase on-scene lighting.

Outside team positioning: second due

The second outside team to arrive moves to bolster the workings of the first ladder company. While the first-due truck is concentrating on areas in, around, and above the fire, the second-due LCC, OV, and ROOF team begin to perform similar tasks, but in areas further distant from the fire.

If the first-due ladder company has thrown ladders and completed VEIS maneuvers on the floor above sides Alpha and Bravo (as they were at greater risk of exposure due to the fire on floor below), the second ladder company outside team should focus on performing the same tactical operations, but concentrate their efforts in the Charlie and Delta sections of the home.

Tool selection should remain the same, largely relying on the use of portable ladders and forcible entry tools that meet the needs of the buildings in your area. Stay alert to the radio traffic of the first arriving units for special requests and possible FE challenges that may require additional FE tools.

Reduced staffing: first- and/or second-due outside team

Short-staffed ladder companies quickly lose members from the outside team. As mentioned, we can only fill spots for which we have people. Taking the total truck crew down to five or four eliminates the ROOF firefighter position and cause the outside team to fall to two members.

This two-member outside team still gives you the ability to have a pair of firefighters that can work together to enter an IDLH and/or VEIS a particular area if needed. However, having just two personnel on the outside slows down the placement of portable ladders and reduce the speed in which the completion other exterior duties occur.

A three-person total staff ladder company forces the LCC to work alone. Even though physically one person, the LCC is still a part of the outside team. While some tasks absolutely require an additional member to team up if needed to enter an IDLH, the single operator can accomplish many of the other tasks at a residential building fire.

Do you have junior or exterior-only firefighters who come to the fire scene with you (fig. 20–22)? They don't have to just stand around taking photos and texting their friends. Put them to work. How can they assist the outside team? How about teaching them to "become one" with portable ladders. Even if they just carry and place the ladders at the level of the windowsill without doing anything else, it will be a help to the cause. Train with your members and show them techniques for safe and effective carrying, self-deployment, and extension of portable ladders. Handling portable ladders is just one of many things individual firefighters can do at residential building fires.

Fig. 20–22. Junior firefighters need to be trained to assist.

Individual firefighters can also work alone to effect coordinated lower floor ventilation by controlling utilities. If they are tasked to do something that requires them to enter an IDLH and use their SCBA, they need to pair up with another member prior to doing so. If the LCC or any firefighter operating sees something that needs to be addressed or that is going to require additional resources, they should advise the IC immediately.

Later arriving ladder companies with reduced staffing may be sent to perform individual tasks as a company. Such items as peaked roof vertical ventilation come to mind. They may also come to the command post and be used to bolster lower-staffed initial companies.

Setting the Plays in Motion

Establishing the two-team concept in your organization may take time, as do most items that cause change in the fire service. One of the best ways we have found to initiate such change is educating your members as to *why* we want them to use this strategy. While we all know it would certainly be easier to just order them to do as they are told, getting them to understand why we are doing what we are doing is integral to program success. You want to get the company to buy in by having your members adapt to utilizing the inside and outside team approach to ladder company work. Preparation and practice will absolutely make you a more efficient team.

In order to accomplish this in the shortest amount of time and yield the greatest margin of success with the two-team truck concept, you have to write your department's own playbook. Start simply. Establish some standard operating guidelines (SOGs) for ladder company actions in the buildings where you most routinely go to fires. Keep in mind your apparatus, your tool compliment, your members, and the ultimate goal of increased safety for your crew and the civilians we are sworn to protect.

The members need to be well practiced and verbal and hands-on tactical training sessions are musts. In addition to the obvious training and drilling schedule, many departments have adopted the SOGs to help the members remember their assignments in creative ways. Many have posted laminated versions of the ladder company assignments throughout the firehouse. They can be blown up on bulletin boards, on a laminated ring for use in the kitchen, on the back of the door in the bathroom stall, put them everywhere (fig. 20–23). Some have even made smaller laminated cards for positional assignments on the physical rig and/or for members to carry with them until it becomes part of muscle memory.

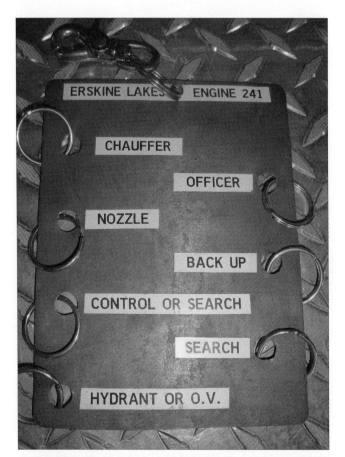

Fig. 20–23. Laminated cards with tool and functional assignments can help remind firefighters what to bring and where to operate.

The Payoff

Do you think that there is no way that you can apply this concept to your fire department? Let me tell you about a fire department that thought the same way initially. A small, rural, bedroom community volunteer fire company in the northwest portion of New Jersey contacted our training group, Traditions Training LLC, inquiring about a truck company, hands-on program for their members (fig. 20–24). When we asked what type of ladder company they had, there was silence. They had no truck in quarters with them.

Fig. 20–24. Erskine Lakes Volunteer Fire Company #1

What they did recognize was that they were going to have to perform truck company functions early on at every fire that they ran in conjunction with engine company movements, regardless of when the neighboring town's ladder company arrived. When we explained and taught them the two-team concept, they realized that they could apply it best to the members responding with the second-due engine company. They normally respond with two engines out of their firehouse for reported dwelling fires.

According to all members of the department, from the fire chief to the newest member, since enacting these riding assignments they absolutely increased their efficiency with all aspects of fire suppression. The two-team concept has allowed for smoother engine company operations and has best utilized the limited staffing that they have on the fireground for ladder company operations.

The increase in fireground efficiency was not just internally noted. While neighboring departments may not have drawn up written commendations documenting the changes they have seen, they have commented aloud that they have noticed a change in this company for the better. These informal accolades are indeed infectious. They are not used to gloat or say they are better than you. It is a fact that they are indeed proud of their reputation. They earned that reputation with the training and practice of their members. It's spreading. It's like a perfect wave, slowly growing and dispersing. As these concepts are applied, they have now been overheard in various firehouse circles and in different departments around their town. Embrace the spread of effectiveness in your area.

The "Why"

How have they shown their members the "why"? They have shown their members the why with positive results from the fires to which they have responded. They have noted an increased level of satisfaction from chief and company officers at incidents. They have noted quicker turnaround times on searches, have held fires in check, and extinguished them more promptly. They have seen a decrease in building damage caused by fire. They have made their fire company better prepared to meet the needs of their citizens and, at the same time, built in another layer in initial accountability.

Operating within the guidelines of the two-team truck will save lives and property at residential fires. It is the best overall usage of staffing for any ladder company, regardless of the total number. Efficient and effective ladder company operations stem from preplanned riding and tool assignments based on your buildings and your staffing for that particular event.

The Backstep

Riding Assignments

Based on this chapter's discussion of individual riding positions on a ladder company apparatus, map out riding assignments for your department. Base these positions on the staffing, tools and pre-arrival assignments, and type of apparatus you respond in to residential building fires. Keep in mind that oftentimes the greatest number of tasks are accomplished by utilizing the two-team approach. Take a stack of 3×5 or 4×6 index cards. Write each individual position on one side and, on the back, the tools and responsibilities. After discussion with the department, these positions may evolve into laminated cards for each spot to act as reminders for members while responding of their role at the incident. It also serves as an excellent training regimen with tool, location, and tactical discussion at each position. The point here is to make it work for you and your department based on variables you can control.

Inside Team

- Officer
- Firefighter 1
- Firefighter 2

Outside Team

- Chauffeur
- Firefighter 1
- Firefighter 2

21 Look, Listen, Feel: Residential Recognition

Fire Senses

Have you ever been at a fire and had to guess at something? Whether it was a direction to search, a time and place to start water in the line, a hydrant to either hit or pass, or deciding what wall to overhaul first, certainly we all have made guesses at fires in our careers. Most of the time we guess correctly, but unfortunately we sometimes do not. Were our actions truly just guesses?

We are of the belief that our fireground guesses are really not guesses at all. Rather, they are quickly assembled hypotheses—educated guesses. The hypotheses that guide our actions come from the blending of your knowledge base, previous experiences at similar fires, and the current information that we are gathering at the fire. These three thoughts—the foundation, the recall, and the recon—blend in our minds prior to our tactical decision making (fig. 21-1).

Fig. 21–1. Tactical operations at residential fires can be influenced by our senses. (Courtesy of Nate Camfiord.)

The ability to quickly make educated decisions on the fireground is not inherently found in every firefighter initially. It is molded and sculpted over time by using our five senses. We refine our decision-making abilities with knowledge, and we can increase our knowledge base through listening to firehouse mentors, reading department guidelines, studying fire books and journals, attending fire service lectures, and conducting realistic hands-on training. In other words, we hone our five senses to adapt to fireground conditions based on what we know. The more we can expand this base, the more it will help in our ability to identify the actions needed and make the correct decisions on the fireground.

How can we use our five senses most keenly when at fires? Obviously, the nature of fire and its blinding and choking smoke causes many of our senses to be darkened, in a very literal sense. But do we really lose the ability to rely on our senses at fires? Not unless we allow ourselves to do so. We argue that on the fireground, our senses are not diminished at all. We see our senses at fires heading to the other end of the spectrum, running at extremely heightened states. As firefighters and fire officers, we need to focus on the ability to maintain control of them. Again, take time to make time. Allow your brain to take an extra second to adjust your senses to the new environment. We are leaving the normalcy of the outside world and headed into what is our truest office, the inside of a burning residential building fire. We need to take a second to let our minds and bodies adjust.

It has been said that if you take away one sense, the remaining ones will become increasingly focused and alert. For instance, a blind person's sense of hearing may be sharper as a result. If we can refocus those heightened senses (since we are used to working with all five of them during the course of a normal day) while at fire operations, it will greatly assist us in doing our fireground tasks.

Through the darkness there comes clarity. Through noisy chaos comes calm. Physical movements become calculated, body and hands moving with purpose, recognition, and direction. We must recognize the necessity of harnessing these heightened senses and allow our brains to digest the information they are receiving. Move with purpose, don't just blindly move for the sake of moving. The ability to focus our fire senses will greatly assist us in keeping ourselves aware, oriented, and alive. It increases our efficiency in all phases of our attack on a residential building fire.

Our Sixth Sense

The importance of analyzing our own human senses while we are in the fire environment is ever apparent. But is there some sort of sixth sense held by certain firefighters and fire officers? I am sure that you could recount many of your own stories here. These would be from past incidents in your careers that reflect an often eerie side of firefighting. Why did that firefighter tell us to move just prior to the window failing? How did that chief know to back the units out of the building just prior to collapse (fig. 21–2)? How can some personnel recognize or perceive events prior to their occurrence? Likewise, why do some others not have it at all, perhaps even after being smacked in the face with similar events a few times?

Fig. 21–2. Senior firefighters can seem to exude a sixth sense at operations. (Courtesy of Nate Camfiord.)

Is it some sort of divine intervention? Have these chosen ones been blessed with inexplicable powers? Not likely. We believe the firefighting sixth sense is a derivative of intuition. It's that hair on the back of your neck rising, that "something just isn't right" feeling. There are going to be times like these when things happen to you, and when they do, listen to this sixth sense. Don't discount it. Be sure to give it its due diligence.

It also comes from our capacity to merge information. For the junior firefighters who have been awakened by such a moment, your knowledge base is growing. Do not be afraid to share what you are seeing with those around you. Senior firefighters and officers often have a better perception of this sixth sense gleaned from their years of experience at seeing fire in action. They have had more time to practice and plan for fires, and they have formed and polished this sixth sense to a masterful craft at fires. They have a thicker recall file, an extended slide carousel of similar events. This leads them to have a greater ability to anticipate what might come next. In conjunction with this is the ability to clearly communicate what they are seeing on the fireground.

This sixth sense might be seen as the culmination of mastery in knowing how your five human senses perform at fires. It is recognition primed decision making at its finest. It doesn't just start up by itself, nor occur at every fire, but it comes from the ongoing absorption of sensory awareness.

Firefighter Senses in Size-Up

As we head to the report of a fire in a residential building, many of our departments have pre-determined reference points that dictate our initial suppression actions and guide our operational tactics as we pull onto the block. When we stated in previous chapters that every department needs guidelines for fire operations, they have to be just that—guidelines. We need built-in flexibility to allow us to complete our tasks. Guidelines provide not just a basic framework for understanding, but also give us latitude to account for events that don't follow the norm.

Our tactical size-up of this particular residential dwelling and the characteristics this fire is exhibiting will either add to or subtract from those aforementioned pre-determined initial guidelines. We have been taught through the years that the incident size-up begins with the receipt of the alarm. The 13 points incorporated in the size-up acronym COAL WAS WEALTH (Construction, Occupancy, Apparatus and personnel, Life hazard, Water supply, Auxiliary appliances, Street conditions, Weather, Exposures, Area, Location and extent of fire, Time, Height) still holds much validity for success.

As we mark on the scene with fire and smoke showing from the first floor at the A and B corner, many of our initial suppression actions and our operational tactics will certainly be dictated by reference points extrapolated from this acronym.

The fire service loves acronyms and phrases to help us recall information quickly in the heat of battle. While the 13 points are always poised in our minds for reference material, they are not always in the forefront as we push the line into a burning house. Let's hope that we are not standing on the front yard of a burning house ticking off the letters in the acronym before we act on our plan. That acronym and the countless others developed in the fire service are for our reference. They add to our foundation for on-scene decision making.

The list is just too long to run through each point in assisting us to making split-second decisions and actions. We must rapidly evaluate the situation at hand; then drawing on that foundation of information (such as those 13 points), we make the action happen. We must commit ourselves to adding those points and what they represent into our memory banks and be able to draw from them to assist us in formulating a plan of action for what we are seeing in front of us at the fire scene.

Fireground intelligence gathering must expand beyond those 13 points. Firefighters and fire officers need to consistently evaluate and re-evaluate conditions encountered. All our senses must be on alert. We must use all our five senses to gather intelligence about the bigger picture, then attempt to validate the smaller tactical pieces to make sure the outcome is what is anticipated. We must absolutely resist the temptation to put the blinders on and run like a moth to a flame in our tactical decision making.

While not discounting the fact that size-up begins with the receipt of the alarm, size-up must be an ongoing, continuous process. It changes as conditions and tactics change. It changes as we enter the IDLH structure. What is the visibility? What are the heat conditions? Where are the victims? Where is the fire area in relation to our entrance point? We must continually update our personal size-up, noting conditions found, reporting actions taken, and requesting additional needs as we travel through the single-family structure.

While we can use those initial snapshots of our arrival on-scene as a reference point, we must resist the temptation to focus on stored or "canned" images captured during the initial on-scene size-up. Size-up is a fluid, ongoing process. While it indeed begins with the receipt of the alarm, it is never really concluded until the fire is out and we are backing into quarters.

Look, Listen, and Feel

We may remember, from somewhere deep in our training past, that we have heard this "look, listen, and feel" verbiage before. It comes from the technique used during CPR. I recall it in the "B" step of the ABCs. Once the airway is opened, we *look* for breathing (chest rise and fall), *listen* for breaths (escaping from the mouth and nose), and *feel* (air against your cheek). We have borrowed these sensory applications from CPR to input them into our continual size-up process within the residential building. The description that follows reveals why these three senses are paramount for success in the fire situation.

Look

As discussed in chapter 7, the dwelling's building style lends visual clues that may assist us in our operations. Visual building landmarks (doors, stairs, window sizes, and layouts) can assist us in identifying floor plans, living vs. storage spaces, location of stairs, and presence or absence of basements. Before we even enter the IDLH, we should be able to visually identify specific features in residential buildings that we can use to assist us to perform at greater efficiency.

As we prepare to enter the structure we should pause, taking time to make time. Take a second to look around. What are we looking for in the residential building? What are we looking with? The obvious answer is that we use our gift of sight for one. The other is through the eye of the thermal imager (TI). TIs and their technological advances in the fire service have certainly increased our ability to see through layers of smoke once inside. While it is a tool that has its own set of limitations, it has certainly increased our visual acuity that was once lost on the interior.

Now that we know what we are looking *with*, what are we looking *for*? That we are looking for victims, fire, and smoke are the obvious answers, but we should also be aware of the home layout, furnishings, and other building construction features that will act as reference points to help us remain orientated in the home. Even though your sight may be diminished due to the smoke condition in the house, use your gloved hands to act as your sense of sight. By touching and then allowing your brain to process what it is that you are feeling, you can visually construct where you are operating in the house. Do not allow yourself to enter the burning home and put blinders on. Look around!

Take every opportunity to look. Once the door is open and we have gained entry and are preparing to move in, look. Many times the smoke and accompanying thermal layer of heat and gasses will lift just to the point that we have a small window of visibility at the floor level. I can distinctly remember the first time I saw this occur.

When I was a new firefighter, I responded as the first-due engine company to a report of a fire in a private dwelling. Upon our arrival, fire was showing from what appeared to be a kitchen (due to the exterior recognition of similar buildings with the smaller, counter height window size adjacent to a sliding glass door) on the first floor in the rear of a small, bungalow style, two-story dwelling. As we flaked out and bled the hoseline on the front porch, I saw the officer open the front door, wait 5 seconds or so, put his head on the porch floor, and peer into the dwelling.

Inquisitively, after the job was over I asked him why he put his head to the floor like that. He explained that once we opened the door, we created an opening to let smoke out. It was just enough to let some smoke lift or "blow," as he put it. We had changed the interior environment by opening that door, insofar as the smoke level rose up off the floor just enough for him to get a quick layout of the front room. He told me he saw where the couches and furniture were placed and that there was carpet on the floor with no victims in plain sight. He also saw the steps going to the second floor. Under the smoke layer he saw the hallway that lead to the kitchen.

His estimation was that we would have a short stretch and a clear path with the hoseline to the fire. He further explained that this phenomenon of the smoke lifting as it did at that particular job does not happen at every fire (due to factors such as humidity, air temperatures, and the volume and density of smoke), but any time it does, we should use the advantage it gives us.

Utilization of the thermal imager is a skill. It is a learned skill, just like many others in our profession. The TI was not designed to take the place of basic fundamental search and rescue techniques. TIs were designed to supplement good techniques, expedite searches, and assist with the overall scanning of rooms. Newer technological advances also have built-in heat register scales and, dependent on brand and model, colorizations to those layers or heat signatures. Features and utilization of TIs is discussed in greater detail in chapter 25.

We prefer the TI to be held by the unit leader. This member, while not necessarily an officer, is managing the search. He or she can use the imager to keep all members of the search safe. I know you are saying that we only have a team of two to conduct our searches, but think about the size of the average bedroom in the residential building in your area. How large are they in reality? Can a team of two search them? Do both members need to be duck walking or crawling into the room in opposite directions until they meet? Or can we have one member conducting the search in a conventional manner, while the other partner stands fast at the bedroom door? This person at the door can be scanning the room for victims and watching the conventional searcher in case of a missed area. The firefighter at the doorway can also relay info such as doors, closets, bunk beds, and windows to the searcher and can also re-scan where you came from as well as scan ahead to where you are going to search next. This reduce duplications of search efforts and leads to quicker and more thorough primary searches.

We believe wholeheartedly in the two-team concept to truck company operation—an inside and outside approach. While we discuss this in greater detail in chapter 20, the *look* concept applies in this section for obvious reasons. Information garnered from exterior firefighting forces can be another ally in our firefight. While units are engaged in operations on the interior, outside team members can relay pertinent information discovered. Communication to team members of the position and location of victims, hazards noted, or fire conditions not initially discovered can ease the burden of the inside team. Any pertinent information should be relayed to the IC and all operating forces.

Because we regularly operate in a visually limited smoke environment, heighten your sense of sight. Stay alert for the glow of the fire, other units' flashlights, and window and door openings you encounter. Keep the blinders off and move with direction and purpose through the residential building.

Listen

Listen. I must say this to my children 10 times a day. Sometimes I wonder if firefighters have the same level of responsiveness. Everyone "hears" things, especially those thrown around the kitchen table, such as "I'm buying the meal tonight" or "I'm thinking about getting married." But on the fireground, with all the associated noise that ensues from our operations, can we really hear anything? Some of our problems in the listening end of communication are technology based. While we touch on that, most of what we discuss in this section focuses on the noises around us, noises that we as firefighters are creating at fires.

Most members today are equipped with some type of portable radio, regardless of brand (fig. 21–3). Whether you call your fireground communication device a handie-talkie, a radio, a portable, or a hoot and holler really doesn't matter. What *does* matter is that you remember to turn it on first. Then you must be certain that you are on the right channel to receive the information. Finally, you must be listening to what is being said (and what is not being said) when it comes over the air. While we discussed the many pitfalls and problems with fire communications today in chapter 4, in this chapter we will focus on the listening end of communications.

Microphone/speaker placement and volume adjustment are the two biggest factors that affect us in actually hearing radio transmissions. The invention of the remote microphones with incorporated speakers dramatically helped firefighters hear fireground audio (fig. 21-4). They are an absolute must for any new firefighter radio setup. If your radio requires you to hold it in your hands and then put it up to your ear to listen and to your mouth to transmit, is that really an effective use of your hands? Older style handheld radios might work for the chief in the street (if that is all that is available), but it is absolutely going to slow down every operation that you conduct when you are in the residential building trying to fight a fire.

Fig. 21–3. Portable radios are imperative for firefighting members. (Courtesy of Nate Camfiord.)

Fig. 21–4. Remote speakers allow for hands-free use. (Courtesy of Nate Camfiord.)

The natural positions for our arms and hands are along our flanks, at our sides. As far as I know, we cannot yet hear radio communications through our hands. There is no doubt that messages will be missed if we are forced to operate and attempt to hold that communication device in our hands (fig. 21–5).

Volume control is the second problem with portable communication devices at fires. It is hard to find a correct volume setting because different brands, models, and even individual radios can have a wide variation. If the HT volume is set at its maximum, speakers can crackle and messages become unreadable. At maximum volume, feedback can become an issue. If you attempt to transmit a message near another firefighter whose volume is at its max, the squeal produced will cover the message. As a method of creating good habits, we prefer to turn our radios on, turn the knob to full volume then back it down a hair. I know that's not a very scientific measure of volume to grasp, but if you were to put it on a scale of 1 to 10, I would guess it would be somewhere around a 7. If the volume is set on low, communications may be lost in ambient fireground noises.

Fig. 21–5. Old style hand-held portable radios

The Screamer. Who among us has not had the distinct displeasure of working with or for a fireground screamer? We have discussed various radio communication problems caused by logistical and technological factors, but we need to try to tame the human errors in our information sharing on the fireground. When the homeowners dial 9-1-1, they have effectively given up control of their lives to us. It is part of our duty to remain calm in the face of fire. We call ourselves firefighters. Why then, when we arrive on scene for a reported building fire, are we screaming over the radio that there is a building fire? Is that not why they called us in the first place? Calm and consistent radio traffic is helpful for all. Now we are not suggesting the stoic and "too cool for the room" radio transmissions that we sometimes hear are the best options either.

We are talking about remaining calm. If you become a screamer at fire, guess what happens? Everyone else has a tendency to be brought up to that amped up level of excitement. It can cause premature and unwarranted fireground decisions to occur. Add the filter of utilizing the SCBA facepiece to the mix and transmissions can quickly become undecipherable for all. An excellent drill to teach a screamer to recognize the downfall of such actions is to record your radio transmissions. If you do not have the ability to review tapes from the communications office and cannot do it from live fires, do it during drills in fire scenarios with a recording device as simple as a smart phone. Let the members replay the transmissions and see just how difficult it is to understand the messages.

We bring the noise. As we move with our firefighting team we make a substantial amount of noise. Silence during the residential building fire will only occur if we make it happen. In our Rapid Intervention (RI) programs we teach a 15-10 approach to silence. This technique provides for 15 seconds of focused movement, followed by 10 seconds of stopping and having the team leader order the crew hold their breath. We are doing this in hopes of hearing noises from a downed firefighter with a focused search, specifically targeting in on the person.

In the residential building, we would encourage a 30-10 search to allow us to focus our senses and just listen. What are we listening for? We are listening for a few things, mainly human life. Obviously we need to be tuned to listen for civilian victims coughing, moaning, crying, and/or screaming. In addition, we should also be listening for the noises that the fire creates (crackling, popping, etc.) and noises from fireground activities that are taking place around us.

Fireground noises, such as breaking glass, the roar of a chain saw, and the flow of water hitting walls can all be all welcome sounds. What is the impact of these noises on our operations? Well, it depends. Let's take, for example, a company assigned to conduct a search in a residential building fire. While the breaking of glass can be a welcome sound if and when ventilation efforts are in order, or if that company is looking to find a way out, it may not be welcome if they are searching above the fire and the first hoseline is not yet in place (fig. 21–6)!

Fig. 21–6. Firefighter clearing a window. (Courtesy of Nate Camfiord.)

For the engine company, noises can also assist us in the ultimate goal to deliver water to the seat of the fire. Practice with your team and listen to the sounds that water makes in residential buildings. Listen for water flow. When you have the line operation, listen for that noise that comes back to you at the nozzle. Think of the hoseline operating ahead of you like a bat sending out a signal and awaiting its return to assist in finding a direction of travel. Sweeping the floor with the hoseline may note holes in flooring, locate steps, and so on. Line operation can also assist us in determining the layout of the floor. Listen to the line hit the wall and then listen to the changes in the sound to find the direction of the hallway. The sounds reverberating back to the crew can give clear indications for movements.

Practice listening for the variations in the noises made from water hitting walls and openings, noting differing sounds with various floor and wall coverings (sheetrock, glass, carpet, and tile), and use them to effectively move the team to the seat of the fire.

For the company officer and incident commander, noises can also aid in accountability to some extent. If we know that we have assigned a firefighter to the roof and we hear the saw in operation, we know that firefighter has reached his objective. Remember, however, that every action we perform on the fireground will cause a reaction on the fire. Sometimes those actions can impact it in a positive direction, other times, often unexpectedly, our actions can adversely impact fire conditions and breathe new life into a fire. Be mindful of the sounds around you and what impact they may have on your current situation within the dwelling.

There is a message that we give to new firefighters: We have been given two ears and one mouth for a reason, because we should listen twice as much as we speak. This is not to discourage them from speaking when appropriate on the fireground. They only need to tune their ears to listen to what is going on around them. We need to apply that to fire operations, as well. Not only must we use our ears to hear messages and trapped civilians, we must truly listen to what is being said across the radio. Again, practice giving and deciphering communications through facepieces, practice giving radio reports with gloves on, etc. We must also condition our sense of hearing to listen to the actions that are being performed around us on the fireground.

Feel

Completing the "look, listen, and feel" approach is the ability to decipher information through our sense of touch. Our feeling, or sense of touch, is instantly limited by the fact that we must operate in fires with gloved hands and must learn to do all tasks on the fireground with them. While we can certainly make excuses for not wearing gloves at fires, they are all completely unacceptable. And when we say gloved hands, we are talking about hands covered with *firefighting* gloves (fig. 21–7).

There has been a perceived uptick in a trend that firefighters are wearing lighter weight gloves into fires because they may afford an increase in the ability to feel. We cannot disagree that a tight fitting pair of leather work gloves gives your hands a greater range of dexterity and flexibility (fig. 21–8). But these gloves *do not* and *cannot* provide adequate protection from the thermal environment in which we operate.

Fig. 21–7. Firefighting gloves

Fig. 21–8. Work gloves

In fact, some man-made, synthetic, mechanic-style gloves, which maybe excellent for auto extrication, are made from components that will literally adhere to your skin under heat conditions. Like shrink-wrap to a boat, they will adhere to your hands when heat is applied. They will add to your pain and discomfort as the ER staff attempt to peel back and separate them from the skin on your hands once you have been burned.

Sure, you may be able to get away with wearing other types of gloves to the next one hundred fires, but you never know when you are going to be heading to "that" fire. It might be the 101st, or perhaps it is the 31st. You, your crew, and your family are going to pay the price when it does happen, and it will. It's just a matter of when. It's hard to do anything with two burned hands. If you want to try to see what it will be like, try to eat with no hands. Or better yet, go to the bathroom with a pair of mittens on your hands and see how easy that task has become. Your loved ones will certainly help you eat, but in the bathroom? That might just cross the line of embarrassment for you and your family.

On the fireground, once your hands are burned, you are now an additional problem that your crew will have to compensate for. Just like the civilians who called us, your team members are depending on you to perform. Your burning hands will not allow you to transmit a message, nor will you be able to extract a victim from the fire area. They will not let you open the nozzle in the fire room, or get more line to make the floor above. If you have a problem with your fire duty gloves, get a new pair. Try as many types and styles as are necessary until you find the ones that are the most comfortable and best allow you to do your job.

If you are a part of a fire department system that issues a specific brand of gloves, as cruddy as you may think that they are, you must learn to use what you have. You must train with the gloves that you have. You must do every task you normally do on the fireground with them in place. If you think there are better options out there for your department, get involved in a PPE committee, or at least pass along your thoughts to someone who is a member. Look forward and get involved. Often the people who purchase the PPE in your department may not be the ones who ever wear it inside a fire (e.g. chief level officers and department finance chairpersons). In being involved, you can help initiate future changes from within.

It's obvious that the aforementioned and largest portion of this section explains the need to recognize and work with gloved hands. When you break it down, even in the feeling that is done through our hands, the sensory receptors in the skin are also responsible for feeling. The skin that covers our hands is just one section of the body's largest organ. The skin's built in heat receptors in other areas of our body can assist us with recognizing facets of the residential building fire. Our body's skin covering, known as the integumentary system, has a complex layered network of blood, tissue, and nerve endings. The skin has as much to do with regulating our internal temperature as it does with our ability to notice changes in our external temperatures. Our skin is a very complex organ.

While exposed skin can begin to burn at temperatures as low as 130–140°F, our focus is on our integumentary system's ability to recognize heat and our brain's interpretations of those heat conditions. We understand that this is in no way a defined science. Each person has different abilities to interpret heat conditions. Also, varying PPE brands and turnout styles are factors in the level of heat transfer to the body. We are certainly not recommending using our bodies as heat thermocouples to identify temperatures in the building fire. We should, however, realize that temperature changes are extremely relevant and are another useful tool for us to be cognizant of at fires.

We know that heat transfer in building fires happens through conduction, convection, and direct heat transfer. If firefighters are fighting the fire and are inside that building on fire, they are without question going to be exposed to most, if not all, of these heat transfer paths. Recognition of the heat conditions around us is another facet of fireground understanding that we can employ to make us work more efficiently on the residential fire.

Our PPE, when worn as designed, has the firefighter nearly fully encapsulated (fig. 21-9). As we look at this photo of a fully dressed out firefighter, we can see no exposed skin visible that could be damaged by fire. In fact, if you look closely, the only skin that is exposed (albeit protected with the SCBA facepiece lens) is around the area of the eyes (fig. 21-10). How can we use our skin to feel if it's encapsulated? Again, we know that we cannot use our skin as a temperature gauge for a true measurement of heat at fires, but in conjunction with our other senses and past knowledge and experiences we can use it to assist us in locating the area of origin. Even through the layers of our protective clothing we can still feel heat, and it may point us in the right direction.

Finding the seat of a residential building fire is not always a simple task. There is never a roadmap to locate its origin. Certainly not every fire will have a clear path to its extinguishment. We must put all our senses together to defeat it. Even though the skin is covered by PPE, it can still detect heat and help point us to the fire area.

Fig. 21-9. Fully encapsulated firefighter. (Courtesy of Nate Camfiord.)

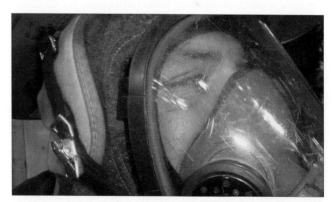

Fig. 21-10. The only skin we have exposed is still under cover of the facepiece.

Heat normally travels in waves. Because of this, heat waves will be emanating from the area of origin and following exhaust paths created in the structure. Using our skin's heat absorption and recognition capabilities, we can often feel the direction from which the heat is coming. This can be utilized in conjunction with your other senses to locate the fire in the house.

Keep your head on a swivel. Take a few seconds as you are searching the building to stop and turn your head from side to side: often the head and face will pick up the heat direction first.

Increasing heat with no visible fire should also send up red flags for operating forces. Rapid increases in temperature without the presence of a charged hoseline allows little room for error. Modern fire loads and the modern components of homes (such as plastics) have reduced the amount of time from incipient fires to flashover. Rapidly increasing heat conditions must sound the alarm for a search team. Retreat, while not a favorite option for a search team, is often the best course of action when faced with these types of rapid heat build-ups.

Reflection

We know that few firefighters or company officers have experienced the perfect fire. In that scenario, every fireground task said and done is accomplished right on point. An introspective self-assessment is something that I know I complete after every fire. After the formal or informal tailboard with the members, I critique my actions at the job. I am normally my toughest critic. In nearly 20 years of fighting fires, I still have yet to come back from a fire and say to myself, "I nailed it . . . and would do it the exact same way again next time!"

Sometimes there are simple things such as, "I should have said that differently on the radio," or, "I should have waited to take that window till the line got here." Sometimes they are life lessons, things you will never, ever forget.

We have to take every residential fire in which we operate as a learning event, as well. We must be consistently updating our slide carousel, adding information that we take in job after job. We need to add it to build the depth of our knowledge base. Be certain to take away something from every fire you go to, and pride yourself that you have the ability to recognize that there is always room for refinement and improvement. Whether personal, company level, or department wide, we can modify and polish our operation for the next fire. We owe that to ourselves, our fellow firefighters, and the citizens we are sworn to protect.

The Backstep

Residential Recognition

What is residential recognition and how can we use it to our advantage on the fireground? Encapsulation in our firefighting PPE modifies the use of our usual five senses. We must practice and work within the confines of those changes. The following are a few tips to attempt mastery of your firefighting senses:

- **Replicate the residential setting.** In your search drills, assemble furnishings from the residential home. Carpeting, wood flooring, and tile allow the searcher to note changes in building features. Include radiators and windows at proper heights (make mock windows on a solid wall with a wooden frame and some Plexiglas). Obtain as many various furnishings as you can to replicate residential rooms (kitchen tables, beds, and couches).

- **Drill blind.** Obscure the trainee's vision. Use wax paper or blackout facepieces. Have members verbally explain to the drill leader what it is that they are seeing, hearing, and feeling around them. This is the way we will be dressed out at the fire. Let's practice the exact same way to create muscle memory.

- **Bring the noise.** Having members searching in silence isn't like reality. Have a recorded track of commonplace fireground noises to add realism. Have the units make radio transmissions to command and have command answer (follow the CAN format, a report that identifies unit conditions, actions, and needs).

- **Record, record, record.** The advent of cell phone and other small digital recording devices makes recording all drills for review a tremendous asset. These recordings must be done with a degree of strict professionalism and under the guise that they will not be used as fodder for jokes at the kitchen table, but for a recipe of company and individual improvement.

22 Ladders, Ladders, Everywhere

You're a combat-ready kind of firefighter. Even though as a volunteer you don't receive a paycheck for your services, you realize all too well that going to fires is no social walkabout. It's tough, it's dangerous, and it can be deadly. You are well read in firefighter strategy and tactics, you take time to practice and train, and you attend lectures and hands-on classes on your own time with your own money just to be better prepared. You're not alone, however. There are others in your department with the same mind set, but there are not as many as you would like. Some of the guys at the station bust your chops about you wanting to make your local departments like the city department 50 miles away. "It's just about being ready all the time," you tell them. But they just don't get it, nor do they get you.

You are the type of firefighter that you want to come to your house, rescue your family, and extinguish a fire should one occur. The firehouse is no coffee club, a flavor of the month for you. For you, it's much deeper than that. You know both the risk and the reward. You aim to be the best that you can be, in and out of the firehouse.

Your latest projects have focused on getting the rigs, tools, and equipment in the same combat-ready mind set that you and your peers have adopted for personal actions in firefighting preparation, practice, and anticipation. The resistance is strong, and traditions run deep. In attempting to effect change, remember to always explain the why.

At 0300 hours the home pager squawks open, breaking the dead silence in your home. Your company is alerted to a response to a structure fire. You know this is going to be something. Honing in on that sixth sense we firefighters have, the time, the "in the area of Elm Street," the tone in the dispatcher's voice, it is all lining up. You spring from the bed, legs falling directly into your neatly placed pants, always positioned in a manner to make them ready to run out. You grab your keys off the night stand, turn back over in the bed, and give your wife a kiss on the cheek.

She groans softly, for she knows what it is you do and what your response means to your community. Somewhere in her mind, if only for a second, she wonders why. As she does every time you head out the door, she clears her throat and says, "Be careful, Jack."

"I'm always careful, babe," you reply, now halfway down the hall. You hit the steps, you're in the car, and at the firehouse in short order.

You find that for this hour, there is a great crew already assembled, five members on Engine Company 22 as she hits the street. While you are a truckie at heart, you are ordered to get on the first piece out. Your volunteer department has set a minimum staffing requirement of a total of four on the engine and four on the truck to leave the bay.

"Let's go now, we roll with what we got," calls out the crusty old lieutenant.

More phone calls come in and the dispatcher tells the chief (who is responding from home) and Engine 22 over the air that additional callers are now reporting people at the windows cut off by fire and getting the address reported as 13 Elm Street (fig. 22–1). You take a second to breathe deeply, wipe that last bit of sleep from your eyes, and listen to your officer as he turns around.

Fig. 22–1. Engine company responding

His face is stern, emotionless, and serious as he turns back to the crew and says, "Slight deviation from the usual, the truck didn't get out yet. We got a report of people trapped, we going to run the engine split!" The engine split is a well-practiced evolution that we can put into play when we know the truck is going to be delayed in getting to the fire. In this area it happens quite often, most of the time in either the middle of the day or the middle of the night. While the split does work best with a full crew of six firefighters, it can work with five or four too.

"Looks like we may be here alone for a bit, there are reports of people at the windows, perfect time for the split guys!"

The lieutenant continues, "Team 22-1, Jamie and Luke, you're with me. Normal engine operations. I'll grab the irons and take a quick lap as you guys stretch the line to the front door. Big Joe, I told county dispatch you are going to drop a line from the hydrant at Cardinal Drive and Woodcock Mountain Road and for Engine 81 to pick up. Joe, give us booster water in the line when you get the pump set, then help Jack set up the portable ladders to get these people. The initial reports said they are trapped on the second floor, unable to pass the fire on the first. You guys will be 22-2 on the radio."

As practiced, the split goes off without a hitch. There is a good amount of fire from two windows and the front door on the first floor, Alpha side, and quadrant one. The fire building is a fairly new-two story, wood frame, lightweight constructed 20×30 colonial with an attached garage on the Delta side.

As you are getting the ladders down from the top ladder rack of the engine, the lieutenant reports victims at the windows on the Charlie side second floor, quadrant Bravo, and they look like they are ready to jump. "Jack, get those ladders here NOW!"

"It's on," Jack says to himself. "Move with purpose here, every second counts." Jack is now moving as fast as he can, half talking aloud, "I know that this is our new engine and primarily responsible for hose and water but, geez, this new ladder rack, Joe! Where is the button to get this thing down? Ah, never mind, I got it! Oh my, it takes forever to hydraulically come down from the top of the rig (fig. 22-2)! It was so much faster to grab the ladders off the side of old Engine 22. Ah, what the heck is this, I got to take the roof ladder off first?" Now tripping over the roof ladder under his feet, Jack struggles to get the 24-foot extension ladder from its mounts off the side of the new apparatus ladder rack system. Jack is a truck guy, and while very comfortable with operating with portable ladders, he never really gave too much attention to the new Engine 22 and how the ladders were placed on her.

Fig. 22–2. Ladders on engine companies are not often given the same attention as ladder company ladders.

These ladders are so new in fact that after the fire Jack thought that this may have been the first time they'd ever come off the rig. The ladder's midpoint is unmarked and, in his haste, Jack chooses a point too far back, an unbalanced place, to put his arm through. Fighting the ladder's desire to comply with gravity for 100 feet or so, he has to stop and reposition it on his shoulder, moving up two rungs.

Once in the rear, Jack sees the male victim crouched on the sill, about to jump. Jack attempts to calm him down once they make eye contact with each other. Smoke is getting darker and darker over the victim's head. Jack throws the ladder efficiently by himself using the building as a brace. With a smack, the ladder hit the siding. Jack grabs the halyard to extend the ladder, but is it jammed! Ah, it's tied around both the bed and fly sections in some sort of unknown type of a knot. Jack struggles to untie the knot while balancing the ladder against the side of the building.

He jumps. Jack hears the bones snap. It makes the hair stand up on the back of his neck. The sound is unforgettable, the sharp crack, in conjunction with a hollow thud of flesh meeting earth, air forced from the lungs like a bellow fanning the flames of a small, smoldering fire. Jack drops the ladder, halyard still tied, as pretty as the day the she arrived on Engine 22. The jumper lives, thankfully. Jack helps carry his mangled body to the now waiting ambulance. Hours, days, weeks, even years of pain and trauma from the event, for the victim *and* Jack, will be etched into their minds forever.

Nothing will keep an occupant from fleeing fire and smoke, whether they're on the 2nd, 5th, or 25th floor. If civilians make it to a window, rescue isn't imminent, and they still feel threatened, they *will* jump. If there is a choice between lying down and dying right there under the window or taking a chance on survival by jumping from a window, they will choose the latter and jump. Civilians and even firefighters will jump when pressed, and those are indisputable facts. In the residential setting we must have our portable ladders ready to work.

Portable ladders have been a staple piece of firefighting equipment on nearly every fire apparatus in history (fig. 22–3). Although portable ladders are historically considered to be a truck company mainstay, most modern engine companies have a compliment of several portable ladders today, as well. While the modern aluminum ladders have replaced older classic wooden varieties in most areas of the country (with the exception of perhaps San Francisco), the utilization of this marvelous firefighting tool hasn't changed. Portable ground ladders (portables) have a long and established history in the fire service. They provide firefighters the ability to traverse up, down, in, and around the residential fire building.

Your Area and Your Portable Ladders

In keeping with the concepts of creating combat-ready equipment, fire department portable ladders certainly deserve our attention. As with most fire department operations, their deployment is extremely time sensitive. The portable ladder coming off a newly purchased piece of apparatus is no doubt functionally operational. However, ladders certainly can be tweaked by firefighters to aid us in our ability to perform with them in a more efficient and effective manner. Save time by making your portable ground ladders ready to work right from the rig.

There are many residential fire situations that can render the ladder company unable to utilize the hydraulic ladder on its rooftop (fig. 22-4). A great majority of residential buildings are set back from the street. They have narrow driveways, trees, and the routine occurrence of electrical wires all clogging access. Any and all of these situations can make life difficult for the ladder company chauffeur (LCC) to find an acceptable position at the fire to utilize the rig-mounted ladder. If this is the case, portable ladders will be heavily relied upon at the residential building fire. However, even in instances when the apparatus's hydraulic ladder is used and operated, it is still just one ladder.

Fig. 22–3. Portable ladders. (Courtesy of Traditions Training, LLC, www.traditionstraining.com.)

Fig. 22–4. Obstacles can prevent the ladder company from using its hydraulic ladder.

Properly positioned portables offer members operating on the fireground various entrance and exit points for our operations. The normal compliment of portable ground ladders vary in lengths and, in our opinion, should be thrown around the fire building until there are none left on the rig (fig. 22–5).

Fig. 22–5. Aluminum fire service portable ladders. (Courtesy of Traditions Training, LLC, www.traditionstraining.com.)

Though modern aluminum portable ladders are found on fire apparatus in varying lengths, there are several common ladder sizes in most departments' inventories. The length and quantity of ladders that your department maintains on the apparatus should tie directly back to your buildings. Examine the ladders on your apparatus and determine if they pair well with what you are doing in relation to your fires. There should be a match of the greatest fire frequency in the buildings for the area in which you serve and the type and length of ladder to best reach all floors of those homes.

If you have primarily one-story ranch homes in your first due and your ladder truck came with a 50-foot Bangor ladder, it is likely gathering dust in that ladder bed. Could you be better served with an additional 24 footer and perhaps a 20-foot ladder in that tray? One might consider replacing it with something more manageable and appropriate to meet the needs of the dwellings in the community. Just because your apparatus came with ladders already on it should not mean they have to stay there forever. Your first due is always changing. We have to adjust our tools and equipment to keep pace.

The fire service is tasked to comply with various recommendations and guidelines set forth by agencies outside of our standard department SOPs and apparatus specifications. The National Fire Protection Administration (NFPA) and Insurance Services Office (ISO) are two such agencies that provide guidelines regarding portable ladder compliments for apparatus, and in both, the footage length has declined over the last few years. NFPA and ISO portable ladders guidelines are just that, guidelines (fig. 22–6). The numbers listed should represent a minimum established requirement base.

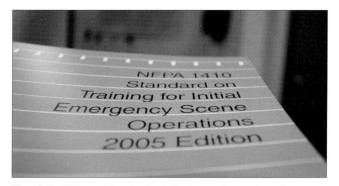

Fig. 22–6. NFPA provides standards for fire service equipment.

According to NFPA standard 1901, a quint apparatus is required to have a minimum of 85 feet of ground ladders. Aerial ladders are required, at a minimum, to carry 115 feet of ground-based portable ladders. The standard goes further to specify one attic ladder, two extension ladders, and two straight ladders with folding roof hooks. NFPA 1931 is the standard that addresses design specifications of portable ladders.

The ISO requires a minimum of 150 feet of portable ladders for quint apparatus and 159 feet of portable ladders for aerial ladder companies to achieve full credit for a laddering compliment. ISO ratings also have specific ladder length requirements for each type of ground ladder.

We have listed, as a point of reference, many of the common portable ladder lengths and ladder weights below. These numbers were extracted loosely from various manufacturers. Because of this, be certain to check your specific brand of ladder on your apparatus for specific details. Again, all weights, closed lengths, and extension ladders, do vary. They have been approximated to compile the list below.

General sizes and types of fire department portable ladders in use:

Extension ladders

35′ extended, closed length 20′, weight 135 lb

28′ extended, closed length 16′, weight 100 lb

25′ extended, closed length 15′, weight 100 lb

24′ extended, closed length 14′, weight 80 lb

16′ extended, closed length 10′, weight 70 lb

Straight ladders

20′: weight 50 lb

20′ (hook) weight 55 lb

12′ (hook) weight 35 lb

Special ladders

14′ A-frame extended, closed length 7′, weight 35 lb

10′ A-frame extended, closed length 6′, weight 22 lb

19′ Little Giant extended, closed length 5′7″, weight 40 lb

15′ Little Giant extended, closed length 4′7″, weight 36 lb

11′ Little Giant extended, closed length 3′7″, weight 32 lb

10′ attic folding extended, closed length 11′, weight 16 lb

Working Load Capacities for portable ladders*

Extension ladders: 27′ to 35′: up to 600 lb

Extension ladders: 26′ and less: up to 500 lb

Straight ladders: up to 500 lb

Folding ladders: up to 300 lb

*Note: This is the maximum load capacity, which includes the firefighter, all associated tools and equipment, victims, and any other weight on the ladder such as a charged hoseline.

Fundamentals of Portable Ladder Construction

The general design of most portable ladders has remained largely unchanged. They still have tips, feet, beams, and rungs. There has been some effort made by manufacturers to reduce the amount of weight of various ladders over the years, but the simplistic yet effective design holds true. To understand the whole we must understand the parts (figs. 22–7 and 22–8).

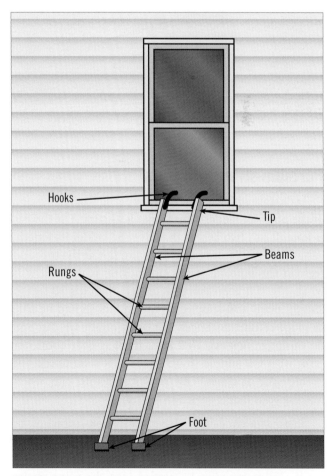

Fig. 22–7. Straight ladder. (Illustration by Matthew Tamillow.)

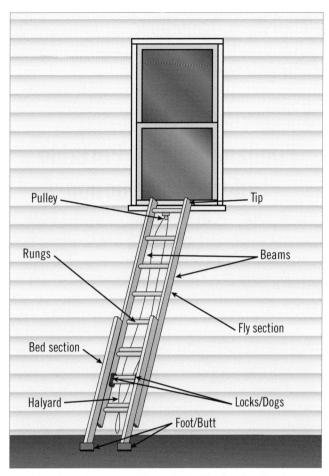

Fig. 22–8. Extension ladder. (Illustration by Matthew Tamillow.)

Portable straight ladders

Let's start with a quick overview of the straight version of the portable ladder. This ladder is as simple as the day is long. It has two beams that give the ladder its length, rungs that span from beam to beam, tips at the top, and feet on the bottom. It has no moving parts, and it is what it is. If it is measured at 20 feet then it's 20 feet in length, plain and simple. All firefighters should be able to judge certain distances, but occasionally we can and do misjudge heights. That said, there is little room for error with the simple straight ladder. It's simplistic but problematic at times. This must come into play when making your ladder selection for use.

Some straight ladders have roof hooks attached to the tip of each ladder beam (fig. 22-9). These retractable spring hooks are stored folded. When opened, these hooks are designed to be nestled into the ridge of the roof. This is not to say a roof ladder can only be used for roof operations. Roof ladders are basically just straight ladders with hooks, and that's it. They can be used like any other straight ladder on the fireground.

Fig. 22–9. Straight ladder roof hooks

Remember when we said 20 feet was 20 feet? If you need only 12 feet of ladder to reach your objective, what do you do with the other 8 feet of ladder? Conversely, if you misjudged your distance and need 22 feet but brought only a 16-foot straight ladder, is there an option to make up the needed difference? Portable ladders when climbed at extreme low angles have a tendency to kick out at the ladder's feet. This is especially true when these ladders are footed on hard surfaces such as concrete or asphalt. On the other hand, attempting to gain extra length by creating an extremely steep angle is just as dangerous. Putting the ladder too tight against the building can cause the ladder to pull away from the structure as you ascend.

Most residential buildings have 8-foot ceilings. Newer homes, however, can be built with the standard 8, 10, 12, or even 14 feet of spacing between the floors. Take into consideration that windowsills are usually three feet above the height of the floor. Judging distances in heights, especially when visual cues are hindered by night operations or smoke conditions, can lead you to make a poor choice. Practice your laddering skills at routine incidents, especially when you have an outside firefighting position. Don't be afraid to take a portable ladder off the rig and bring it to your objective. It's good practice. We are certainly not advocating causing undue damages or smashing out every window on your next run, but make your ladder selection, bring it to the point where you would be utilizing it, and see how well you did.

Remember, especially when taking portable ladders to the rear or sides of the structure, that there may be differences in topography that are unknown. As more homes are built on less available land in your town or city, the tendency is to build homes on lots that were previously undesirable due to the terrain (fig. 22–10). Walkout basements and building setbacks are other factors that may influence your choice of the ladder you decide to bring around back. Listen on your radio for the report from the officer who took a lap. Listen for any noteworthy concerns mentioned in the rear or from unviewed sides of the fire building. What you see of the front of the fire building from your position on the rear of the rig may be vastly different when you make your way around to the rear of the property.

Fig. 22–10. Topography differences from front to rear can influence ladder choices.

One of the reasons I mention this is because it happened to me. This exact situation caught me off guard. A simple enough job, I had the outside vent (OV) position in the ladder company for a fire in a two-story, peaked roof, semi-attached private dwelling fire. Getting off the rig with nothing showing in the front and the chauffeur unable to set up the tower ladder (TL) to the fire building due to low-slung electrical wires, my position assisting with the initial TL setup was unwarranted. I was to head to the rear. Quickly sizing up the front of the dwelling and those up and down on the block, I estimated that taking into consideration the ½ basement windows and seven or eight front steps, a 20-foot straight ladder should set me up nicely to get to the second-floor rear bedrooms. If I needed to throw a ladder to the front windows, the 20-footer would be an exact fit. This ought to be an easy one, or so I thought.

As I rounded the Alpha/Bravo corner I had that sinking feeling start to creep into the pit of my stomach. There were two 5-foot high chain link fences between the fire building and where I was, the kind just high enough to be a pain in the rear, literally! I transmitted over the radio that I would be delayed in getting to my position. Tools over, ladder over, me over, and again, tools, ladder, me. The process was only complicated by two angry barking dogs nipping at my heels (which the owner did loosely attempt to bring inside). As I stopped to catch my breath, the next problem became apparent. There were *three* full stories in the rear! You can see where this is headed.

I immediately got back on the radio to the second-to-arrive OV and told him to bring the 28-foot extension ladder. My assumption from the front could not have been any more wrong. My 20-foot straight ladder was going to leave me well short. At the highest point to which I safely could climb, all I could complete was ventilation of the top floor windows after communicating with my officer.

What if a civilian was at one of those windows in the rear? What if one of our guys got stuck and needed to get out? These were just a few of the thoughts that raced through my head after clearing the window, making the window opening into a door. Watching the smoke drift from the opening, I paused for half a second collecting my thoughts, attempting to find a way to recover from my ladder length miscue. I noted a picnic table in the adjoining yard. In my head I thought, if I needed to, I could drag and throw that over the fence. Perhaps I could set the ladder up on top of that table and grab a few gawking civilians to foot it until reinforcements arrived if an emergency situation presented itself.

Recover. Show me a firefighter who has performed flawlessly and to the letter of the SOP at every fire and I will show you a bold-faced liar. While we try to avoid errors at all costs, we can and do make mistakes, have miscues, missteps, whatever you want to call them. We must learn to use them as motivation to not make them again. Studying the miscues of others could perhaps prevent us from making those same mistakes ourselves. They can be great learning tools to study over the long term. On the fireground, however, there is no time to throw your hands up, standing slack jawed staring at the situation you are facing. The mettle of the character of the firefighter is forged in how he or she adapts, recovers, and learns from a mistake. Don't languish over what is done. Accept the situation, address it, and move on, obviously in the shortest amount of time possible.

As the second to arrive OV made the building corner, I gave him a hand setting up his extension ladder. By the time we were able to get the ladder properly positioned, the fire was largely extinguished and the primary searches completed. Once we entered the home via the ladder and made contact with our bosses, we assisted in opening up in the fire area.

The take-away point to the above story is that straight ladders are what they are. A 20-foot ladder is 20 feet; no more, no less. They do not offer the flexibility in terms of length options that you find in extension ladders. If you have a choice, especially when taking portable ladders to places that you have yet to be able to verify a height with your own size-up, take the one ladder with the greatest variety of options, the appropriate portable extension ladder.

Portable extension ladders

Portable extension ladders are comprised of two or more sections. The base is referred to as the bed section and the movable sections as fly sections. There is a rope (normally a braided sisal type) that permits the fly section to be raised from the bed section to the necessary height. Refer to the list provided earlier in this chapter for nested (retracted) and extended lengths for common extension ladders. Ladder locks, often referred to as "dogs," are what locks the portable ladder fly section(s) in place as they are extended from the bed section. Anytime you are setting up and climbing an extension ladder, check to be certain that the ladder locks are fully engaged. These locks, when engaged correctly, bite onto and around the rungs of the bed section. The more fly sections that we have on the portable ladder, the more ladder locks we will have. Be sure to keep an eye on these, because the locking mechanism is what keeps the ladder extended. Extension ladders offer firefighters the greatest degree of flexibility when judging heights and should be utilized in most instances to achieve the greatest level of safety.

In the opening of this chapter we mentioned pre-tying your halyards. The halyard is the piece of rope that goes through a pulley atop the bed section that is used to extend the fly section of the portable ladder. Most, if not all portable ladders come from the factory with the loose end of the halyard tied to the two sections. The bed section is literally tied up to the fly section (fig. 22–11).

Fig. 22–11. Sisal halyard found on extension ladders binding the two sections together must be untied before the ladder is able to extend.

When a life hangs in the balance, the extra time it may take to untie this knot is absolutely avoidable and unacceptable. The rope likely has a memory and has been sitting out in the elements, making it even harder to untie. Do you think this photo was taken in Nowhereville? No, this picture was taken in an operating FDNY firehouse, on an in-service, front-line, single engine company (no truck in quarters). They had recently received a new piece of apparatus and the working members had no idea that the ladder was tied this way until it was pointed out to them. The halyard, as you can plainly see, would need to be united prior to being able to be extend the ladder.

Take the loose end of the halyard and tie it to the one of the lower rungs on the bed section of the ladder (fig. 22–12). Make sure that the rope remains taut when doing so. Now when the ladder is removed and brought to the objective, the only thing you need to do is pull down on the halyard to extend the ladder into position. Once it is in position, and if time permits, you can certainly take the excess halyard and tie off the ladder through the rungs with a clove hitch or other approved knot.

Fig. 22–12. Tie halyard rope taut to fly section of ladder. This may require periodic tightening as the rope may stretch over time and use.

Maneuvering with Portable Ladders

There are many options for carrying portable ladders from the apparatus to the fire building. A quick review of the average weights of a varying brands and types of extension ladders shows us that there aren't many that weigh less than 100 pounds. While we aren't normally hiking ladders for miles, the weight is a factor we cannot ignore. Compound the actual ladder weight with the added compliment of tools and firefighting equipment that we are tasked to bring to the fire, and you can see how quickly even the fittest members can be overwhelmed. We must find ways to work smarter, not harder, to get these ladders into position.

We are all taught multi-person ladder carries in our entry-level fire school (fig. 22–13). We are shown methods for transporting portable ladders, specifically the extension variety, with two, three, and four persons. These numbers of carrying firefighters make the task seem easy. They are wonderful for an initial introduction in how to move portable ladders from place to place, but have their only real application on the training ground where hands are plentiful. At initial operations, even in the biggest city fire departments, it's just not going to happen that way. Please show me a department that is able to commit four persons to carry one ladder, and if you do, I will show you a department that does not manage its initial resources to their best advantage.

Fig. 22–13. Four-firefighter carry of portable ladder

24-foot, 25-foot, 28-foot, and even 35-foot extension ladders can be maneuvered effectively by one firefighter who is well practiced (fig. 22–14). This should become routine for your firefighters. Practice with portable ladders does not require a great deal of materials nor a specific space. We have the ladders, we have the tools, and we have the PPE needed with us at all times. As with most combat-ready drills, when practicing with portable ladders it is imperative to practice as we play. The meaning is to use all the tools that we would normally bring with us to the fire. Practice bringing the combination of your firefighting tools along with the portable ladders. There are many ways to do this. We discuss three methods for single firefighter portable ladder and tool carries. Practice what works best for you, dependent on your portable ladder configuration and tool assignments.

Fig. 22–14. One firefighter carrying tools and portable ladder

Low shoulder carry

Perhaps the most common portable ladder carry is the low shoulder carry. Utilizing this technique, the firefighter rests one beam of the ladder on his or her shoulder, places one arm through the rungs, and grasps the rung with one hand. Most firefighters gravitate to make the shoulder/arm holding the ladder their most dominant one. A compliment of hand tools can be carried in the opposite hand.

Figure 22-15 shows one firefighter carrying the extension ladder on one shoulder with the 6-foot metal hook resting on the lower beam of the ladder, leaving the other hand to carry the Halligan.

Fig. 22–15. One firefighter carrying tools and portable ladder in low shoulder carry

Some fire companies have taken this concept one step further, affixing a tool to the ladder itself (fig. 22-16). In this photo, you can see where this slight modification allows a 6-foot metal hook to be tied in to the fly section of the 24-foot extension ladder. The tip of the hook is inserted into the tip cap of the fly section of the ladder and the shaft of the hook is secured with a length of Velcro strapping to a rung further down on the fly section.

Fig. 22–16. 6-foot metal hook attached to fly section of portable ladder. (Courtesy of Traditions Training, LLC, www.traditionstraining.com.)

High shoulder carry

The high shoulder carry for portable ladders takes considerable practice to master (fig. 22-17). Utilizing this method places the lowest beam on top of your shoulder. To control the ladder in this position, both hands are normally required to hold the rungs. Hand placement using the high shoulder carry does not leave a hand free, nor does it allow great flexibility to carry additional hand tools. While it is certainly an option for ladder movement, we consider it to be less advantageous than the low shoulder or drag methods. Additionally, unlike the low shoulder carry, placing the lowest beam of the ladder on top of your shoulder can limit some vision as you maneuver. Obviously, prior to any ladder movement your travel path should be scoured for anything overhead, such as electrical lines, low tree branches, and other obstructions.

Fig. 22–17. High shoulder carry of portable ladder

Fig. 22–18. In short-staffed companies, heavier portable ladders can certainly be dragged into position by one firefighter.

For all shoulder carries of extension ladders, attempt to place the fly section out away from your body. The "why" is explained in the deployment section of this chapter.

The ladder drag

Members with smaller physiques or any member tasked to move ladders of heavy weights (such as the 35-footer) can run into problems when limited staffing is a factor on your fireground (fig. 22-18). Additionally, any member physically carrying portable extension ladders for long distances can be quickly worn out. Throw in elements such as snow, ice, and unsure footing, and the difficulty levels continue to increase. There is no reason to fight the ladder! Fire service-rated portable ladders can certainly be dragged to their destination. Place all necessary tools on the bed of the ladder and away you go! Might you have to clean the mud off the ladder butts when the job is over? Sure, you might, but you made your objective and should have more energy left to accomplish other fireground tasks.

Multiple ladders? How can one firefighter bring multiple ladders to the fire building in one trip? Have you ever tried to carry two ladders, one on each shoulder? It is not very easy, nor does it leave you any room to carry tools or an option to protect yourself should you lose your footing or balance. Simply stated, combine and drag (fig. 22-19). This works best with one straight ladder with roof hooks and one extension type ladder. We have seen two methods that can help you work smarter. Both involve first extending the roof hooks.

Fig. 22–19. Multiple ladders in one drag

The first method involves hooking the roof ladder on top of the extension ladder, using the roof hooks to hook over the top of the rungs of the extension ladder (fig. 22-20). The second method involves placing the extension ladder on top of the roof ladder. Place the extended roof hooks in the upward position and place the extension ladder on top of the roof ladder. The roof ladder hooks will capture the butt of the extension ladder as it nestles down into place, resting at the hooks (fig. 22-21). For either type of multiple ladder carry, be certain to lift with your legs, grab through both sets of ladder rungs, and drag the ladders to position. Hand tools such as hooks, Halligans, and saws can be placed on top of the ladder and come along for the ride. This ladder drag set up can also be used in wintery weather by throwing a small tarp over the rungs to transport hose, tools, salvage materials, etc.

Fig. 22–20. Place the roof hooks over the end of the extension ladders.

Fig. 22–21. Slide the extension ladder into the notch created by the deployed roof hooks.

Ladder Markings

We mark everything in the fire service. Using permanent markers, paint pens, spray paint, colored electrical tape, and other materials, nearly everything we personally own and most of the tools and equipment that can come off our apparatus have some kind of markings. Why do we so heavily mark our fire service possessions? The obvious reason is so they don't go missing, are borrowed by someone, or wind up with other fire companies. We mark them so that the tool gets back to where it belongs.

Ladders, while marked for similar purposes—so that we know what rig they came from and can make sure they are returned correctly—should also be marked in other ways. Additional markings on portable ladders can aid us in our quest to increase efficiency and operate in the safest way possible on the fireground.

As we were just discussing carrying ladders, one such marking that can easily be done to increase operational efficiency is to mark the balance point of your portable ladders, whether they are straight or extension (fig. 22–22). As previously discussed with the rationale behind the pre-tied halyard, marking the balance point is a time saving step when seconds count. A ladder's balance point can be determined by finding the point at which the ladder equalizes (think balanced scales) on the member's shoulder.

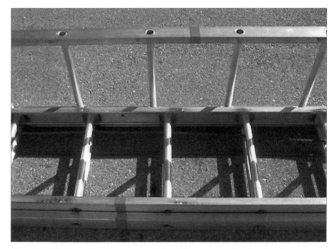

Fig. 22–22. Mark the balance point with a piece of reflective tape.

There should be no reason why you should have to guess where this point is for your ladders. If you grab your ladder at a position too far back it will cause the tip of the ladder to plunge down, too far forward and the butt will drag. Either position will likely cause you to have to stop, put the ladder down, and reposition it. This will delay you from reaching your position in the most expeditious manner possible. We strongly recommend using reflective tape to increase visibility for nighttime operations.

Reflective marking tape can also be utilized on the ladder tips. One fire service company we routinely call on for innovative ideas is Firehouse Pride (www.firehousepride.com). They have recently begun to market adhesive style ladder wraps (fig. 22–23). These highly reflective ladder tip wraps can replace the spray painted tips of your ladders. Think of the safety factor when reflective material is added to your portable ladders that are placed at a window or above a roofline. They sell various colors to match your department's needs and can customize department apparatus name and company numbers into the design of the wrap (for instance, "Ladder 39"). These reflective markings can also be placed near the butt end of the ladders for easy recognition of the ladder's length and additional company name identifiers to be seen as they are stored in the ladder beds on the apparatus.

Fig. 22–23. Mark the tips of the ladder with a piece of reflective tape or paint. (Courtesy of Firehouse Pride, www.firehousepride.com.)

Placement Order of Preference

We have chosen our ladder length and moved it from the apparatus to the residential building. Now it is time to get it into position against the fire building. Where do we start placing ladders at the residential fire if no obvious rescues are initially required? Additional ladders will come as the incident progresses, but where do we start? If we are placing ladders largely for our access and egress, we need start our placement with purpose. Ideally, we would like to see at least one ladder on every side of the building and at least one ladder to each floor, but where to begin?

1. The fire floor

Initial laddering operations should begin by creating an additional egress point with a portable ladder placed to a window on the fire floor (fig. 22–24). This is imperative on upper-floor fires. Most residential structures have only one set of stairs, which limits our ability to go up and go down in the home. We know from review of several LODD and close call reports that the stairs can quickly become clogged with those who want a piece of the action and can be cut off from operating forces with rapidly extending fire.

Fig. 22–24. Ladder placed to adjacent window on fire floor. (Courtesy of Kentland Volunteer Fire Department, www.kentland33.com.)

Specific placement, as in which window to choose, will largely fall back to your knowledge of your buildings. For instance, most second floors in colonial style dwellings have rooms (bedrooms and bathrooms) that run off a center-type hallway that normally follows the ridgeline of the home. Placing a ladder to a window in the fire room may not be your first choice as members will not likely be *in* the fire room until the hoseline is moving in. Perhaps one room over would be better served initially to act as that egress point.

Do your homework and review your building styles. Think about why you would or would not place a ladder to each particular window. Corners of buildings can often be covered with one ladder when there is window on side Alpha and side Bravo that service the same room. Another consideration is to avoid placing ladders directly in front of entry doors. Members moving hoselines, tools, and equipment may cause the ladder to be dislodged and can impede hoseline movement and overall progress. Remember, we are setting these ladders up for *us*. We must think how and where the interior forces are operating and plan our placement accordingly. A well placed portable ladder can quickly become a lifesaving tool to anyone who needs it.

2. The floor above the fire

At first-floor fires, the floor above is the first place we should aim to throw our portable ladders. Outside of the immediate fire area, this can be one of the most dangerous places to operate. Avoid placing the ladder to a window directly over the fire area on the first floor if at all possible. Just because the fire isn't currently coming out that first-floor window now doesn't mean it won't in the future. This could place anyone operating on that ladder or anyone who may need to come down it later in a seriously precarious position.

In dwellings that have multiple stories, after the first ladder is positioned to the fire floor, the next ladder placement should be to the floor above (fig. 22–25). Remember that if you need a 24-foot ladder to reach the fire floor, you will likely need the 35-foot ladder to reach the window on the floor above. As stated in the previous paragraph, the floor above can quickly become untenable, forcing members to seek refuge in those rooms off the main hallway. In many instances the interior stairs can quickly be compromised. The ladder can be placed for coordinated ventilation, an exit point for members on the floor above, and possible vent, enter, and search (VEIS) techniques (reviewed at length in chapter 24).

The perils from residential open interior stairs are well documented for firefighters, and these are very common in the residential setting. The point at which fire can cut a team off from a safe retreat to the stairs can happen in seconds. If the stairs brought you up, and this route is now compromised and is no longer viable, what's next? Portable ladders may be our only way out!

3. Top floor

Our next point to place portable ladders is the top floor (fig. 22–26). We should aim to coordinate ventilation efforts to provide an exit path for smoke and also, condition dependent, VEIS operations. Again, the commonly recognized open interior stairs will allow the top floor to quickly fill with the products of combustion. Ladders thrown to the top floor can be used as a platform for ventilation and the possibility of vent, enter, and search.

Fig. 22–26. Ladder placed to window on top floor. (Courtesy of Kentland Volunteer Fire Department, www.kentland33.com.)

4. All floors in between

Once we have enough ladders to cover our primary areas of responsibility (the fire floor, the floor above, and the top floor), when resources arrive we should look to fill in the floors in between. We should have ladders against the building to cover all floors of the structure. A portable ladder to every floor ensures a secondary means of egress from every level of the home. If radio traffic permits, a quick transmission as to their placement would be helpful.

Fig. 22–25. Ladder placed to window on floor above fire. (Courtesy of Kentland Volunteer Fire Department, www.kentland33.com.)

5. To the roof (if flat roof move to priority #4)

We discuss the pros and cons of placing operating members on the roof at length in chapter 23. Peaked-roof residential dwelling fires are best extinguished from inside and from below. If roof work is applicable, for instance the presence of a flat roof on a residential dwelling, we should indeed ladder to the roof. Again, in peaked-roof residential dwellings our preference is that portable ladders be placed to all sides of the building and secure at least one to every floor before laddering the roof.

Deploying Ladders For Use

Once we have decided the appropriately sized ladder, carried it to the fire building, and determined where we want to set it up, how do we best utilize our minimal resources to get that ladder up and against the building? While our generic fire academies have again given us three- and four-firefighter raises (with a safety spotter, of course), this is not the reality of today's fireground staffing. We must utilize other options to make up for the lack of initial staffing that most of us face.

In this section we discuss the one- and two-firefighter ladder throws. Remember, we are talking about throwing and placing portable ladders, not throwing and ventilating or conducting VEIS. Ventilation today requires the strictest discipline and coordination. Entry into an IDLH requires two persons. Your most junior and even exterior-only firefighters should be taught techniques to place ladders by themselves. Again, we are not discounting ladder fundamentals taught to us in the academy, but we must refine our deployment options with the realities of actual fireground staffing.

One-firefighter portable ladder placement

While it is not optimal to work alone anywhere on the fireground, throwing ladders can be the one exception. There is little need for team coordination if it's just you. Not often does available staffing allow us to hear, "Prepare to lift. Lift! Prepare to raise. Raise!" We have all seen or been a part of a ladder evolution where you go left and the assisting ladder carrier goes right; they go forward and you go back.

As a single firefighter making ladder decisions on the outside, you really aren't doing anything for *you* initially. What you are doing is providing a lifesaving service to the people inside. You are giving them another way out. While you may not get the "salt" of being in the job, your role is undeniably important. Ask anyone who has had to rapidly leave a dwelling, coming out a window onto a portable ladder, if they were thankful it was there. The answer will be self-explanatory. To many, throwing ladders is a thankless position. Sure, perhaps most times. But that one time, whether it is one in a hundred or one in million, your ladder placement role will be lifesaving and absolutely invaluable.

In limited staffing situations we must use the building as our second firefighter. There are two methods we discuss for single-firefighter deployment. Both are not difficult conceptually and should be easy evolutions to master with the right amount of practice and training. Limited staffing fire companies can train members to throw all of their apparatus' portable ladders, up to and including the 35-foot extension ladder, with these methods. Method 1, "dip and dive," takes the portable ladder right from the shoulder carry, in one movement, to the building. Method 2, "beam slide," begins with placing the ladder down perpendicular to the dwelling. The member then lifts, gets under, and pushes and slides along the beams, moving the ladder up into a vertical position. We break down each method into several simple steps.

Method 1: dip and drive

- Carrying the ladder with the fly section out away from your body, determine your line of approach to the building. You will look to set the ladder to one side of the window, the windward side if possible so that smoke will not factor into your ability to judge position (fig. 22–27).

- Note any overhead obstructions on your route (electrical lines, cables, tree branches, etc.).

- As you near the fire building:
  - Drop hand tools off to the side as you near your objective.
  - Just prior to reaching the building, flatten the ladder on your shoulder (fly section now up) keeping one hand firmly grasped on the rung (fig. 22–28).

- In one fluid motion, *dip* the butt down and *drive* the ladder into where the ground and building interface. At the same time, punch up with the hand on the rung and drive the ladder against the building (fig. 22–29).
- Secure the ladder to the building, standing off to the side with your knee/leg. You may want/need to pull the base out a few inches from the building for added stability.
- Pull halyard to desired height.
- Roll ladder into position at the level of the windowsill.

Fig. 22–28. Get ladder flat, dip butt down and drive toward building/ground interface.

Fig. 22–27. Line up your approach.

Fig. 22–29. Punch up, press ladder into building, and secure.

Method 2: beam slide

- Carrying the ladder with the fly section out, away from your body, determine your line of approach to the building. You will look to set the ladder to one side of the window. If possible, again go to the windward side so that smoke will not factor into your ability to judge position.

- Note any overhead obstructions on your route (electrical lines, cables, tree branches, etc.).

- As you near the fire building:
  - Place the ladder on the ground, fly up, with the butt of the ladder at the ground/building interface.
  - Go to the ladder tip and, lifting with the legs, move ladder up to eye height.
  - Duck your body under ladder, holding the rails, and slide/push ladder up with momentum to vertical (fig. 22-30).
  - Secure the ladder to the building, standing off to side with your knee/leg. You may want/need to pull the base out a few inches from the building for added stability.
  - Pull halyard to desired height (fig. 22-31).
  - Roll ladder into position at the level of the windowsill.

- If you are unable to reach the building/ground interface with the butt of the ladder due to obstructions (bushes, fences, etc.), girth-hitch your looped piece of webbing five or six rungs from the butt end on the fly section, run the webbing along the top of the fly section, over the butt end and back under the bed section of the ladder. Complete the raise as above, and as soon as able, step on webbing that will present itself under the bed section as you raise the ladder. This will now act as the building wall does, bracing the ladder and limiting butt kick-out.

Fig. 22–30. Grasp ladder at tip along rungs, push and slide until against building.

Fig. 22–31. Grab halyard and extend to desired height.

Two-firefighter portable ladder placement

Honestly, when two firefighters are assigned to outside positions at the residential fire, we would expect for each of them to bring their own ladder. They should be trained and practiced to throw all sizes individually. Obviously, if they need assistance they should let each other know and not keep it a secret. We must communicate. They should coordinate with each other as to their ladder placement as they approach the structure. Communications is key to avoid duplication. Let each other know who is going to place which ladder in which position (i.e., which side and floor). Each firefighter, if familiar and trained with carries and/or drags and single deployment, in one trip from the rig to the fire building, should be able to place a ladder to cover the fire floor and the floor above instantly. Later arriving companies, as needed, can bolster this initial compliment of ground ladders.

Ladder Climbing Angles

While firefighters are creative, tool savvy, and packrats to some degree by nature, few carry protractors in their turnout gear. Determining the ultimate perfect textbook climbing angle can be hard to guesstimate in the heat of battle. How do you determine what's good? The old buffalo's rule of thumb was always to stand on the lowest rung, put your arms out straight, yeah, that's good enough. Another climbing angle determination uses one quarter of the working length (20 feet of a 24-foot extension ladder used to reach objective = 20 feet of working length). In this instance, to obtain a proper climbing angle, your ladder's base should be 5 feet from the building.

Ultimately, when all is said and done, you are looking for a proper climbing angle of about 70 degrees. Some manufacturers have helped us out, placing a decal on the side of the ladder's beam showing us the proper angle (fig. 22–32). Take into consideration the climbing angle when determining the length of ladder needed to reach your destination. Too steep of an angle may cause the ladder's tip to pull away from the building. Angles that are too low have been known to kick out at the base.

When footing ladders, it is our recommendation to stay on the outside of the ladder, facing the building and the ladder (fig. 22-33). Take a boxer's stance, placing one foot on the bottom rung and your hands on the beams of the ladder. This vantage point allows you to see what is going on with the member above you on the ladder. If you are underneath the ladder, unless you strain to look up, you have limited visual awareness of what that member is doing on the ladder above you.

Fig. 22–32. Ladder climbing angle stickers

standing outside, not in an IDLH environment, with a facepiece fully affixed, breathing down valuable cylinder air nowhere near the fire or products of combustion. As discussed in chapter 5, there is a finite amount of air in your cylinder. Once it's gone, it's gone.

Masking up quickly is another combat-ready skill that we have mentioned earlier in the book. It will prevent you from succumbing to the temptation to be a yard breather. Either practice masking up while at the ladder's tip or at the bottom of the ladder, leaving your regulator out, popping it in at the top. Until you are ready to enter the IDLH at the top of the ladder, you shouldn't really need to breathe down your cylinder and be "on-air" while setting up your ladder for use (fig. 22–34).

Fig. 22–33. Foot ladders from the front to keep eyes on the building and member climbing.

Fig. 22–34. Firefighter masks up quickly at the top of the ladder. (Courtesy of Traditions Training, LLC, www.traditionstraining.com.)

Working On/Off Portable Ladders

When working at fires with portable ladders, first and foremost have all your PPE in place at the ready. Unless immediately headed to operate in an IDLH environment, you will not need to use your SCBA, but it must be on your back and at the ready at all times. Someone may appear at a window at a moment's notice that requires your help. Do not be a "yard breather"—someone who is

When climbing portable ladders, keep your feet slightly out to either side of the rungs as you ascend and keep your hands on the beams. In bad weather keep your feet deeper into the rungs and farther to the outside where the rungs meet the beams for a more secure footing. When traversing ladders with hand tools, how should we carry them? Well, we don't have to carry all our tools with us at all times. It's very easy to hook a tool onto several rungs above, climb empty handed, then retrieve the tool, move it up, and so on. If climbing with several tools, hook what you can, but leave one hand free for your secure grip. Slide the other hand holding the tool along the beam of the ladder as you move upward (fig. 22–35).

Fig. 22–35. Carrying tools on ladders

Many departments issue their members ladder belts or safety harnesses with hooks attached built into their personal protective equipment. These should be utilized (hooked into a rung) as needed to add a measure of security when working for a period of time from a portable ladder. We should not routinely operate laterally far beyond the confines of our ladder. If you need to reach far to the left or right for something, get down and reposition the ladder or have someone bring an additional ladder to get it.

When drilling with portable ground ladders, at a minimum you must have helmets and gloves in place.

How many runs a year do we respond to for civilians who have fallen off ladders? It can happen to us, too. If you are going to be at any level above the ground, wear your helmet. Ladders can develop burs that lead to splinters and razor-like cuts in our hands. Not only wear your gloves, but also as with all training, wear your firefighting gloves. Practice the way we will play!

Inspection, Care, Maintenance

Chapter 6 of *NFPA 1932* specifies recommendations for inspection and maintenance of the portable ladder. We do not spend time discussing the intricacies of the inspection list here; just know that your ladders should be inspected visually on a monthly basis and after each use. Damage to portable ladders can occur from fire operations, drills, or even just in placement on the apparatus. Be certain that your ladders are kept secured to your apparatus and do not bounce around against the rig.

A clean tool is a happy tool, and portable ladders are no exception. Keep your ladders free and clear of fire debris and clean as needed with careful focus on the extension ladder's moving parts. Pulleys, ladder locks, and fly section channels are prone to collecting debris. Refer to your ladder's manufacturer for specific lubrication information, but remember that ladders do occasionally need a few drops of oil on pulleys, roof hooks, and so forth.

Tip In/Tip Out

An excellent combat-ready portable ladder tip was brought to our attention by fellow Traditions Training instructor Joe Brown. Joe is a DCFD tiller man assigned to a ladder company and captain at the Kentland Volunteer Fire Department assigned to TL-33. Joe has assisted in having his crews set up their ladder bed on the apparatus to have a mirrored supply of ladders. Half the ladders have the tips of the ladders facing in to the ladder trays toward the cab of the rig, and the other half have the ladder's tips facing out at the rear of the apparatus (fig. 22–36). Why does this have any bearing on efficiency when we make our ladder selection?

Fig. 22–36. Portable ladders in bed, half tip in and half tip out. (Courtesy of Kentland Volunteer Fire Department, www.kentland33.com.)

Let's do a quick review of your response area and focus on the suburban/urban mix. As homes are now being built, they are tight on space. There are no palatial front, rear, and side yards. The community is maximizing homes on minimal land. You have narrow spaces between dwellings, tight backyards, fenced-in alleys, and so on. This is where this "tip in/tip out" methodology for your ladder storage is best applied.

Let's say we are throwing ladders to the front of the fire building. Which ladder would you select, tip in or tip out? Well, with the tip in, as you pull the ladder, it comes out butt first. This will lead you, butt first, toward your objective. Most of the time, this is our preference in the way we travel with portable ladders. The firefighter can then take that ladder, place the butt end right into the building's base, and raise it up into use. But, what if we had to take that same ladder down a narrow alley to use in the rear of the house?

If we were to carry the ladder butt first into the rear yard through a narrow side yard into a tight backyard, we would need to somehow spin the ladder around so that the butt of the ladder faces the rear of the building. If instead we grab the tip out ladder, we would be in much better shape. Again, knowing your buildings is paramount. With the tip out, we would be heading tip first toward the building. With this approach, we can simply take the tip straight back past the objective, stop, turn to face the building, and the butt of the ladder is now in position facing the building!

Keep It Simple

Portable ladders want to be used at fires (fig. 22–37). They want to get out of their ladder racks and be used, see the light of day, be near the heat and smoke, and help you do your job more quickly and easily. They live to be thrown against residential building fires. It makes them happy. They love to be carried or dragged into place by one firefighter. They want to be placed at the level of the residential windowsill for best overall operations. They provide access and egress into the home from points elsewhere. The correct placement of portable ladders can add an additional layer of safety for us and decrease the time it takes to search for and save civilians. They are simple yet effective tools in our fire attack arsenal. Firefighters can quickly and easily grow comfortable around portable ladders, they just have to love them a little. Don't make laddering the building harder than it has to be. Get the ladders off the rig and become a laddering pro in your department.

Fig. 22–37. Portable ladders in operation at a residential building fire. (Courtesy of Kentland Volunteer Fire Department, www.kentland33.com.)

The Backstep

Laddering Drill

As we discussed, nearly every rig has a portable ladder on it these days. They require little imagination to set up and deploy. Obviously we know that you cannot just pull up in front of a residential home as you are driving around your area, stop, and throw ladders. But if you have a blank wall at the firehouse or the local school, you can easily make windows everywhere! With the company out of sight, take a piece of chalk to the blank brick or wooden wall and draw windows at varying heights. Then have the members get on the rig, disembark, and head to the building with a ladder to match the markings that you placed on the wall. Need more windows? Need differing heights? Get more chalk! When finished, take the water can and, presto, the markings disappear (figs. 22–38 and 22–39)!

Fig. 22–39. Members practice throwing ladders with full PPE.

Fig. 22–38. Make chalk windows at varying heights.

23 To the Roof?

The title of this chapter, asked in the form of a question, is not an accident (fig. 23–1). "Is anyone headed to the roof?" is a question that is regularly heard at residential building fires. For some departments, every fire gets a hole cut in the roof, but why? What is it that roof ventilation accomplishes?

Fig. 23–1. Firefighters operate on a residential roof. (Courtesy of Chuck Ryan.)

What you may need to ask yourself is, "Do we *need* to send someone to the roof?" Is a cut in the roof a top priority in this operation? How else may we ventilate to remove the built up smoke and heat from this dwelling? Does your department cut roofs just because they always have? This chapter discusses best practices with respect to ventilation and residential building fires. For every action we take on the dwelling, the fire will have a reaction. Perhaps nothing influences fire conditions more than ventilation efforts.

Let's recap the concept of ventilation in a nutshell before we move on to specifics for tactical considerations at residential fires. Ventilation can be described as the systematic removal of smoke and fire's toxic by-products from the confines of the fire building through building openings. Sometimes we create all the openings; sometimes they may already be opened up when we get there. It is our job as firefighters to control existing conditions and to coordinate and communicate further ventilation efforts at fires.

Ventilation

There are generally two forms of fire department controlled ventilation efforts: vertical and horizontal. Vertical ventilation requires an exhaust opening to be created and opened at or near the highest point of the structure (normally the roof level). This facilitates the movement of smoke and by-products of combustion up and out of the fire area. Horizontal ventilation results from members creating horizontal openings (normally through doors and windows) on each level of the home. It may occur on the fire floor and/or on the floors above, depending on conditions found.

The efficiency of vertical and horizontal ventilation efforts can be increased by two methods, both provided by fire department members. One method to speed up ventilation is to place mechanical fans in and around the dwelling. Another simple yet effective way to clear an area of smoke is through hydraulic (hoseline) assistance.

Ventilation Advantages

The mechanical method to aid fire operations with smoke movement comes in various forms of ventilation fans. Fans can be a positive pressure ventilation type (PPV) or a simple propped or hanging exhaust fan (fig. 23–2). They can be gas or electrically powered. Some PPV fans can move upwards of 20,000 CFM (cubic feet per minute) of air. They can be strategically placed to assist our removal of smoke from various areas in the dwelling.

Fig. 23–2. Firefighters operate with ventilation fans.

There is, however, an ongoing debate in the fire service similar to the apparently never-ending "smooth bore vs. fog nozzle" fight as to when, where, and if PPV fans should be set up. Others wonder if they should be utilized at all. This debate reaches a fevered pitch, especially when discussing the role of PPV during fire attack. When used correctly, coordinated with the fire attack team that has control of the fire area with a charged hoseline in place and fire room door control, they work well. PPV fans can assist in clearing the home quickly of smoke, aiding in searches, limiting smoke damage, and pushing the fire's extinguished by-products out of the home.

Conversely, when used incorrectly, PPV fans create the same effects that are produced by wind driven fires. When you set up and operate a PPV fan, you are effectively charging that structure with fresh air, pumping it into the dwelling. This can be an enormous problem. As discussed in chapter 13, most fires today are oxygen regulated. Please take a moment to go back and review chapter 13 for what happens when we saturate a starving fire with air. The ensuing events are indeed dramatic and often deadly.

It has been our experience that PPV fans and ventilation exhaust fans should not be used in fire attack, but in the salvage and overhaul portion of the fire. However, extreme caution must still be exercised with fans, especially the PPV, at this and any stage in an operation. You are, in effect, turning on a wind creating device! Any small, unexposed pocket of hidden fire can quickly show itself once our fans are running and fresh air is being introduced to the environment.

Hydraulic ventilation is another method that we can use to assist in smoke removal once the bulk of the fire has been knocked down. It can assist in opening up and final extinguishment (fig. 23–3). While operating the hoselines, moving water through the nozzle creates a Venturi effect, drawing air in from behind it. The pressure from the escaping water draws air and smoke in from behind it and pushes it forward. Our job is to maximize the Venturi effect to draw and push the smoke out an exhaust port with the nozzle.

Fig. 23–3. Firefighters vent area with hoseline stream out window. (Courtesy of Kentland Volunteer Fire Department, www.kentland33.com.)

A glass free, cleaned out window makes an excellent opening to practice hydraulic ventilation. Have the nozzle firefighters position themselves, with SCBA in place since they will be drawing all the products of combustion past them, about 3–5 feet back from the window. Open and adjust the nozzle pattern to nearly fill the opening of the exhaust port. Smooth or fog nozzles work equally well to accomplish this task. For smooth bore nozzles just crack the bail back slightly to produce a broke stream, or use a straight stream and rotate the line in a clockwise motion out of the opening.

Ventilation, either horizontal or vertical, is done for specific reasons: either to facilitate smoke movement to sustain human life (vent for search, or VFS) or to create a exhaustive relief point to channel smoke and fire from the structure (vent for extinguishment, or VFE). Ventilation is so much more than just running around the dwelling and breaking windows or getting on the roof and indiscriminately cutting holes.

Fig. 23–4. Vent for life includes vent, enter, and search. (Courtesy of Traditions Training, LLC, www.traditionstraining.com.)

Ventilation for extinguishment (VFE) is a ventilation effort designed to be applied using primarily horizontal ventilation techniques . Vertical ventilation in the VFE scenario is discussed later in the chapter. This VFE strategy is set in motion in cases where a ventilation position can be secured in or near the fire area opposite the operation of the hoseline. Utilizing the VFE method, a firefighter creates a ventilation opening that acts as an exhaust port for fire, smoke, and steam to pass out of the structure. As a best practice, this is not completed until the hoseline is advancing and is ready to hit the fire area (fig. 23–5).

Why We Vent

"Vent for search" seems self explanatory, but it's often misapplied under the guise of "search." Again, we have to remember that every action we take on the fireground will have an effect on the fire. Venting for search may occur without the initial protection of the hoseline. While searching, ladder companies' inside teams may commence VFS maneuvers as they move through the dwelling while doing their primary search.

Searching without the protection of the hoseline is certainly hazardous. Breaking windows indiscriminately as firefighters move, however, can and will increase the overall level of danger. While removing smoke horizontally by taking out windows as we search the dwelling may help trapped civilians and perhaps momentarily increase visibility in search, if a hoseline is not in place, there is an inherent risk of drawing fire to our position. Or, equally as dangerous, we may create a negative change in the fire environment. Without a hoseline at the ready, this may increase the intensity in the fire area. It can increase fire growth by the addition of an intake/exhaust opening that was created. VEIS maneuvers also fall under VFS and have ventilation effects at residential building fires (fig. 23–4). VEIS is discussed in greater detail in chapter 24.

Fig. 23–5. Firefighter taking window opposite hoseline advance. (Courtesy of Nate Camfiord.)

Members assigned to the outside team of the ladder company are in the best position to accomplish this. It *must*—we cannot stress this enough—*must* be coordinated with the advance and operation of the hoseline. Firefighting members assigned to the ladder company's outside team must utilize sound judgment regarding when and where to apply VFE techniques. In all cases of ventilation of the fire floor, the first-due ladder company officer or inside team leader should coordinate and control such efforts inside and outside the fire building.

Staffing and Ventilation

We are fighting fires in the age of limited staffing. Whether you're riding a rig in a career department that has fallen victim to staffing reductions or a volunteer department that is desperately trying to keep members coming to the firehouse, it is a reality for all of us. Firefighting is not a business that can "do more with less." With less, we can do less. With less staffing it will take us longer to complete our duties. While we cannot use it as an excuse for poor performance, we must prepare, practice, and anticipate the problems that short staffing will present for us at operations. It is what it is. We have to make sure that we let others around us know that working with less will place firefighters and civilians in greater danger. We cannot place blame anywhere in particular, that is just the reality of the nationwide trend toward reduction in staffing.

When it comes to performing ventilation at the residential building fire, we must make decisions and manage our resources wisely. We must play the hand we are dealt and use the firefighters we have for maximum benefit.

We discussed the effectiveness of the two-team truck in chapter 20. While the ladder company has many tasks to accomplish on the fireground, the two-team truck concept pairs firefighters to address interior and exterior horizontal ventilation. Even with short-staffed trucks (remembering that a minimum effective truck company has three members), mission-critical tasks will be covered. However, as the chapter title indicates, we must think about ventilation, specifically peaked-roof ventilation, and whether or not we need to place our initial ventilation emphasis in this area.

The Residential Roof

In many departments, every fire in a residential building gets a hole in the roof. Is this really the best practice and utilization of your short-staffed crew? What else might we be able to accomplish around the fire building without having two members married to one roof cutting operation? While firefighting is inherently dangerous, roof work takes it to the next level. Think for a moment about how many runs we respond to yearly for construction workers and homeowners who have fallen from roofs. I know my roof, the shape it's in, how old it is, and the last time the shingles were replaced (fig. 23–6). Do you know anything about the roof you are putting your firefighters on?

Fig. 23–6. The safety of a residential roof is largely unknown to responders. (Courtesy of Nate Camfiord.)

Roof Construction

Do you know if the roof you are working is constructed of lightweight wooden trusses held together with gusset plates (fig. 23–7)? Is it made from disintegrating layers of water-soaked or delaminated plywood, is it treated for fire resistance or gummed-up pressboard? Do your firefighters know the difference between a good roof and a bad roof and what exactly comprises a "spongy" feeling?

What other factors can we not control? Is it snowing or raining? Has thick morning dew settled on the 30-year-old moss-slicked shingles? Oh, we nearly forgot about us! Did we forget to remind you that your firefighter will be using a chain or rotary power saw on this roof? More than likely, your member will be shrouded in a smoke and fire condition and dressed in nearly 70 pounds of protective gear. Each addition (shoddy construction, age, condition, and environmental factors) multiplies the dangers exponentially. We could go on, but I hope you are getting our point.

Fig. 23–8. What's in your attic?

Fires in attic spaces are best extinguished from below. Members with hooks and hoses with water are best suited to extinguish the fire via the top floor. Another roof operation we have all seen is a residential fire with half the roof burned through, and yet two firefighters are 3 feet away cutting an additional hole. It's also common to see handlines and master streams being shot into those openings from the ground. We call it "sky protection."

Fig. 23–7. What is the support structure of the roof comprised of?

What's In Your Attic?

There is an argument regarding cutting holes in peaked roof residential buildings when the fire is in the attic (fig. 23-8). To that we say, what worse conditions are you now putting your firefighters atop? Take in the previously mentioned roof hazards and compound them by placing a fire directly below them, eating away what little support they have left. Let's think about why we are opening up. What are we saving in this situation? While certain building types have attic space that is converted to living areas, most attics are filled with things people do not even want in their houses.

Our attics are filled with the kids' outgrown clothes, holiday decor, and the sweaters that we keep getting from the aunt who still thinks we are 12 years old. Are we risking the welfare of our members to save these things? Think about what you have in *your* attic.

Horizontal vs. Vertical Ventilation

Let's be clear, we are not saying that there will never be a time or place when roof cutting is required. There are very few absolutes in the fire service. What we are saying is that in the initial stages of the fire operation, especially when staffing is limited (which seems to be true about every fire these days), departments would be better served to choose horizontal ventilation first.

Horizontal ventilation, coordinated with the inside operating units, is quicker to perform than sending members to the roof. Horizontal ventilation can also be accomplished with one person, perhaps even an exterior-only junior firefighter under the direction of the ladder company's outside team. Instead of adding an additional layer of danger by sending members to perform roof work, horizontal ventilation actually adds an extra layer of safety. How is this, you ask? Well, the majority of horizontal ventilation is performed on portable ladders. When a portable ladder is left at a window, it becomes a point of potential egress for us (fig. 23–9).

When given the okay to horizontally ventilate a window, we must not merely check the box when doing so (fig. 23–10). When we say vent the window, what we really mean is clear the window. All efforts should be made to remove the entire window—glass, sash, curtains, and all. We would like to see the window morph into a doorway (fig. 23–11). Anything you leave at the window will be an obstruction. We want nothing to hinder us from the ventilation effects and firefighters getting in or out of the space we create.

Fig. 23–10. A window left in this state does little to help an exiting firefighter.

Fig. 23–9. A well placed ladder becomes an additional avenue to exit the home. (Courtesy of Traditions Training, LLC, www.traditionstraining.com.)

Fig. 23–11. Make the window into a door. (Courtesy of Roger Steger.)

As extra units arrive and staffing levels increase, if roof work is still warranted, address it with extreme caution. Even the FDNY, which is often thought of an organization that operates only in high-rise buildings, responds to thousands of working fires in peaked roofed residential buildings. Also, the FDNY in terms of staffing, has the luxury of operating with a total of six company members on its trucks. In my experience working in the outer boroughs (Bronx, Queens, Brooklyn, Staten Island) in peaked roofed structures, some attached, some stand-alone, not once ever did I see a peaked roof being cut during the attack phase of the fire.

In fact, the dedicated FDNY ladder company roof firefighter who you would assume always goes to the roof actually teams up with the outside vent firefighter when arriving at a fire in a peaked roof dwelling. They do not go to the roof at all. Together, the roof firefighter and outside vent firefighter are tasked to VEIS areas remote from the fire via portable ladders. The roof firefighter does not even give the roof a second thought in their operational plan once it is determined to be of the peaked variety.

Experience in fighting fires in residential buildings has shown the FDNY that roof ventilation, while having its place, is not an initial consideration to be addressed. So even with six ladder company personnel on each truck and a second ladder normally only 1 minute behind them, they do not send members onto peaked roofs. Why is this?

We agree with the decision because horizontal ventilation takes fewer members to accomplish. In most instances, it is normally completed faster than vertical ventilation. Horizontal ventilation via portable ladders makes us more successful overall in the fire operation. Horizontal ventilation incorporates the vent, but also may be used to enter for search. It also leaves that portable ladder in place, providing potential avenues of egress for interior operating crews.

Roof Positioning

If it is determined, after all initial ladder company fireground duties are completed, that roof operations are warranted, there should be an order of preference in accomplishing it (fig. 23–12). The order should give operating firefighters the greatest measure of safety that we can provide for this dangerous task. First, prior to starting any power saw, any and all natural roof openings should be removed. This includes items such as skylights, gable end vents, and perhaps removal of any attic fan type ductwork.

Fig. 23–12. Peaked roof residential building

If at all possible, the natural openings and any initial roof cutting work should be completed from the safety of a tower ladder (TL) bucket (fig. 23–13). This is the safest platform from which firefighters can operate. Placement should have the TL with the turntable on the corner of the home with the ability to reach two sides of the structure. The boom to the roof should be placed over the corner of the building, up to the ridgeline. As you may imagine, the TL bucket is also the FDNY's first choice for roof cutting operations at the pitched roof dwelling.

Fig. 23–13. Tower ladder bucket platform. (Courtesy of Nate Camford.)

The written evolution that describes the FDNY's TL peaked roof cutting operation literally calls for the firefighters to tie themselves to the device. Members are to make a bowline in the bucket with a length of rope tied to a substantial section of the TL basket. This loop is created with a section of rope that reaches the outer edge of the bucket. Then they are to clip their personal harness into that loop prior to operating. They are to cut and work the saw with one foot outside on the roof and one foot in the bucket. There is also a safety rope tied off to the saw and lashed to the basket or controlled by a backup firefighter in the TL bucket.

If no tower ladders are present, roof cutting should commence from inside the safety of the bed of the aerial ladder (fig. 23–14). The member should be placed as close to the peak of the roof as possible. The valley of the roofline is the second choice. If neither TL nor aerial ladder can be placed on the roof, roof work should be completed with the utilization of a combination of portable and roof ladders.

Fig. 23–14. Tip of aerial ladder. (Courtesy of Nate Camfiord.)

Making the Cut

When making cuts on the peaked roof itself, try to have the wind at your back if at all possible. Also, we also recommend roof firefighters wear all of their PPE, including SCBA, for roof operations. If your department does not have a policy on rooftop PPE, make one now. If the IC has made the decision to send members to the roof, the commander has done so with the expectations that there is smoke and fire that need to be vented from that area. Have your mask on and breathe air as you make your cuts. Remember why are you making that cut. You are expecting smoke and possibly fire to come out. If you make your cuts and nothing comes out, guess what? One of two things happened. Either you cut in the wrong place or, more likely, you didn't need to be on that roof cutting it in the first place!

Let's talk for a moment about the tools that are used in most rooftop cutting operations (fig. 23–15). Chain or rotary disc saws with carbide-tipped blades are most common in the majority of fire departments. Both of these two tools require great care, regular maintenance, and competency to use correctly and safely. These saws, like fires, have no consciousness. They are powerful tools that will make quick work of roofing material, human flesh, and bunker gear. One thing not to overlook when assembling your tool selection is a set of irons. They can be driven into the roof boards and used as footholds. If there is a failure of the mechanical saw at the roof level, they can be used to open the roof (fig. 23–16). There are many great videos online that show the effectiveness of using the flat striking backside surface of the axe to break through older roofing plywood. The striking end is used in place of the cutting end as the sharp edge tends to get wedged into roofing materials. The backside of the axe bashes clean through the boards and requires less effort per stroke for removal.

Fig. 23–15. Chain and rotary saws for roof operations

Fig. 23–16. Set of irons used on the roof. (Courtesy of Nate Camfiord.)

So, your saws are working at the job today, but you and your partner bring a set of irons with your roof hook just in case. With all of your PPE and SCBA in place, the cutting member should start the initial cut as close to the ridgeline of the rooftop as possible. If the fire is near a particular end of the building, then place the cut slightly closer to that end of the structure, but keep in mind the building's configuration. The reasons for this are twofold. One, "near the ridge" has to do with making your cut at the highest point (which I hope is self explanatory). Two, most peaked roof residential buildings have center hallways that follow the direction of the main gable. After we make the cut and open the attic space, use the 6- or 8-foot roof hook to push down the ceiling below. Depending on the pitch of the roof and depth of the attic, varying length roof hooks may be needed. We should then be able to catch and relieve conditions in that hallway, and assist in venting the entire top floor.

There are probably more styles for making roof cuts than there are brands of wood cutting saws. As with most fire department operations, the style of cut your department chooses to teach should be based on the buildings in your area. Because of this, we discuss a few options based on roof pitch including low pitch, steep pitch, and flat.

Depending on the pitch of the roof, one of the most effective styles of roof cut we have seen is the "hinge" or "louver" cut (fig. 23–17). On low-pitched roofs, this cut seems to be the quickest and easiest method for opening up a roof hole on a residential home. It does, however, make the members move a bit farther out onto the roof itself and may not be ideal if fire has control of the attic space or roof stability is in question.

Fig. 23–17. Louver cut. (Courtesy of Nate Camfiord.)

Place the first horizontal cut parallel to the ridgeline, approximately 4 to 5 feet in length. While doing so, note the position of the roof rafters if possible. We will want to make our subsequent vertical cuts on either side of these rafters to hinge or louver the cut on the rafter itself. Make your second cut vertically down from the ridge toward the ground, starting near your initial horizontal ridgeline cut. Make this second cut approximately 3 feet downward. Repeat this downward cut adjacent to the first, moving along the initial horizontal ridgeline cut. Place each downward vertical cut on either side of the roof rafters to create a hinge. Once the downward vertical cuts are completed, usually four or five, marry all of them together with a final horizontal cut. Be certain to cut an overlap where two cutting lines meet. If done properly, this louver can be opened with your hook by pushing down on the upwind side and pulling up on the downwind side. The benefits of this cut are not just the louver that you created, which limits the amount of pulling needed on the roof material, but also should prevent any debris from falling through on the members who may be operating below.

On steep pitched roofs (fig. 23–18) use extreme caution if attempting to cut the roof. Again, do this from the safety of the ladders as previously mentioned. Do not overextend yourself while making these cuts.

Fig. 23–18. Steep pitch roof

flat roof, go get those vertical openings. Remember to notify the inside team via portable radio prior to breaking skylight glass. After you make the radio transmission, attempt to break out a small piece of glass first, pause momentarily, then take out the remaining skylight. The sound of the breaking glass may allow anyone who missed the radio transmission to get off the stairs before more glass starts raining down.

Fig. 23–19. Flat roof residential building

Fig. 23–20. Skylight over stairs

The Flat Roof Residential Dwelling

Flat roof residential dwellings are obviously very different from the peaked roof variety (fig. 23-19). In instances of fire in flat roofed private dwellings, there is a greater need to place a member on the roof. Flat roofs, in comparison with the pitched styles, are stable platforms from which we can do roof work. Most of the time flat roofed dwellings are constructed with a skylight and/or scuttle cover. Normally the skylight is positioned directly over the top of the main staircase (fig. 23-20). The scuttle cover may be at the stairs, in a closet in the hall, or in a rear bedroom. These vertical ventilation ports are some of the most important ones we can get open quickly.

When we get on a roof, whether flat or pitched, we should always be looking for multiple routes of egress. Before you even step on the roof, you should be visually scouring it looking for a few different ways to get off (rear fire escapes, secondary ladders, etc.). Once on the

Many skylights also have a Plexiglas draft stop located a few feet down from the skylight, usually the same depth as the cockloft, that must be removed to facilitate ventilation. Any tarred-over skylights or scuttle covers should be cause for a saw to be called to the roof immediately. Once things on the roof are opened up and venting, check the perimeter of the building for victims and locations of fire to report back to the IC and the inside crews.

Cutting the Flat Roof

Cutting on the flat roofed residential building normally occurs for fires on the top floor. It is done based on the premise that doing so will help relieve smoke and heat pressures that are pushing down on the crews operating below. Again, cutting the roof itself should not be initiated until all natural openings are used and the victim perimeter search is completed. As there is no high point on a flat roof, cutting should commence directly over the fire area. If you are able to discern the fire location from the building or room layout, try to make your cut catch the fire room and the hallway or adjacent room, if possible. Make your cut a "7-9-8," which is an approximately 4×6-foot "coffin-style" cut. This roof hole can be further extended if necessary (fig. 23–21).

Fig. 23–21. Cutting the flat roof residential building. (Courtesy of Traditions Training, LLC, www.traditionstraining.com.)

The Backstep

Modern Residential Ventilation Practices

While the focus of this chapter relates to the effects that reduced staffing has on prioritizing ventilation in horizontal vs. vertical methods, there are a few recent studies that warrant discussion and review by all members. Ventilation is an often unpredictable fireground operation. Take the time to review the studies outlined in chapter 13. Find out why we ventilate the way we do. Take time to become better versed in the dynamic processes that come with various forms of ventilation. As you will come to find out, most fires in today's dwellings are oxygen regulated. Ventilation not only relieves smoke, heat, and the products of combustion, but also brings in fresh air.

Ventilation is so much more than just cutting holes in roofs. With many tasks to accomplish and crew sizes diminishing, fire departments need to take a hard look at where members are best placed for the initial fire attack. The early placement of members in positions to vertically ventilate peaked roof structures is often not the most prudent use of resources. Utilization of horizontal ventilation techniques dovetails nicely with the concepts found in the two-team ladder company outlined in chapter 20.

Horizontal ventilation minimizes risks while maximizing rewards. Portable ladder placement for firefighter access and egress, coupled with the potential to locate victims when VEIS is applicable, are great advantages. If you are ordered to the roof, be aware of your surroundings. Utilize whatever safety advantages are available (full PPE, TL bucket, roof ladder, guide, etc.) and make your cuts. Open that roof, drop that ceiling, and get off that peak! You are far more valuable to the rest of your crew coming down and getting other things accomplished rather than standing and admiring your cut.

24 Probability in Search: VEIS

We are in the lifesaving business. There is perhaps no greater reward found in our profession, no greater level of satisfaction that arises, than that we get from saving a human life. Ask any firefighter to describe how he or she felt after taking part in a lifesaving event. It is the epitome of what we do as firefighters and fire officers.

As firefighters we are well trained to save lives in many ways. We save lives with our emergency medical first responder skills (EMS), our vehicle rescue extrication techniques, swift-water skills, and so on. Across the world, firefighters save countless lives in all facets of fire service response disciplines.

While many of our "saves" come from first responder EMS incidents, many others cause firefighters and fire officers to put themselves into dangerous situations. At fires the dangers are everywhere. Structural firefighting puts our members in IDLH environments every time. What clues and cues can we employ to find fire victims and increase their chances for survival while minimizing our own risk? In this chapter we discuss how you can utilize a search technique known as vent, enter, isolate, and search (VEIS) to quickly investigate highly probable victim areas and limit our operational exposure.

Firefighters are lauded by the public as heroes (fig. 24–1). Politicians and civilian groups give speeches at events touting our bravery and courage, recounting our actions as if they were there. The reality, of course, is that the only group who *really* knows what we truly do is our brothers and sisters on the job. We know the inherently dangerous risks that come with firefighting. "Firefighters—Saving Lives" isn't just a slogan for a T-shirt, nor should it be taken for granted as routine by any stretch of the imagination.

With the multitude of caveats that the residential fire throws at us, it can seem to stack the deck against us. Locating a fire victim during search isn't an easy task. It is, however, what we should have been ultimately trained to do. A successful civilian rescue takes teamwork from engine, truck, rescue, and command personnel. It requires dedication, perseverance, and sometimes, a little bit of luck. It can seem like we are playing Russian roulette with six bullets in a six-shot revolver, but occasionally there is a misfire and we "make a grab."

Fig. 24–1. Meritorious acts noted at a medal day ceremony. (Courtesy of Michael Stothers.)

Let's talk for a minute about search probability. What exactly does this concept mean? Focus on the probability end of that last statement for a moment. What does probability have to do with firefighting? Probability is a mathematical branch that deals largely with the recording of uncertain events. Can we draw a parallel, in that searching for fire victims is largely an uncertain event as well? Unless we receive specific information during the initial response or when arriving on the scene, search is an absolute uncertainty. Occasionally we are alerted to a particular area to search for trapped occupants in the home like, "My mom is in that room." Most often, the reports that we routinely receive of people trapped are rather vague and uncertain in direction. How can we locate these victims more quickly? Where should we begin?

What can we do as firefighters to better the odds? How can we make educated fireground decisions to increase our probability in finding civilian victims in the residential building fire? In this chapter we will talk about a specific tactical skill that may just increase our odds of finding civilians more quickly. Time saved in our line of work can equate to lives saved.

In chapter 25 we also speak about the primary search and how it should begin at the greatest area of danger, the immediate fire area, and work outward from that point. We also discuss in chapter 20 how the two-team concept of inside and outside ladder company teams is best suited to attack the fire, allowing the inside team to locate, confine, and control the fire while the outside team throws portable ladders, ventilates when ordered, and conducts applicable VEIS operations.

While the ladder company outside team is in the best position to initially VEIS the residential building fire, they certainly should not be the only one capable of doing so (fig. 24–2). All members of the department should be trained to understand the concept, not just recite what the acronym stands for. So many times while instructing we hear from our students, "We VEIS at every fire we go to. We vent, we enter, and we search, don't we?"

Fig. 24–2. Instructor drilling on the concepts of VEIS. (Courtesy of Traditions Training, LLC, www.traditionstraining.com.)

That is not the case. VEIS is not applicable in all situations or for every residential building fire. VEIS is a specific tactical operation. When done correctly it can pinpoint target-rich areas for victims to increase their chances for survivability and decrease our time to get to them. The concept of vent, enter, isolate, and search (VEIS) is not a new operational technique. It has been taught around the country for years and has yielded several civilian rescues on multiple occasions. It is a technique that has certain applications for search success at the residential building fire. VEIS, in a single statement, is a targeted search that is based on probabilities.

While we dissect the steps to this operational technique in its entirety throughout this chapter, we need to understand the core concepts behind the actions. It is not applicable to every residential fire or every room in the home. Nor is it without other safety parameters and limitations. VEIS actions at residential building fires are normally conducted when the selection is made of a room in the home to be searched based on the probability of civilian victims being present in a particular area. It is a quick yet thorough search of a particular room.

VEIS Tactical Skill Review

Safety

We must first start with a discussion about our safety. There are critical safety concerns when conducting a VEIS operation. First, the IC and team leader of the company performing the operation must be advised via radio when and where the activity is to be performed and also when it has been completed (fig. 24–3). Members will be entering an IDLH environment, many times above the reported fire area. Team members must be paired into a team of two prior to entering the structure from the outside. Further, members must fully understand the concept that this is a room-targeted probability search and not deviate from it.

Some fire departments shy away from the concept of VEIS, stating that it places their members in a position that is overly dangerous. Think about the concepts involving interior search, are they not a great deal more dangerous? By properly training, practicing, and implementing a residential building fire VEIS maneuver, we would argue that members would likely be more protected, better accounted for, and more able to conduct a safer search. Think about the interior residential primary search. Instead of traversing the home from room to room, largely into the unknown, members are conducting focused searches of specific areas with a team of two notifying command before and after search.

Fig. 24–3. Member notifies command prior to commencing VEIS operations.

Positioning

The first component of a successful VEIS operation happens well before the first letter of the acronym "V" for ventilate. How can that be possible? Well, as in many of our fire service acronyms, we shorten the steps or leave out parts of the whole evolution in our quest to make something easier to remember. One of the critical components in initially conducting a VEIS operation is found within the building construction arena. Building construction? How is that, you ask?

When we think about probabilities, we are basing them on the rooms people are likely to be (fig. 24–4). We need to know, based on the buildings in our first-due areas, the likely layouts of each style of home. This will serve to guide us in our quest to increase our odds. We should be pros at knowing the dwelling layouts to which we respond. Our knowledge of the buildings combined with the time of day the alarm is received will further narrow the target. For example, when responding to fires at night, most occupants will be sleeping. Where are the likely sleeping quarters for the dwelling? Another example may be an open window remote from the fire in the dead of winter. This may indicate an area where a civilian was trying to get help or exit prior to our arrival.

Fig. 24–4. Knowing your buildings can increase your probability of finding victims.

Once we have decided on a geographic area to access, the room is entered from the exterior of the building, providing a quick and simple entry point. This is normally conducted from the exterior with a portable ladder. Take a moment to review chapter 22 to guide you in the selection of the correct size, carrying form, and placement of the portable ladder. All portable ladders placed against the fire building should be left at the level of the windowsill (fig. 24–5). This allows the member the easiest access and egress to and from the window. Again, refer back to your home's layout, as one-story ranch style homes can be excellent candidates for VEIS as well, perhaps without even requiring a ladder.

Fig. 24–5. Portable ladders are best placed at the level of the sill.

When working in your team of two, to best maximize time we must have excellent communication between the members. One member of a well-practiced team should be able to anticipate the moves of the other. In most instances, one well-trained member is all that is required to throw and set the ladder into the building (see chapter 22).

While we will operate in a team of two, we do not need both members in the room to search. The firefighter who sets up the ladder becomes a safety beacon for the member who conducts the room search. If the team has a thermal imager, the outside member should have it, not the searching firefighter. Once the ladder is set and the searching member is ascending, the second firefighter should be ready to follow the first to the top of the ladder once the first member is inside.

As one firefighter is setting up the ladder, the second firefighter should fully don SCBA with the regulator left initially unattached, grab the hand tools, and be ready to immediately scale the ladder upon placement. At the top of the ladder, the member should click in the regulator and be ready themselves to take the window.

Vent

The ventilation in a VEIS operation should not be construed to mean the same thing as our typical fire department ventilation performed to remove smoke and heat from the dwelling (fig. 24–6). While it is true that in conducting a VEIS we will be entering through a window and therefore creating a ventilation opening, we don't want the impact of such an activity to affect the fire. The reality is that we aren't trying to create a flow path for the fire to exhaust byproducts of combustion. Actually, we are not looking to ventilate anything other than the conditions that may be found in that particular room. Here is where the science of fire behavior discussed in chapter 13 can and will be a factor. A quick review tells us that every ventilation opening that we create in the building will have some effect on the fire conditions.

How does that relate to VEIS? Controlled and coordinated ventilation is a must at residential fires. Venting prior to VEIS is no different. If we receive permission to VEIS, take the window with the tip of the ladder, reposition it at the sill, then get masked up and ascend. Let's say this takes 60 seconds in total. A quick review of several UL and NIST ventilation videos shows how much 60 seconds of a positive flow path can change fire conditions. When we are positioned at the window, prior to breaking it, we must be cognizant that there is a chance that conditions may rapidly change. We want to interrupt and stop this potential flow path as quickly as possible.

you need to exit the room in haste, you do not want any obstacles in your path of egress. Turn the window into a door.

Before committing to entering the room, take a look at what you see. Is there smoke? If there is good fire and no smoke shows initially, chances are that the door to the room is in the closed position. This is good. If the smoke condition is turbid, thick, and pressurized, this should be an indication that the door is open and your initial vent opening has created that exit flow-path we discussed earlier. Attempt to landmark the room with any visual acuity that the initial vent opening produced.

From your position at the top of the ladder, take the hook and place it into the room with the hook end resting on the sill. This will become one of your landmarks for your return to the window. Use the Halligan tool to first sweep the floor under the window. Many victims are found near exit points, such as under windows and behind doors. Once the floor is swept, use the tool to sound the floor by slamming the tool into it to check for the floor's presence and initial stability.

Firefighters need to be proficient and practiced in entering windows from portable ladders. Countless times in our training we have seen firefighters dive head first into the room from the windowsill or awkwardly back themselves off the ladder into the room. "Controlling the sill" is a technique we teach our firefighters to use when entering the room. The first step is to climb to the top of the ladder. Keeping your head on the outside of the building, put one shoulder into the side of the window jamb, place the other foot inside, and sit onto the sill itself. At this point we are straddling the windowsill, one foot in and one foot out on the ladder (fig. 24–7). If we need to get out at this point we can quickly do so. To get into the building from this position, sit back and dip your head into the building. Place your opposite shoulder behind the window jamb and slide in and down into the room with the outside leg (that was on the ladder) following you (fig. 24–8). Be aware that sill heights vary from home to home.

Once the initial firefighter makes a controlled entry into the room, the second firefighter (who initially placed the ladder) climbs up to the tip to act as a beacon for the searching firefighter. The second firefighter should use the thermal imaging camera to guide the searching firefighter.

Fig. 24–6. Member using 6-foot metal Halligan hook to clear window

Enter

We recommend a six-foot metal hook and Halligan bar as the hand tools of choice for VEIS operations. When fully encapsulated in your PPE (facepiece and all), use the 6-foot hook's reach to strike downward toward the center of the window sash. Trim out the remaining glass shards as much as possible. When taking the window, from either the portable ladder or the ground, clear out the entire window, sash and all. Remove any obstructions like curtains or blinds that may hang you up. If

Fig. 24–7. Member controls the sill prior to entry

Fig. 24–8. Member controls entry into the room from the sill

Isolate and search

We have successfully notified command and placed our ladder, cleared the window, and entered the room. Here is where our isolation and search begins. Initially, what are we searching for? We are searching to find the door from the room to the rest of the house. If we encounter a victim while searching for the door, that's great, but we cannot just stop there. We must get to the door to isolate our area and stop the potential flow path. We should issue a radio transmission notifying command that we found a civilian and request additional assistance if needed, but we still need to find and close that door. If we focus on the victim and don't close the door, more smoke and fire may be headed our way.

Remember, we must stop the flow path and eliminate the exhaust or intake port.

What is the quickest way to find the location of the door? Many homes are constructed in a manner which places the room door in line with the placement of the window. This allows for air to move through the room (fig. 24-9). Keep that in mind as you search for the door. Also, by this time the second firefighter should be atop the ladder watching the searcher and helping to direct the search. Listen for these instructions, as they should be able to aid you in finding the door that much faster. Perhaps you will note an increased heat signature coming from that area as you scan the room. Once at the door, perform a quick sweep of the public hallway for fire conditions or any victims, then close the door to the room you are searching.

(by closing the door) is a critically important step to curb the impact of ventilation effects on the fire, the VES acronym has never been fully descriptive. We have always taught our firefighters to find the door as the number one priority, then sweep the hall, close the door, and search the room (fig. 24-10). In fact, most fire department acronyms leave several bits of information out because they merely serve as a guide for operations, not a full and complete recall of all the considerations. That said, we feel that if it helps your firefighters recall this particular step, so be it. In light of new scientific modeling, we have put it in there.

Fig. 24-10. Firefighter sweeps the hall and then closes the door.

Fig. 24-9. Many times windows are in line with the door.

There is a school of thought to add a letter "I" to the older VES equation, forming a new expanded acronym VEIS: vent, enter, isolate, and search, which we have used throughout this text. We feel that since isolation

Once we have secured the door in the closed position, we can continue our quick but thorough primary search of the room and the room *only*. We do not use VEIS to search the entire floor, additional rooms, or other spaces. If you are in a bedroom, most have closets. Some even have bathrooms, so be sure to search those areas. Again, the second firefighter on the ladder with the TI can aid you and watch as you quickly move through the room. The second firefighter can scan the room, helping you as you move through by saying things like, "There's a bed. Get behind it, under it. Okay, 2 feet ahead, looks like a closet with sliding doors on your left." As you move through, the other person should be monitoring conditions in the room and making sure that you cover the entire area effectively.

If no victims are located, complete your search and return to the window from which you entered. If you stayed in contact with one wall all the way around the room, you should wind up back where you started. Remember, you will have a fellow firefighter at that window who can make noises to help guide you back (bang the hook on the sill, talk, or yell). Notify command and your team leader when you are out of the building. From here, we prepare to move to the next room based on the probability of finding survivable victims.

One point to note here is that we have left the door closed during this entire search. Once the search is complete, but before leaving the room (depending on radio traffic and status of the operating hoseline), you should check in with the IC to see if they would like the door reopened for ventilation purposes before you leave. This may be likely if fire extinguishment is underway, but that isn't your call. That is for the IC to make. If there is any doubt, leave the door closed and move on.

If you discover civilians while conducting your search, you need to get them out as quickly and efficiently as possible. Victim removal is one important facet of the overall VEIS operation that isn't normally covered in VEIS training. When do we normally stop the VEIS evolution? Well, most of the time it is when the searching firefighter has performed a VEIS of the room, found a victim, and brought the victim to the window. That is where the drill ends. We need to conduct our drills of the VEIS skill set to completion with the victim removed to the ground and all searching firefighters out of the building.

If a victim is quickly located, removal via an interior exit, while the easier and preferred method, is also probably the route most contaminated. Firefighters can get the victim out via the portable ladder, but it requires practice and skill. Many times the most difficult aspect of extricating an unconscious person comes when we have to lift the lifeless victim up onto a windowsill (fig. 24–11). This is another potential use for that 6-foot metal hook we left on the windowsill earlier.

Fig. 24–11. Firefighters operating at a windowsill. (Courtesy of Traditions Training, LLC, www.traditionstraining.com.)

With victims who are larger or otherwise difficult to manage, drape the body onto the shaft of the tool head toward the window. Remember that the hook end is affixed to the sill. Lean the victim's torso to rest on the shaft and place their legs on either side of the shaft. In coordination with the outside member lifting and pulling on the victim's shoulders, the inside firefighter reaches through the crotch, grabbing the tool and lifting it upward. This effectively hinges the victim onto the horizontal plane with the fulcrum being the windowsill, which places the body at the height of the sill. Once at the level of the sill, we can reposition the victim for placement onto the ladder for removal to the ground (fig. 24–12).

from ladder firefighter to searching firefighter until low on air and/or conditions on the interior change to allow different tactics to be used.

Fig. 24–12. Victim lifted to sill with the use of a 6-foot hook

Leapfrog rooms?

VEIS can be done over and over, room by room, throughout the structure until the fire is extinguished to allow safe passage for members operating in the interior. In the team of two we teach a "leapfrog" method to save time in moving from room to room. It takes practice to make it efficient, but so does every action we perform.

Here is an example of a team of efficient firefighters performing VEIS around the second floor of a residential structure with a fire that has initially compromised the stairway leading to the second floor. The members have completed an initial VEIS operation on floor two's Alpha/Bravo corner. As the first searching firefighter (called FF1 in this case) returns from the VEIS room search, the ladder firefighter (FF2) climbs down the ladder. FF2 makes the notifications to command of the completion of the initial VEIS and asks permission to VEIS floor two's Bravo/Charlie corner, which is the next room based on probability. FF1 comes out the window onto the ladder and carefully drops tools to the ground. FF2 retrieves as FF1 comes down the ladder. FF1, now ready, moves the ladder by rolling it into position and foots it for FF2 to become the searching firefighter. FF1 takes the TI from FF2 and becomes the ladder firefighter. For all intents and purposes, they leapfrog positions

The Backstep

VEIS—Who, What, Where, When, and Why

While in the context of this book we consistently attempt to help you explain the "why" of what we do, VEIS deserves a bit more attention as we see it as a lifesaving firefighting tactic that is often misapplied, misused, and misunderstood.

Who should conduct VEIS?

Any firefighter, when paired with a partner trained in VEIS techniques, can conduct the maneuver at fires. Initially, the best suited members at residential fire situations will be those from the first and second to arrive ladder companies, specifically their outside teams. Each team member should bring a portable ladder and a compliment of hand tools when disembarking the apparatus. While carrying portable ladders toward the fire building, members should be mindful of their placement against the home. Ladder placement should be in areas of the greatest danger to our operating members and to areas of high probability for locating civilian victims.

Outside team members conducting VEIS will be operating remotely from their officer and must be equipped with a portable radio. Firefighters must notify the IC and the ladder company's inside team leader as to where and when they are headed to perform VEIS maneuvers. Here is an example of such a transmission.

> "Ladder 22 OV to Ladder 22. I am in position to VEIS the 2nd floor on the Alpha/Bravo corner."
>
> "13 OV to Command. Did you copy my last transmission?"

What do we VEIS?

We shall VEIS particular rooms of the fire building based on the probability that civilian victims will be located within them. Knowing the interior layout from your routine responses helps you master your knowledge of residential structures. This increases the chances of picking the right rooms based on time of day and other factors. As stated earlier, bedrooms, especially at night, are often the initial targets for VEIS maneuvers.

While historically we think of VEIS as a technique conducted from portable ladders to upper floors, VEIS can even be performed on first-floor fires. We can VEIS any probable areas in the home regardless of height. The key to VEIS is that while we would have covered all the rooms of the home eventually, taking an inside (typical search operations) and outside approach (VEIS), we can get to our highest probability rooms faster and more efficiently. In VEIS, we normally enter the home by what we call non-traditional means, like going through windows.

Where to VEIS?

While slightly repetitive, teams need to locate the best place to work. While we have discussed working from a portable ladder in great detail, we certainly should not overlook using the building's built-in characteristics to our advantage to assist us in our VEIS options. A porch roof or rear building setback makes a great platform for us to use for access in VEIS maneuvers (fig. 24–13). Take, for instance, a good fire on the first floor and a VEIS operation being conducted on the second floor. Observe how one ladder to this porch roof allows the entire front of the building to be a candidate for VEIS.

Fig. 24–13. A house with a front porch that can be used for VEIS

Again, knowing your buildings is the key. Many center hall colonials have windows in the center that are open to the foyer below, so be sure to sweep and check the floor prior to entry and control the sill before you make your final move to enter.

When to VEIS?

We entertain the notion to VEIS based on probabilities that persons are in the home and, specifically, in the rooms that we are choosing to VEIS. We do so as a team of two radio-equipped firefighters. We VEIS when the fastest and most efficient manner to reach said persons would be from the outside, a direct route into that room. We do so to perform an immediate targeted search of that area. We only VEIS this specific area after notifying and receiving permission from the incident commander.

Almost as important as knowing when a room or a fire building would be an excellent candidate for VEIS is knowing when it isn't. When would we not VEIS a room or a specific area of the home? What would constitute a no-go to VEIS? We do not conduct a VEIS operation into a room that is engulfed in fire. Our turnout gear is not a proximity suit. We should not conduct a VEIS maneuver into a room that is in a rollover state or headed in that direction. We must carefully weigh the smoke and fire conditions that appear at the window, and in the room once we take it. We must let it blow for a second or two, see what is happening, then make our next move to close that door.

Why do we VEIS?

When residents call 9-1-1 and summon our response, they have effectively given up. They have given up control over the situation before them. They call for us knowing full well that property, possessions, and possibly their own lives are now no longer in their hands. With our response, they have invited us to provide influence, control, and order to what is occurring.

It is up to responding firefighters to search for and find our citizens. We must locate them and remove them as quickly as we can. How can we initially narrow the focus of our primary searches to quickly assess areas based on the probability that victims will be located there? How can we use a two-pronged attack—interior and exterior—to reach them in time? How can we do this as safely as possible, reducing the possibility of members freelancing? Create a department SOP that delineates the actions required and the specific tools necessary to complete the task. Prepare, practice, and anticipate your next VEIS evolution. Train and explain the why behind VEIS. Make it part of your combat-readiness mantra. If you do, VEIS will save lives in your community.

25 We Save Lives

It's 11:30 p.m. and you sit in the house watch. Overall it was busy day, and you have settled into the semi-comfortable yet half broken, stuffingless recliner. Just as you are glancing up to catch the end of your local news at 11, the firehouse alarm printer spits to life. "Engine, ladder, and chief go," you announce over the PA. "Everyone goes first due, for the house on fire." It's a job, and it's in the basement.

When you arrive there is a civilian, smoke-stained and soot-covered, standing on the lawn (fig. 25–1). She tells the chief that she was alone in the house and got herself out of the basement. Thankfully, before completely overcome by the products of combustion, she was somehow awakened and escaped, dazed but largely unharmed. The chief relays to all units that the house is reported to be unoccupied according the occupant of the dwelling.

Fig. 25–1. Civilian talking to firefighters outside the dwelling

The home is a semi-attached (duplex type) 20×30 two-story, peaked roof, single-family dwelling. Engine and ladder companies get off the rigs and have begun to operate from their SOGs for basement fires. The incident commander isn't quite as eager as the members masking up. He can see that this fire has a good hold on the basement, and it's starting to poke through into the first floor.

The engine brings the charged line into the first floor to protect the interior operating units and find the basement stairs. The ladder company officer and engine boss quickly realize that they are not going to be making it down the stairs because there is too much fire. They decide that they are going to hold the fire in check here, hold the stairs, and get a quick search completed. They communicate these facts to the IC and he concurs. The IC alerts the second-due engine company to coordinate fire attack at the basement level with the first engine now at the top of the stairs. There is an outside entrance. The second engine will team up with the second to arrive truck to make entry, search, and bring the line into the basement from the outside.

In the short time between these orders being delivered by the IC and the crews carrying them out, this fire continues to grow. Watching the fire conditions, the chief is starting to have second thoughts about the status of the companies operating on the first and second floors. This fire needs to be knocked down, and fast. These companies are literally sitting above the fire on the first floor's beams like steaks on a barbecue grill. For a moment, he contemplates backing them all out of the house until the fire is knocked down. It is empty, right?

After finding the stairs and coming to the rationale that they were not going to make the basement, the ladder company did a quick primary search of the first floor. Hearing the second due truck assigned to the basement, they knew that they were required to search all the floors above.

As conditions continue to deteriorate, the ladder company inside team pushes on and up to search the second floor. What is this? It can't be. The chief said everyone was out! Horrified yet amazed, the ladder company gives the radio report, "Chief, we got a kid on the second floor, bringing him out now."

Both the chief and the civilian on the lawn are in shock, the woman in hysterics. An astonishing discovery: a semi-conscious seven-year-old child found in a bedroom on the second floor (fig. 25–2). The chief turns to the lady in the yard, and yes, she is the child's mother. She is now in full-on hysterics, and rightfully so. How is it that she said she was alone? Did she purposely deceive us?

Fig. 25–2. Firefighters remove fire victim. (Courtesy of Kentland Volunteer Fire Department, www.kentland33.com.)

Like many households, the relationships within this one were complicated. The woman was divorced, having joint custody of this child with his father. The boy was supposed to be with the father until the following morning. The dad figured that he would just drop him off late the night before instead of early the next morning. Due to the time of night, he apparently took the boy straight to his bed since he had keys and had let himself into the home. He thought he was doing the right thing by not waking the sleeping mother, who was in the basement. She never knew. She found out her son was in the home when the members carried him out, unconscious.

The story you just read is not fiction, it's fact. Though it reads like a script, the reality is that a life was saved that night. Thankfully this child survived and made a full recovery from the smoke inhalation he suffered after a short hospital stay. How many stories can be told that read in a similar fashion? What if the ladder company didn't do that search? Would the story have the same outcome? I think not. Can you imagine living with the idea that you didn't do a search because someone told us everyone was out of the house? Should we listen to that someone?

There has been great debate in the fire service recently regarding if and when to search dwellings for occupants. I can tell you this, if my house was on fire I would expect the fire department to search it. I can assure you that our citizens expect the same. I would expect that my neighborhood firefighters are physically, mentally, and tactically prepared to do their jobs. The firehouse may have a clubhouse feel with jocularity among the members between runs, but when that bell goes off and we are heading to a fire, you had better be able to perform. Your search decisions are life and death ones. Those for whom you are responding are the citizens who entrust that we will ultimately save them and remove them from harm.

You had better be practiced and prepared to search the house. That's what you signed up for. If you are not ready, two things should happen. Make yourself and your team better prepared and practiced, or move along and find a new vocation where lives are not at stake. On our job, seconds count. There are no do-overs or second takes. If things go wrong, people *will* die. Sometimes it's just that simple.

Misinformation

We have all received erroneous information on the fireground from bystanders and do-gooders. In fact, any and all information that we garner from onlookers and homeowners alike needs to be verified by our fire department members. I'll relate a quick story here.

We had a run for an odor of smoke in a private dwelling. We arrived at the address and the reported building owner met us in the street saying, "Everything is fine, no problems, I was just doing some work, you know, renovations on the first floor."

The temptation may be to say, "Okay, you're the building owner, you say everything is fine, it's fine," and walk back to the rig. Our philosophy should always be, "That's great, but let's go see. Show us exactly where you were and what you were doing." We did just that. We discovered that the owner was removing glued-down floor tiles with a roofer's torch. This was on the first floor, inside the building!

This was a two-family private dwelling with occupants in an apartment above. They were the ones who called the fire department, but that was not specified on the response ticket. The owner lived on the first floor and had storage in the basement. On that second floor, aside from being half banked down with the smoke condition, there were carbon monoxide (CO) readings of nearly 80 ppm (parts per million)!

Further inspection of the first floor revealed that fire had gotten into the walls at two different places where the torch was used to remove the tiles that abutted the wall. A precautionary line was stretched and the ladder company quickly opened up the wall area and knocked down the fire with the water can. Imagine what might have been the outcome if we had walked away without further investigation. Always verify the information that you receive from outside sources, including other public service agencies such as the police.

I love the police. I have many family members and friends who are police officers. I certainly do not want to do their jobs in that while fire can be unpredictable sometimes, humans are unpredictable *most* of the time. I joke with most of my police officer friends that it seems many of them like to "play" firefighter sometimes. They seem to gravitate to fire hydrants. If you can't find a plug, just look for where their police car is parked. However, they will rarely admit it (fig. 25–3). It's not that common the other way around. We've met very few firefighters who "buff" police incidents. We don't normally take the fire trucks over to hang out at the police station. We trust and respect each other, but operationally it is only to a point, right? While we both serve the public, we seem to take different paths to get there. That is because, in the end, we have different jobs to do.

Fig. 25–3. Police car on hydrant. (Courtesy of Firehouse Pride, www.firehousepride.com.)

That being said, could you ever imagine a time that you would ask a police officer to give you his gun? Try it. Tell him or her that you just want to hold it for a little while. It's not going to happen, and I should know, I have tried. The level of trust and responsibility between that officer and you is just not there. How is it, then, that we can be so trusting of the police and other agencies with human life when it comes to fires?

For the police to advance a hoseline or exclaim beyond a reasonable doubt that no one is inside a home is beyond the scope of their training. Similarly, you holding an officer's pistol is beyond the scope of yours. We *must* determine the presence or absence of human life in the residential building.

Search Realities

When confronted by the city council, family members, friends, and neighbors after a tragic fire death and they ask you why you didn't search that house, can you in good conscience state, "The police on the scene said no one was in there." No other public agency is going to take responsibility and accountability for a missing civilian left to die in a fire except the fire department. That is exclusively our job. This is where we make educated decisions and yes, occasional sacrifices. The report *U.S. Fire Administration Firefighter Fatalities in the United States in 2010*, released September 2011, notes 3 of the 87 line-of-duty deaths in 2010 were noted as "search and rescue." We must check our egos and stand firm behind our vast training and knowledge base. We *will* determine that there is no one inside.

Yes, we have committed to search for someone that was reported trapped only to find out the person had self-evacuated prior to our arrival. Yes, we have searched for people who were never in the home in the first place. Have we discovered the opposite? Have we found persons who were inside when all were reported out? Have we found persons in vacant buildings when they were reported as vacant? The answer is, of course, yes. This is the most critical part of our job, saving lives. We are in the life saving business! Buildings are reported empty and the searches for civilians are complete and negative when the fire department's units say so!

Need to Delay

This is not to say that searches must be conducted immediately in all dwellings at all times, regardless of conditions found (fig. 25-4). We must recognize the limitations of our equipment. We must understand and maintain the situations in which we find ourselves through education, training, and experience. We must utilize good judgment. We must master the aggressive mentality that is explored in chapter 6. We cannot rush foolhardily into our search just for the sake of searching, but rather we should utilize our fire senses to continually evaluate the situation ahead as we move through to search the dwelling.

Fig. 25–4. Fire at the front door. (Courtesy of Kentland Volunteer Fire Department, www.kentland33.com.)

"Ladder 191 to command, primary searches on the fire floor delayed due to heavy body of fire. Chief, we got to wait till the engine knocks it down to get into the back bedrooms, okay?" Have you ever had to make a transmission like this? Has there been a time when a residential search was delayed? Of course there has. There are many times when fireground situations may warrant a delay in the start and/or completion of a search.

Our PPE, while getting better and better, is not a proximity suit. Because of this, we cannot search rooms that are currently full of fire. We are not provided with any encapsulating body armor that can protect us from imminent or occurring collapses. Fire conditions, structural stability, and Collyer's mansion–like clutter conditions, are some indications that may be cause for delay, just to name a few. This doesn't mean we aren't going to search, only that the search may take a little longer than normal.

Search Generalities

When searching for life in the residential dwelling, we need to do so based on the primary and secondary search. Let's quickly review these two search avenues as they apply to the residential structure:

Primary search: A rapid search for life before and during actions to bring the fire under control. It is a quick once-over of the entire accessible area, with an emphasis on checking the most likely locations victims may be found. The primary search is conducted with a delicate balance between fast and thorough.

Secondary search: A painfully thorough search to ensure that there is no possibility that a fire victim remains in the building. A company that did not conduct the initial primary search should carry out the secondary search, if possible.

Where Are They

First and foremost, when we search residential dwellings for civilian victims, we should be keenly aware of likely locations that they are found. Although it may seem like a rather basic statement, it warrants discussion since our quest in finding victims is conducted quickly yet thoroughly. Again, knowledge of residential building layouts, particularly those found in your response area, greatly enhances your ability to quickly move through the dwelling while searching for victims.

Let's talk a moment about behavior traits that we have seen as it relates to escaping from fire. If awakened, adults do not normally lay down to die when faced with fire. They normally attempt to flee their home using the routine exit paths, ones with which they are most familiar. Children under duress can panic and, in fact, may retreat into hiding. They normally do not attempt to exit through routes that put them in danger (smoke and heat). They may try to shield themselves from the fire's byproducts in the space in which they find themselves. These factors should be incorporated into your search training.

In either case, adults and children paralyzed by smoke do not normally make it very far in the structure. An overcome, unconscious, or semi-conscious state renders the victim unable to move, dropping them to the lowest point within their surroundings (the floor, on a bed, couch, etc.).

Adult fire victims can be generally located along the exit paths from the home. While attempting to flee, many may be overcome by the products of combustion. Key locations along exit routes to check for adult victims include behind doors. Be sure to check behind every door you go through or open. Doors have a sneaky way of hiding things. They can cover up an unconscious person behind them, and when opened fully they may block an additional room or hallway to points further in the dwelling. If the door seems stuck or semi-obstructed, the cause may well be a victim behind it.

Many times adult victims are found in hallways, overcome while moving in the path of their escape. This is one of the reasons why we conduct our firefighting operations from the front door, quickly checking and then protecting the main path of egress from the home.

Another likely position to find victims is directly under or near windows of the home. If you are entering a room from a portable ladder or porch roof, sweep under the window with your tool or hand before you sound the floor. Check for a victim before you check the stability. You may have an unconscious victim located there.

Nothing gets a firefighter's adrenaline coursing faster than a radio report of children trapped by fire. Children are not supposed to die. Parents are never supposed to bury their children. That is not the natural progression. Nothing is as precious as the innocence of a child. While all fire deaths are traumatic for operating members, the death of a child can trigger particularly difficult emotional responses in firefighters. If you have ever experienced the death of a child on the job, it is not easy to forget. Be certain to utilize the support network that your firehouse and your department provide for both peer and professional counseling and critical incident stress debriefing and management (CISD, CISM).

Kids are kids. They only know what we teach them when it comes to fires and fire safety. If you have children, take the time to remind them of your family plan for emergencies in the home. If you are at the firehouse and have a school class visit, be certain to stress to the children the importance of not hiding in fires to the kids. Take every opportunity to teach family and friends about the importance of fire safety and having a fire safety escape plan from their homes.

Children are expert hiders (fig. 25–5). They play games like hide-and-go-seek, trying to out-hide other kids and adults alike. They can get their bodies into positions and places that adults couldn't dream of, and this isn't necessarily good for searching firefighters. That said, closets are common hiding places for any child. Any time my kids and I play hide-and-seek, invariably they are hiding in the depths of a closet. Unfortunately, during fires children have the same affinity for hiding in them. Be certain to search closets with great care.

Fig. 25–5. Child hiding in a closet

While it's not an easy task to go through anyone's closet when you can see perfectly clearly, it's even more difficult in a smoke condition. If indications are that children are missing, this is a place you *absolutely* want to check thoroughly. Most of the time in the majority of residential homes, if you come across a bed you are more than likely in a bedroom. This should cause you to be prepared to find a closet and be diligent searching it. Most, but not all bedrooms in private dwellings will have some sort of closet, so it is something we must be sure to look for.

According to the 2006 U.S. Fire Administration/ National Fire Data Center study titled *Fire in the Older Adult* (FA-300) published by the U.S. Department of Homeland Security, based on 2002 NFIRS data nearly 40% of all adult residential fire deaths occurred while the victims were sleeping. In addition, the majority of fire victims who succumbed to fire were located in bedrooms (53.3%).

Where in the bedrooms are victims located? The easy answer is on the bed, right? Well, that is true, but adult and child victims can be found in, on, and under the bed. What is the first thing a searching firefighters should do when they discover a mattress? The answer is to stand up. Why? You need to be certain to check for the presence of bunk beds (fig. 25–6). Quickly stand up and raise your arm over your head to see if there is framing for an additional mattress.

Fig. 25–6. Bunk beds

An upper bunk is often overlooked, as there is tendency is to just get on the mattress and search the first one you find. Bunk beds are becoming more and more common in children's rooms. Sometimes they may not even have a lower bunk. The lower bunk may be removed and furniture or a desk placed in its spot. If you think you are in a bedroom and cannot find a bed, start thinking bunk bed. Conversely, if you come across a seemingly random post (usually 4×4 or similar) in the middle of a small room, it may have a bunk above it.

While you may find a victim in the middle of the bed, chances are it's not going to be that easy. In all bedding configurations, be certain to give particular attention to the space between the mattress and the wall, if so positioned. This nook seems to be a very cozy place in which to find someone nestled (fig. 25–7). It may be easily overlooked if you don't get up on the mattress and search it specifically. Our search technique for beds is to get on the bed, not just sweep your arm over the top of it. The primary quick search technique is to feel the mattress, stand up and check for a bunk bed, get up on the mattress, check all voids and nooks where the bed meets the wall, and sweep under the bed when the top search is completed.

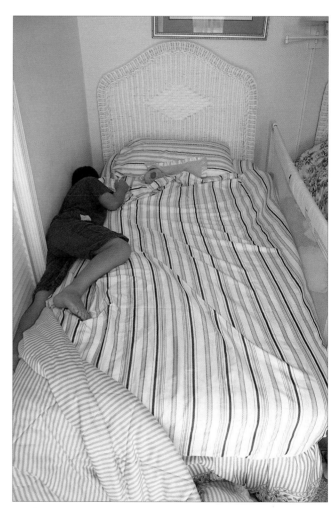

Fig. 25–7. The space where the wall meets the bed is where many fire victims are found.

Bathrooms are other places where trapped persons tend to gravitate when cut off from escape. As stated in the aforementioned U.S. Fire Administration study, the second most common place to find fire casualties is the bathroom, accounting for 10% of victims. One thing is certain: no matter how big or how small a residential home may be, every house has at least one bathroom. No primary search is complete until the bathroom is located and searched. Be sure to check the tub! During a fire, something about the human mind can cause a person to gravitate toward bathrooms because of the presence of water, and they may specifically be found in the tub.

Firefighter Search Training

How are your firefighters taught to search from their indoctrination into the fire department? Searching in fire and smoke conditions are not natural actions. On the training ground we spend hours and hours throwing ladders and pulling hoselines. We rack and re-rack thousands of feet of hose, show forcible entry techniques, and practice saw work on roof props, but how much time is dedicated to learning about the search? If life safety is supposed to be our number one priority, why isn't there greater focus on search in most training academies? If fireground decisions on where, when, and how to search are so difficult, why do we not address them in greater detail here?

Like fine wine, individual search techniques get better with time. They evolve over one's career, as experience and increased comfort tends to aid efficiency. However, can we wait for experience? Should we? Increasing our member's comfort level in conducting searches can be accelerated through realistic training. Realistic training includes practicing a task as we will perform it on the fireground. Build your rooms as close to a real home as you can. Request donations for used furniture as training props. We owe it to the young firefighter to make it as real as possible. Searching empty rooms with four concrete walls isn't the best way we can accomplish search training.

Once we have created a "real world environment" to teach our search skills, then comes repetition. Repetition builds consistency. Consistency can build in a level of comfort. Consistency and comfort lead to mastery. While we don't expect all new firefighters to have mastered search skills, we certainly want them to be comfortable with them and have consistency in the task. While on-the-job training and practice at real fires significantly reduces fatalities in most areas of the world, realistic repetitive training is really the only tool we have left. Having a firefighting team that can search is vital to what we do. More civilians die in residential buildings than any other! We are talking about the lives and safety of our constituents.

When conducting these repetitive and realistic search drills, especially with novice firefighters, there is no need for trickery. We are attempting to hone skills and create good search habits. Initially, place mannequins in areas of your training facility where they are likely to be found in real homes. Put a child in the closet. Wedge an adult between the bed and the wall (fig. 25–8).

Fig. 25–8. Conduct realistic search drills. (Courtesy of Traditions Training, LLC, www.traditionstraining.com.)

We've been in training scenarios where instructors have literally lifted up large three-cushion couches and then placed them on top of the victim mannequin, completely covering them. Do you know what that teaches firefighters when you then lambast them later for missing a victim? Flip over and toss around every movable item they come across! The way you work the junior firefighters on the training ground is the way they will work on the fire floor. Create good search habits for them now and break bad ones that you notice early.

Bad habits? Can we introduce bad search habits in the training academy? There is a tendency to initiate bad search techniques in the start of a firefighter's career, so we need to find a way to re-evaluate the way we are teaching firefighters to search. We must start teaching and refining search techniques with our new firefighters early on, even before they leave the academy.

While many new firefighters will understandably have issues with confidence in searching at the start, we must begin to promote confidence early in their training. Why is it that we still have new firefighters hold the boot of the firefighter in front of them while they search? Then, the next firefighter down the line holds their boot, and so on down the line. Have you ever watched this? It's comical, yet disturbing to see. The first firefighter in the line literally acts like the lead elephant in a circus, with each firefighter's elephant trunk (hand) holding the firefighter's elephant tail (boot) of the one in front (fig. 25–9)! The first firefighter crawls head down as fast as possible until the wall is hit. The next person rams into the first stopped firefighter, and so on, like an interstate motor vehicle pile-up or lemmings dropping off a cliff.

What are we trying to accomplish here? All that we are successfully teaching them to search is a 3-foot space along every exterior wall. Usually, a firefighter elephant team of two follows this large team search technique. We need to approach the way we are teaching them to search in a very different manner. Perhaps one solution is that we just need to give new firefighters more SCBA time before we address search techniques. More time and comfort with the SCBA itself may ease some inhibitions in search.

Once the new firefighters have grown more accustomed to the SCBA, plan to progress slowly. Have your firefighters search a few rooms on the training ground with open, clean air and light, no SCBA in place. Take them through in groups of two, showing them what we want them to search for. Show them, tell them, and teach them. Next have them go through the evolution with lights on and the SCBA in place. Again, teach, tell, and show. The progression can then continue with a few evolutions with flashlights and SCBA. Move then to darkness and SCBA without lights, then wax paper in the facepiece, and so on. Continue then to smoke and heat. I think you get the point.

We cannot just throw our newest firefighters into a smoke condition and say, "Okay, grab a leg and go!" We need to give them plenty of exposure to the techniques and teach them the mental, physical, and communicative properties that will give them confidence with their search habits from the start.

Body Positioning in Search

Most firefighters are still being taught to search on their hands and knees, on all fours, in a crawling position for movement. While most of us learned to crawl before we learned to walk, we will argue that this method does equate to the best tactical advantages while searching.

We prefer to teach a modified duck walk while searching. This technique uses one outstretched leg in front and the other leg kneeling, under the body's mass. Our training with this position has shown us increases in tactical advantages for the searcher. We call the on all fours the "head-down search" and the modified duck walk the "head-up search."

Fig. 25–9. Are we still teaching search like circus elephants?

First of all, instead of utilizing your arms to hold your body weight as in head-down search, the modified duck walk leaves both hands free to actually search the room. Your arms will not be as fatigued from holding up your body weight with each crawling movement you make. Now some may say that this head-up position raises the firefighter's profile during the search, causing a more vertically aligned body profile. Yes, that is true. Take a look at the side-by-side photos to see the slight increase in the searcher's profile. However, the advantages of this head-up search position are multiple.

Observe what that vertical alignment does for the firefighter's ability to monitor conditions ahead (figs. 25-10 and 25-11). No longer are firefighters fighting themselves to look ahead, straining their necks, trying to manage and control that interface of helmet and SCBA riding up their backs. Also, when searching in the head-down position, the weight of the SCBA and your upper body are in a forward stance. If there is a set of unrecognized steps or a burned through section of flooring, the natural tendency (based on the current weight distribution) would tend to drive the firefighter forward into that area. With the head-up search, the majority of the weight is distributed toward the rear of the searching member. This, coupled with the presence of the outstretched forward leg, may lead to earlier recognition and the ability to sit back on the SCBA and back out to avoid the situation ahead.

Fig. 25–11. Head-up modified search position

While we have mentioned thermal imagers (TIs) in other sections of the book, we will revisit it in search, as it is a great assistant to our efforts. Regardless of what brand TI you utilize, training with it and understanding its abilities and limitations are key to maximum success with its operation.

Let's share a few simple tips for utilization of the TI in the residential building fire (fig. 25–12). But first, let us mention a TI disclaimer. The TI is a tool, an expensive and delicate tool. It's not unbreakable, nor is it infallible. It's like any other tool that we have in our arsenal. It has its strengths and weaknesses. While it can assist us in search, it is an electronic, battery-powered tool that can fail. We cannot rely upon it to operate. As such, we must continue to remain oriented utilizing good search habits we have created for ourselves.

Fig. 25–10. Standard all fours searching technique

Fig. 25–12. Thermal imager

We have found it best to have the TI carried by the officer or senior firefighter in the search team when searching the residential dwelling. Think about those

team of two searches, one member with the TI and one without. Do both members need to enter every room we find to search? Most of our experiences are in residential dwellings with rooms that are 12×12 on average. Can't one member quickly search that room? While one member is searching, the officer or senior firefighter can be scanning the room with the camera, verbalizing closets, beds, and obstacles. That officer or senior firefighter should position him- or herself in the threshold of the door, monitoring both the searching firefighter and conditions in the hallway. This TI position can also scan forward to pick the next place to continue the search.

Another TI tip comes with a simple rotation of the camera. Most TI viewfinders are oriented in the landscape position. This landscape position, when scanning the floor of a dwelling, often does not capture the heat signatures found at the level of the ceiling (fig. 25–13). Simply tilting the camera 90 degrees to the side into a portrait orientation will normally allow the viewfinder to pick up floor-to-ceiling images.

Fig. 25–13. Rotate the position of the thermal imager.

Have you even gone to one of those "free" vacation giveaways? You know, the ones that are free except you are required to listen to the sales pitch at the beginning or end of the trip? If you haven't, we are sure you know someone who has. A great way to get some additional training on thermal imagers is to sign up for a free vacation trip. A what, you ask? What we are saying is, give a friendly call to your local TI vendors. Arrange for them to come to the firehouse and bring their new TI products. Most of them will come right to your firehouse, demonstrate the new features, and answer any questions you have at your convenience.

Victim Packaging

"Ladder 73 to Command."

"Go ahead 73."

"Chief, 10-45 first floor in the rear," the muffled voice of the Ladder 73's officer comes over the handie-talkie. The radio code 10-45 is the FDNY's radio signal given by operating units when they locate a civilian fire victim. The signal transmitted by the finding firefighters and verbally received by the incident commander does not get this civilian out of harm's way. It is up to firefighters to devise the best methods and strategize exit options to get this civilian out of the dwelling and into the arms of emergency medical services (EMS).

Before we delve into tactical considerations for victim removal from the residential building, think for a minute about the resources that you have responding on your typical dwelling fire. Do you have an EMS ambulance or medic unit assigned to every fire (fig. 25–14)? If not, why not? At a minimum, your department should arrange to have an EMS unit assigned to every working fire, and the earlier they can be assigned to the incident the better. They will be in place for any civilian victims and, equally importantly, for our members. Train with your local EMS members to ensure that they are ready to help one of us if we are injured. Topics such as fireground ambulance positioning, EMS member staging location (near the command post), and bunker gear removal techniques are just a few topics on which we can train together with our partners in the lifesaving business.

Fig. 25–14. Medical personnel on the fireground

Okay, back to the fire floor. Ladder 73 has an adult victim that needs to be removed from the fire building. First, let's talk about packaging techniques in the residential structure for civilians. Unconscious civilian victims, especially those who are large, can be difficult to move. The amount of packaging you will have to do will be directly based on how far you must travel with this person.

Fight the instinct to just grab an arm or leg and head out. That is useless unless you are right inside or near an exit point. The body's limbs are difficult to maintain a positive grip upon, even with our firefighting gloves. This especially comes into play if the victim has at all been exposed to heat and has suffered burns of any degree. Burned flesh tends to slough off. The removal of the outer layer of skin causes the body to become extremely slippery. Grabbing a section of clothing may initially work, but this can quickly tear and come off while attempting to move the victim. The best practice for removal of civilian victims is to take hold of the torso (fig. 25–15). Depending on the size of the victim, it may be a one-firefighter drag with the other leading the path to exit. In the case where there is a large victim, it may be a side-by-side, under the arm tag team. If either member has a section of tubular webbing, this can greatly increase the speed at which we can remove the victim.

Utilizing tubular webbing can greatly assist you in victim removal (fig. 25-16). But again, its speed and success in application will depend on your level of training with it in these situations. If you are proficient with your hitches and knots, it may take no time at all, but you must be prepared to do it. Perhaps it's just the handcuff knot, or a simple girth hitch that is needed. An even better and sturdier option is a hasty harness that you may be able to secure to the victim. You and your partner will have to judge the distance you need to travel, the difficulty of that distance (up/down steps), and time needed to apply the removal aid to see if it is warranted for each situation.

Fig. 25–16. Tubular webbing tied in a loop can be a multifaceted tool for firefighters.

We have certainly seen firefighters and fire officers call for other items such as backboards and stoke-type baskets to aid in victim removals. Most of the time there is little room for such devices to be maneuvered inside the residential setting, with or without a smoke and heat condition. These tools can, however, be placed at the point of egress to quickly and expeditiously move the victim to EMS care.

Victim Removal

In removing civilian victims from fire building we will refer to a pneumonic device that was taught to FDNY firefighters. It lists an order of preference for removal that is easy to remember: "I Have Five Little Rats."

Fig. 25–15. Firefighters "wrap up" a fire victim.

> *I = Interior*: Removal via the interior from the residential dwelling is normally the fastest and most direct method to get civilian victims out.

This avenue is also normally protected by the presence of the hoseline and affords the best protection for firefighters and civilians alike.

Have = Horizontal: Horizontal movements of civilians should be considered if initial removal attempts from the interior are precluded by fire operations in the dwelling, especially if an interior removal would put them in unnecessary danger. Many times occupants can be shielded in areas (rooms or adjacent apartments) horizontally remote from the fire until avenues in the interior become more tolerable.

Five = Fire escapes: While not found in all residential buildings, there are occasions where fire escapes may be an avenue to remove victims to safety. The removal will certainly prove more difficult, however it is an option that can be explored.

Little = Ladders: Portable and apparatus-mounted ladders are mainly designed for firefighters use in entering and exiting dwellings. Attempting to navigate a victim, conscious or unconscious, onto a ladder is no easy task and certainly needs to be practiced regularly with members.

Rats = Rope: While lifesaving rope evolutions at peaked roof residential dwellings are rarely seen, there may be instances on flat roof dwellings where portable ladders may not reach the rear or sides of the dwellings, causing a civilian roof-rope rescue to ensue. Placing firefighters "on-rope" to rescue victims at windows increases the dangers to both parties.

The Backstep

We Save Lives

Firefighters save lives. It is at the top of the proverbial hierarchy of our duties. We owe it to the civilians who entrust us with their safety to be prepared, practiced, and anticipating our next search to save a life. Utilizing the tips and techniques discussed in this chapter should aid you in refining your search skills for the residential dwelling and all other types of dwellings. We do not have the luxury to wait for time on the job to gain fireground search experience. We must focus and sharpen our skills at every opportunity, from checking building layouts on our routine responses to well planned, realistic search training.

At several points in the chapter we discussed the "why" in why we search. We implore you now to review fact and psychology behind the "where" we search. Saving lives is the overriding mission, as such we must never stop learning. Review the reports from the USFA and other agencies that show statistical residential fire data, points in the home where fires originate, where victims are found, and so on. Study information that extrapolates physiological and psychological "fight or flight" instincts relating to civilians in fire. Time, as it relates to fires, is not on the side of the civilian. The faster we locate and remove them, the better their chances for survival.

Index

A

access and egress. *See also* doors; windows
    in basement fires, 100–101, 112–113
accountability
    in line-of-duty deaths, 87
    in riding assignments, 149–150
actions
    in CAN reports, 34, 37
    of firefighters, 139–140
addresses, 29
    in on-scene reports, 89–90
aerial ladders, 280
after-fire reflection, 40–41, 248
Agassi, Andre, 3
aggressive
    anticipation for, 59–61
    definition of, 53
    as duty, 56, 59
    experience in, 60
    fire companies as, 60
    fire fighting as, 53
    fires as, 53
    how to be, 55–56
    practice for, 59, 61
    term use of, 53, 60
    training for, 53–55, 59
    understanding of, 60
aggressive preparation, 61
    information for, 57
    injuries and, 57
    personal protective equipment in, 58
    physical fitness for, 57–58
    plans from, 56–57
    routine in, 58
    skills for, 59
    tools in, 58
aggressive training, 53
    discipline in, 54
    double standards in, 54
    energy harnessing in, 55
    habits in, 55
    hoselines in, 56
    on-the-job learning and, 54–55
    situational awareness in, 59
    team in, 59
    time in, 55
air recognition drill, 52
always going to fires, 8
apartment houses, xiv
Aristotle, xiii, 192
Armstrong, Neil, 31
assisted living facilities, 17–18
attics, 277

attitude
    control of, 13
    effects of, 5–6
axe, 21

B

backdraft, 136
Backdraft (film), 131
backup
    alternative for, 153
    as backbone, 159
    blind side for, 159–160
    body position of, 161–162
    hoseline turns for, 161
    intricacies of, 160–164
    kink elimination by, 160
    mathematics for, 163–164
    nozzle reaction for, 163–164
    obstacles for, 160–161
    pace facilitation by, 162
    position of, 154
    shortest route for, 161
    snake hoseline by, 161–162
    understanding about, 164–166
    vigilance of, 163
balance point, of ladders, 262
Bangor ladders, 254
bars, window, 125
    attachment of, 204–205
    communication about, 204
    forcible entry of, 203–205
    tools for, 204–205
    ventilation and, 205
basement fires, 126
    access and egress in, 100–101, 112–113
    assisting at, 112
    basement checks for, 114–115
    collapse in, 113
    communication about, 105, 112
    delay in, 114
    engine company operations at, 117–120
    without entrance, 119–120
    examples of, 100–101, 111–112
    fire travel in, 114
    one-hoseline attack for, 119
    plan for, 114–116
    recognition of, 114
    renovations and, 122
    searches in, 297–298
    spiral stairs in, 100–101
    stairs in, 100–101, 115–116
    standard operating procedures/guidelines for, 118–120

without 360-degree check, 121–123
truck companies at, 116–117
two-hoseline attack for, 118–119
victim removal from, 297–298
basements
ceilings of, 113
cellars compared to, 112
definition of, 112
material of, 112–113
mechanicals in, 113
spiral stairs to, 100–101
use of, 113
walkout, 112
beam slide ladder deployment, 267
bedrooms, 29
blind, 249
blind side, 159–160
body position
of backup, 161–162
for searches, 304–306
self-contained breathing apparatus and, 305
brand recognition, 22
Brannigan, Francis L., 65, 78
breathing. *See also* self-contained breathing apparatus
breath holding in, 45
emergency and, 48–49, 105
of oxygen, 44–45
Brooks, Thomas, 121
Brown, Joe, 270
Broxterman, Captain, 122–123
building construction, 19, 65, 71, 241
appreciation of, 83
assembly line for, 81
attention to, 78
calls and, 66
of Cape Cod home, 66–67
of colonial home, 67, 91
design in, 80, 82
developments of, 80–81
engineered wood in, 73
of estate homes, 68–69
examination of, 80–83
first floor in, 83
foundation in, 83
glulam in, 73
green in, 82
guts of, 66–69
hardware store for, 86
learning about, 18–19, 72, 74, 83, 287
of McMansion homes, 68–69
methods of, 137–139
new trends in, 72, 78, 81
newspaper on, 18–19, 74
old compared to new, 72, 81
outside of, 82
profits in, 78, 81
of rancher/rambler homes, 68
renovations in, 84–85
representative to, 81, 86
of roofs, 276–277
sharing information about, 78–79
site managers of, 81
of split foyer homes, 68
of split level homes, 67–68
structural insulating sheathing in, 73–74
tactical operations from, 69
vinyl siding on, 72–73
Building Construction for the Fire Service (Brannigan), 65

building frames
balloon frame, 69–70
lightweight construction in, 71
ordinary frames, 70
platform frame, 70–71
building officials
codes from, 80
relationships with, 79–80
building side, 89–90
building size and type, 90–91
bunker gear, 7

C

call types, 27–28
can firefighter, 41–42
CAN reports, 34, 37
can/hook firefighter
goal of, 225
opening up by, 226
tools for, 225–226
Cape Cod homes
building construction of, 66–67
stairs in, 106
cellars, 112
changes
for combat-ready fire apparatus, 20
in forcible entry, 193–194
of habits, xii–xiii
rigid flexibility, xii
chauffeur, 151, 177
ladder company, 228–230
children, 300–302
citizens, 17–18, 59
COAL WAS WEALTH (Construction, Occupancy, Apparatus and personnel, Life hazard, Water supply, Auxiliary appliances, Street conditions, Weather, Exposures, Area, Location and extent of fire, Time, Height) (13 points), 241
cockloft explosion, 132–133
colonial homes
building construction of, 67, 91
stairs in, 106
combat readiness, xi. *See also* aggressive
differences and, 8
injuries and, 10
oath for, 10
Traditions Training, LLC, 7
weak links in, 11
combat-ready fire apparatus
approach of, 15–16
area structures for, 16–17
blueprint for, 20
brand recognition for, 22
change for, 20
citizens and, 17–18
construction evaluation for, 19
decisions about, 16
equipment book for, 23
for fireground, 15
Halligan bar as, 21
marrying tools for, 20–21
newspaper for, 18–19
pre-plans for, 19
real estate developments and, 19

strategies for, 18
training for, 18, 21
commercial buildings, 26–27
commitment, 4–5
communication
 about basement fires, 105, 112
 clarity of, 31–32, 37
 dispatchers in, 25, 33
 drill for, 36–37
 from first-arriving engine, 169
 ideal example of, 32
 leadership and, 37
 Lego drill and, 36–37
 Mayday in, 34–35
 nozzle obstruction and, 186
 from outside team first due, 234
 about stairs, 116
 tasks or, 33–34
 terminology in, 145
 of 360-degree check, 129–130
 truck company assignments without, 217
 2 1/2-inch hoseline and, 158
 for vent, enter, isolate, and search, 288
 about window bars, 204
communication policy, 31–32
 CAN reports in, 34
 development of, 33–34
 for firefighter in trouble, 34
 rationale for, 33
communications, training for, 36–37
community relationships, 77–86
company officers
 engine officer, 151–152
 Halligan bar and, 200, 224
 tools for, 224
 truck company assignment of, 223–224
compassion, 211
complacency, 8, 25, 28
computer-aided dispatch system (CAD), 33
conditions, actions, needs (CAN), 34, 37
confidence, 304
Conroy, Patty, 121
construction evaluation, 19
construction methods, 137–139
control, 155
 of attitude, 13
 confine, control, extinguish, 206–208
 of doors, 196–197
 of electricity, 213–214
 of fire senses, 239–240
 of natural gas, 212–213
 of stairs, 102
 of 2 1/2-inch hoseline, 157
 of water, 214
coordination
 of hoselines, 117–120
 of ventilation, 142
courage, personal, 13, 285
CPR, 241
Croker, Edward F., 4
crosslay, 178
cylinders, 42–43, 47–48, 52

D

dead load lines, 177
DeNiro, Robert, 131
Department of Homeland Security, 154, 302
depth
 of door locks, 195
 on Halligan bar, 199
depth marker, 21
"Deputy Chief's Wife," 34
detached residential buildings, xiv
dip and drive ladder deployment, 265–266
dispatchers, 25, 33
dogs, 127
door locks, 29
 depth of, 195
 force for, 200
 gap for, 198
 Halligan bar for, 196
 hardware stores and, 194
 Hyrda-Ram for, 195
 irons for, 195
 set for, 198–199
 shock for, 197
doors, 125
 control of, 196–197
 directional swing of, 194
 on firehouse, 78
 sidelights on, 194, 196
 on stairs, 104–105
 in vent, enter, isolate, and search, 290–292
 ventilation from, 197
 as victim locations, 301
 window placement and, 291
doors locks, 79
duck walk, 305–306
duty, 12. *See also* line-of-duty deaths
 aggressive as, 56, 59
Dwight, Timothy, 67

E

electricity control, 213–214
emergencies
 breathing and, 48–49, 105
 9-1-1, 5
emergency medical services (EMS), 306
energy
 efficiency of, 82
 harnessing of, 55
engine company operations
 at basement fires, 117–120
 first-arriving engine, 32, 61, 92, 169
 ladders on, 252
 second-arriving engine, 32, 61
 third-arriving engine, 32, 61
engine compared to truck companies
 chores for, 192
 at firehouse, 191
 goals of, 214
 independence of, 193
 parallels of, 192
 rivalry between, 192
 roles in, 191–192, 221

engine officer, 151–152
engine split, 252
engineered wood, 73
engineers, hydraulic, 176
enter, in vent, enter, isolate, and search, 289–290
Ermey, R. Lee, 183
estate homes
    building construction of, 68–69
    hoseline for, 84
    renovations in, 84–85
    setback of, 17
explosion, CO, 132–133

F

facepiece, 43, 50
    removal policy on, 211
fans
    positive pressure ventilation, 208–209
    for ventilation, 208–209, 274
FDNY, 55, 98, 164, 167, 258
    Firemen's Memorial, xi, 56, 77
    Rescue 2, 6–7, 19–20
    residential building fires for, 278–279
    self-contained breathing apparatus airway checks for, 46
    tools for, 20
    urgent reports for, 34–35
feel
    gloves and, 245–248
    heat transfer and, 247
    heat waves, 247
    rapid heat increase, 248
    skin for, 246–247
fire
    attributes of, 131
    floor above, 101, 113
    fuel composition for, 139
    locating seat of, 100–101
    love of, 131
    oxygen and, 132–134
    phases of, 134–136
    smoke without, 92
    tetrahedron of, 132
    triangle on, 131–132
    in windows, 92, 127
Fire Exposures of Fire Fighter Self-Contained Breathing Apparatus Facepiece Lenses, 50
fire extinguishers, 207–208, 225
Fire in the Older Adult, 302–303
fire senses
    control of, 239–240
    decision making through, 239–240
    educated guesses from, 239
    experience for, 240
    feel as, 245–248
    guidelines and, 240–241
    heightened states with, 239–240
    intuition in, 240
    listen as, 243–245
    look as, 241–243
    reflection on, 248
    seniority in, 240
    in size-up, 240–241
    13 points and, 241
firefighter assist search team (FAST), 41

firefighters
    actions of, 139–140
    family of, 4–5, 111–112, 251
    football compared to, 65, 149, 220–221
    military compared to, 6, 77, 85, 88, 144
    mistakes by, 257
    volunteer, 5, 7–8, 78–79, 251
fireground stressors, 36–37
firehouse. *See also* engine compared to truck companies; truck companies
    doors on, 78
    engine compared to truck companies at, 191
    as home, 77
Firehouse Pride, 262–263
first due
    inside team, 226–227
    outside team, 233–235
    reduced staff inside team, 228
    reduced staff outside team, 234–235
first ladder, 61
first-arriving engine, 32, 61, 92
    communication from, 169
first-due truck, 32
flameover, 136
flashover, 136
flat roofs, 282–283
flow testing, 176
focus, xiii
fog nozzles
    inspection of, 184
    reaction of, 163
football, 65, 149, 220–221
forcible entry, 224–225
    changes in, 193–194
    convention of, 196–200
    force in, 200
    gap in, 198
    of gates, 202–203
    hardware stores and, 194
    set in, 198–199
    shock in, 197
    smoke and, 92–93
    by truck companies, 193–205
    as ventilation, 92–93
    of window bars, 203–205
    of window guards, 201–202
    of windows, 200–202
forcible entry firefighter, 225
    tools for, 224
forward lay, 167–171
fourth engine, 61
Franklin, Ben, 46
Fredericks, Andrew, 164
fuel, 132
fuel composition, 139
Full Metal Jacket (film), 183

G

gallonage selector, 184–185
gates, 202–203
gloves, 9, 245–248
glulam, 73

H

habits
- in aggressive training, 55
- bad, 304
- change of, xii–xiii
- definition of, 8
- excellence as, 192
- for gear, 9–10
- mistakes as, 9
- training for, 304

Halligan bar, 21
- company officers and, 200, 224
- depth on, 199
- for door locks, 196
- for force, 200
- for gap, 198
- for set, 198–199
- for shock, 197
- thickness of, 198–199
- for windows, 201–202

hardware stores
- for building construction, 86
- door locks and, 194

head-down search, 304–305
heads up display (HUD), 39
heads-up search, 304–305
heat, 132
heat increase, rapid, 248
heat transfer, 247
heat waves, 247
heavy rescue, 61
helmet cameras, 3–4
helmets, 42–43
heroism, 285
high shoulder carry, 260–261
high-rise rack, 177–178
history, xi–xii
home security, 79, 194
honor, 12–13
Horizontal Fire Ventilation Experiments in Townhouses, 92, 143–144
horizontal ventilation, 233, 273, 277–279, 284
- ventilation for extinguishment for, 275

horizontal victim removal, 308

hoseline stretch
- dead load lines in, 177
- flow testing for, 176
- high-rise rack in, 177–178
- long line stretch for, 173, 175–179
- obstacles for, 173–174
- pony sleeves in, 178–179
- pre-connected hoseline, 173, 175, 177
- preparation for, 174–175
- pump pressures and, 176
- racking for, 174
- setback for, 175

hoselines, 126
- in aggressive training, 56
- alley stretch for, 161
- coordination of, 117–120
- kink elimination of, 160
- kinks in, 160
- for McMansions, 84
- nozzle operator and, 152–153
- obstacles for, 153–154, 160–161
- pre-connected, 173, 175, 177
- push method for, 109–110
- for residential building, 17–18
- searches without, 275
- snake of, 154
- for spiral stairs, 101
- on stairs, 97–101, 109–110
- understanding of, 176

hydrants, 167–170
Hydra-Ram tool, 23, 195, 204–205
hydraulic engineers, 176
hydraulic ladders, 253
hydraulic ventilation, 274–275

I

identity, 22
immediately dangerous to life and health (IDLH), 44–45
injuries
- aggressive preparation and, 57
- combat readiness and, 10
- from portable ladders, 270

inside team
- can/hook firefighter, 225–226
- company officer, 223–224
- first due, 226–227
- forcible entry firefighter, 224–225
- reduced staff first-and/or second due, 228
- second due, 227–228

instructors, 21
- equipment book for, 23

Insurance Services Office (ISO), 254
integrity, 13
International Building Code (IBC), 105, 109
Internet, 3–4, 57
intuition, 240
the irons, 21
isolate and search, 290–293. *See also* vent, enter, isolate, and search

J

jack-of-all-trades, xiii
jokes, 3–4
journals, trade, 94

K

Kennedy, John F., 12
kink elimination, 160
kitchens, 29
- as seat of fires, 206

Kolenda, Mark, 121

L

ladder company. *See* truck companies
ladder company chauffeur
    duties of, 229–230
    as go-getter, 230
    ladders for, 229–230
    status of, 228
    truck location and, 229
ladder deployment
    beam slide for, 267
    dip and drive for, 265–266
    one-firefighter portable ladder placement in, 265–267
    two-firefighter portable ladder placement in, 268
ladder drag, 261–262
ladders. *See also* portable extension ladders
    aerial, 280
    balance point of, 262
    Bangor, 254
    climbing angles for, 268–269
    delay of, 252–253
    on engine company operations, 252
    entering windows from, 289–290
    help with, 234
    hydraulic, 253
    jumping without, 252–254
    for ladder company chauffeur, 229–230
    markings for, 262–263
    obstacles for, 253
    removal of, 252
    roof, 256, 261–262
    special, 255
    for truck companies, 209–210
    for vent, enter, isolate, and search, 287–288
    vent firefighter and, 231–232
    for victim removal, 292–293, 308
ladders placement order of preference
    for fire floor, 263–264
    for floor above fire, 264
    for between floors, 264
    to roof, 265
    for top floor, 264
laying out
    crosslay for, 178
    decision for, 168, 171
    excuses about, 168
    forward lay in, 167–171
    hydrant location for, 169–170
    with hydrants, 167–170
    mission in, 170–171
    opportunities for, 168–169
    parking for, 169–170
    police hindrance and, 169, 299
    racking hose and, 168–169
    reverse lay in, 167–169, 171
    route for, 169
    for rural setting, 169
    silence in, 170
    split lay in, 167–168, 170–171
    for suburban setting, 167–168
    for water delivery, 167–168
    water supply plan for, 171
Layman, Lloyd, 193
LDRSHIP, 12–13
leadership
    balance of, 11–12
    communication and, 37
    of engine officer, 151–152
    excellence in, 11
    outside team without, 228
    reading for, 14
    traits of, 11–13
Lego communication drill, 36–37
life, 44–45, 241, 309
    ventilation for, 208
lifesaving, 300–303
line-of-duty deaths (LODDs), xii–xiv, 126
    accountability in, 87
    expectations of, 7
    purpose from, xi
    360-degree check and, 121–123
    weak links in, 11
listen
    as fire sense, 243–245
    noise, 244–245, 249
    radios, 243–244
    Rapid Intervention programs for, 244–245
    to screamers, 244
locks. *See also* door locks
    on portable extension ladders, 258
look
    as fire sense, 241–243
    objects of, 242
    opportunities to, 242
    under smoke, 242
    thermal imager to, 114, 210, 241–242, 292, 305–306
    two-team concept to, 242–243
    visual landmarks, 241
low shoulder carry, 260
loyalty, 12
Lund, Peter, 6–8, 13, 167

M

markings, for ladders, 262–263
Martin, Nicholas, 224
mastery, xiii
Mayday
    communications of, 34–35
    self-contained breathing apparatus air emergencies as, 48–49
McCormack, Ray, 55
McMansion homes, 16
    building construction of, 68–69
    hoseline for, 84
    renovations in, 84–85
    setback of, 17
medical assists, 27–28, 306
mental demands, 45
    fireground stressors as, 36–37
military
    firefighters compared to, 6, 77, 85, 88, 144
    Marine Corps, 144
    preparation of, 16
mission
    arrival and, 174
    in laying out, 170–171
    in riding assignments, 149
    water delivery as, 167–168
monument, xi, 56, 77
Mora, William R., 45
multiple dwelling (MD), 97–98

N

National Association of Home Builders, 16
National Fire Protection Association (NFPA), xiv, 254
National Fire Protection Handbook, 17th Edition, 163
National Institute for Occupational Safety and Health (NIOSH), xiii–xiv
natural gas control, 212–213
NBPA 1404: Standard for a Fire Service Respiratory Protection Training, 50
near miss reports, xiii–xiv
needs, in CAN reports, 34, 37
never going to fires, 8
newspaper
    on building construction, 18–19, 74
    for combat-ready fire apparatus, 18–19
    real estate section, 75
NFPA 1983: Standard on Open-Circuit Self-Contained Breathing Apparatus (SCBA) for Emergency Services, 49–50
NFPA 2010: Fire Estimates, 65
9-1-1, 5, 89–90, 93
NIST study, 143–144
noise, 244–245, 249
nozzle inspection, 182
    bales in, 184–185
    flush function in, 183–184
    of fog nozzles, 184
    gallonage selector in, 184–185
    malfunction in, 184
    of smooth bore nozzle, 185
    washers in, 184
nozzle operator, 152–153
nozzle reaction
    assist for, 164
    for backup, 163–164
    review on, 165–166
    scale for, 165–166
nozzles
    creed for, 183
    obstruction of, 186–187
    policy for, 181
    protection of, 182–183
    storage of, 182

O

1 3/4 inch hoseline, 84
one-hoseline attack, 119
on-scene reports
    actions for water supply in, 92
    address in, 89–90
    building side in, 89–90
    building size and type in, 90–91
    clarity in, 88
    concision of, 88
    conditions and location in, 91–92
    consistency in, 88–89
    details in, 87–89
    line-of-duty-deaths in, 87
    pertinence in, 88–89
    purpose of, 92–93
    review for, 94–95
    templates for, 88–90

operational manuals, 26
organization levels, 12
outside team
    duties of, 228
    first due, 233–235
    ladder company chauffeur, 228–230
    without leadership, 228
    reduced staff first-and/or second due, 234–235
    roof firefighter, 232–233
    second due, 234–235
    vent firefighter, 231–232
outside ventilation (OV), 115
overhaul, 210–211
oxygen, 126–127, 131, 134
    backdraft and, 136
    breathing of, 44–45
    explosion from, 132–133

P

Patton, George, 12
peaked roofs, 279–281
personal alert safety system, in self-contained breathing apparatus, 48
personal protective equipment (PPE), 7
    in aggressive preparation, 58
    gloves, 245–248
    for roof ventilation, 280
physical demands, self-contained breathing apparatus and, 44–45
physical fitness training, 44–45
Pike, Albert, xiv
police
    as hindrance, 169, 299
    misinformation from, 299
policies
    for communications, 31–34
    for nozzles, 181
    public image and, 4
    on removal of facepiece, 211
    on self-contained breathing apparatus, 181, 211
    for two-team concept, 218–219
pony sleeves, 178–179
portable extension ladders
    locks on, 258
    maneuvering of, 259–262
    pre-tying, 258–259
    sizes and types of, 255
portable ladders, 230, 234
    care of, 270
    carrying practice of, 259
    carrying tools on, 269–270
    design of, 255
    drill with, 272
    falling from, 270
    footing on, 269
    full use of, 254
    guidelines for, 254
    high shoulder carry with, 260–261
    injuries from, 270
    inspection of, 270
    ladder drag with, 261–262
    low shoulder carry with, 260
    maneuvering with, 259–262
    masking up on, 269
    multi-person carries with, 259

multiple ladder carry of, 261
review for, 272
simplicity for, 271
tip in/tip out for, 270–271
working load capacities for, 255
working on/off, 269–270
portable straight ladders, 258
angles and, 256
parts of, 256
sizes and types of, 255
typography and, 256–257
positioning, in vent, enter, isolate, and search, 287–288
positive pressure ventilation (PPV), 208–209, 274
positivity, 5
power lines, 128
pre-connected hoseline, 173, 175, 177
preparation. *See also* aggressive preparation
of military, 16
pre-plans
on bedrooms, 29
for combat-ready fire apparatus, 19
for commercial buildings, 26–27
opportunities for, 27–29, 79
for routine, 26–27
on stairs, 29
testing about, 29
Preventing Deaths and Injuries of Fire Fighters Working Above Fire-Damaged Floors, 113
pride, 21–22
primary searches, 300
private dwelling (PD), xiv
stairs at, 100–101
professionalism, 12–13
public image, xii
jokes and, 3–4
policy and, 4
recordings for, 3–4
technology for, 3–5
pump pressures, 176
push method, for hoselines, 109–110

R

racking, 168–169, 174
radios, 243–244
rancher/rambler homes, 68
stairs in, 107
rapid heat increase, 248
Rapid Intervention (RI) programs, 244–245
rapid intervention team (RIT), 32, 41
real estate developments, 75
combat-ready fire apparatus and, 19
recordings, 3–4, 249
reduced staff truck company, 222
first-and/or second due inside team, 228
first-and/or second due outside team, 234–235
reignition, 210
from natural gas, 212
Reilly, Kevin J., 45
Reilly Emergency Breathing Technique (R-EBT), 45
relationships
with building officials, 79–80
with community, 77–86
renovations
basement fires and, 122

in building construction, 84–85
stairs from, 101, 108
rescue, 61, 205, 300–303
FDNY Rescue 2, 6–7, 19–20
residential building
apartment houses, xiv
assisted living facilities, 17–18
hoseline for, 17–18
McMansions, 16–17
pre-plans for, 26–27
setback of, 17, 97, 173–174
size of, 16–17
volunteer firefighters and, 78–79
residential building fires, 15
for FDNY, 278–279
percent of, 206
seat of fires in, 206
training for, 25–26
2006–2010, xii, 72
residential drills, 249
residential recognition, 249
respect, 12
reverse lay, 167–169, 171
reviews, 13. *See also* VEIS tactical skill review
combat-ready equipment book, 23
community relationships in, 86
cover drill in, 94–95
on fire behavior, 145–146
hoseline movement on stairs in, 109–110
leadership reading in, 14
Lego communication drill, 36–37
newspaper real estate section in, 75
on nozzle obstruction, 186–187
on nozzle reaction, 165–166
for on-scene reports, 94–95
for portable ladders, 272
prepare, practice, anticipate in, 61
pre-plans as, 29
residential recognition, 249
on self-contained breathing apparatus, 52
360-degree check in, 130
on truck company assignments, 237
on 2 1/2-inch hoseline, 156–158
of vent, enter, isolate, and search, 294–295
on ventilation, 284
Ricci, Frank, 45
riding assignments
accountability in, 149–150
additions to, 155
backup, 153–154, 159–166
chauffeur, 151, 228–230
control, 155
engine officer, 151–152
mission in, 149
nozzle operator, 152–153
training for, 150–151
understanding, 155
Riley, Richard, 6–7, 13, 56
rollover, 136
roof firefighter, 232–233
roof ladders, 256, 261–262
roof ventilation
advantages of, 274–275
for flat roofs, 282–283
hinge or louver cut for, 281
horizontal, 273
making the cut for, 280–282
personal protective equipment for, 280
positioning for, 279–280

reasons for, 275–276
skylights and, 282–283
tower ladder bucket for, 279–280
vertical, 273
roofs, 276
attics and, 277
building construction of, 276–277
construction of, 276–277
cutting tools for, 280–281
ladders placement order of preference to, 265
peaked, 279–281
steep pitched, 281–282
rope, for victim removal, 308
routine
in aggressive preparation, 58
complacency and, 25, 28
definition of, 28
dispatch and, 25
impossibility of, 173
operational manuals and, 26
pre-plans for, 26–27
understanding of, 25–26
rural setting, 169

S

safety, 48
in VEIS tactical skill review, 286–287
in vent, enter, isolate, and search, 286–287
salvage, 211
Santayana, George, xi, 42
saws, 280–281
science
NIST study, 143–144
UL study of, 141–143
on ventilation, 145
screamers, 87, 200, 244
searches, 41, 242, 290–293. *See also* 360-degree check; vent, enter, isolate, and search
in basement fires, 297–298
body position for, 304–306
delays in, 300
generalities about, 300
head-down, 304–305
head-up, 304–305
without hoselines, 275
necessity for, 297–299
primary, 300
secondary, 300
training for, 303–304
truck companies related to, 215
second due inside team, 227–228
second ladder, 61
second-arriving engine, 32, 61
secondary searches, 300
second-due truck, 32
security, home, 79
hardware stores and, 194
self-contained breathing apparatus (SCBA), 36
air conservation and, 47–48
air emergencies with, 48–49
air recognition drill for, 52
air status of, 46
attachment points on, 40
body position and, 305
check for, 46
chronic exposure and, 47
comfort level for, 39
cylinders and, 42–43, 47–48, 52
entanglement of, 48
facepiece of, 43, 50
failure to use, 49
failures in, 49–50
free air in, 47–48
functional operation test for, 46
hips and, 40
liabilities and, 45
lost air from, 48
Mayday and, 48–49
mental demands and, 45
personal alert safety system in, 48
physical demands and, 44–45
policies on, 181, 211
presetting straps for, 43
preventive maintenance of, 50
reliability of, 49
reviews on, 52
seconds with, 43–44
self knowledge and, 39, 50–51
shoulder straps on, 40
storage of, 182
story about, 41–42
straps and, 40–43
time for, 39, 44
training in, 44, 48, 50, 304
upgrades of, 50
use of, 47
waist straps in, 40–43
service, selfless, 12
setback
of estate homes, 17
for hoseline stretch, 175
of residential building, 17, 97, 173–174
single room occupancies (SROs), 113
single-family dwelling (SFD), xiv
as battleground, 65
situational awareness, 59–60
size-up
fire senses in, 240–241
as process, 241
13 points for, 241
skylights, 282–283
smoke
analysis of, 126–127
carcinogens in, 45
without fire, 92
forcible entry and, 92–93
lack of, 142–143
look under, 242
throughout private dwelling, 100–101
tunneling of, 142
in vent, enter, isolate, and search, 289
smooth bore nozzle, 185
special ladders, 255
spiral stairs, 100–101
split foyer homes, 68
stairs in, 107
split lay, 167–168, 170–171
split level homes, 67–68
stairs in, 106–107
stairs
alternatives to, 102
in basement fires, 100–101, 115–116
breathing emergency on, 105
in Cape Cod homes, 106

clues about, 108
in colonial homes, 106
communication about, 116
components of, 104
control of, 102
deadliness of, 102
design and layout of, 103–106
discipline on, 105–106
doors on, 104–105
enclosure of, 103–104
fire at top of, 116–117
fire impact on, 102
firefighters on, 105–106
hoselines for, 97–101, 109–110
landing in, 104
location of, 115–116
no fire at top of, 117
open staircase in, 103, 105, 137
open tread design in, 105, 137
pre-plans on, 29
at private dwelling, 100–101
push method for, 109–110
in rancher/rambler homes, 107
from renovations, 101, 108
rise of, 109
rope stretch for, 99
second line for, 102
size of, 109
spiral, 100–101
in split foyer homes, 107
in split level homes, 106–107
straight run of, 103
use of, 102
variations and, 108
well hole in, 97–100
width of, 105
windows and, 99
standard operating procedures/guidelines (SOPs/SOGs), 111, 115
for basement fires, 118–120
for two-team concept, 235
status, 5
steep pitched roofs, 281–282
stressors, fireground, 36–37
structural insulating sheathing (SIS), 73–74
Structural Stability of Engineered Lumber in Fire Conditions, 113
structured insulated panels (SIP), 73, 83
success, xi
swimming pools, 128

T

team, 13. *See also specific teams*
excellence in, 11
role in, 5–6, 11, 200, 244
two-team concept, 218–219, 221–223, 235–236, 242–243, 288
for vent, enter, isolate, and search, 231
technology, 3–5
terminology, 145
thermal imager (TI), 114, 210, 241–242, 292
rotation of, 306
training with, 305–306
Thiel, Adam K., 49
third-arriving engine, 32, 61

13 points, 241
360-degree check, 114
basement fires without, 121–123
building status in, 125, 130
communication of, 129–130
distractions to, 122–123
environmental conditions in, 124–126, 130
exposures in, 128, 130
fire location and conditions in, 126–127, 130
hazards in, 127–128, 130
information gathering in, 129
line-of-duty-deaths and, 121–123
occupants for, 123
resources in, 128, 130
in review, 130
success for, 123
topography in, 123–124, 130
victims and, 124, 130
of windows, 124–125
of yard, 124
time, 13, 241
in aggressive training, 55
for self-contained breathing apparatus, 39, 44
Timex watches, 22
tools, 3–4, 6, 20–21. *See also* combat-ready fire apparatus
in aggressive preparation, 58
for can/hook firefighter, 225–226
for cutting roofs, 280–281
equipment book for, 23
for forcible entry firefighter, 224
Hydra-Ram, 23, 204–205
saws, 280–281
truck company assignments of, 219–220
for vent, enter, isolate, and search, 289
for window bars, 204–205
topography, 29
in 360-degree check, 123–124, 130
tower ladder (TL) bucket, 279–280
townhouses, 143
Traditions Training LLC, 236
training, 7, 236, 251
aggressive, 53–56, 59
for combat-ready fire apparatus, 18, 21
for communications, 36–37
for confidence, 304
consistency in, 303
critique on, 145
data review in, 145–146
with emergency medical services, 306
excellence and, 11
for habits, 304
physical fitness, 44–45
realism for, 303
for residential building fires, 25–26
for riding assignments, 150–151
for searches, 303–304
in self-contained breathing apparatus, 44, 48, 50, 304
with thermal imager, 305–306
topics for, 150
on ventilation, 145
for zero visibility, 185–187
training tools, 6. *See also* combat readiness
Internet for, 3–4
truck companies, 32. *See also* engine compared to truck companies
at basement fires, 116–117
confine, control, extinguish for, 206–208
electricity control by, 213–214
fire extinguishers for, 207–208, 225

forcible entry by, 193–205
functions of, 193
incident stabilization for, 215
ladders for, 209–210
lives and, 215
location of seat of fire by, 205–206
natural gas control by, 212–213
overhaul by, 210–211
principles for, 215
property conservation for, 215
reduced staff in, 222, 228, 234–235
salvage by, 211
search related to, 215
success with, 236
utility control by, 212–214
water control by, 214
truck company assignments
can/hook firefighter, 225–226
chart for, 222
without communication, 217–218
company officer, 223–224
delays in, 217–218
drill on, 237
flexibility in, 218
football compared to, 220–221
forcible entry firefighter, 224–225
guidelines for, 221–223
inside team first due, 226–227
inside team in, 223–228
inside team second due, 227–228
lack of, 217
names for, 219
outside team first due, 233–234
outside team in, 228–235
outside team ladder company chauffeur, 228–230
outside team roof firefighter, 232–233
outside team second due, 234–235
outside team vent firefighter, 231–232
posting of, 219, 222–223, 235
preplanning for, 219–220
reduced staff, 222
reduced staff first-and/or second due inside team, 228
reduced staff first-and/or second due outside team, 234–235
responsibility in, 219–220
review on, 237
teams for, 218
tools in, 219–220
two-team concept for, 221–223
trust, 5
Twain, Mark, 33
2 1/2-inch hoseline, 84
advancing with, 157–158
communication and, 158
control of, 157
as discipline, 156
equipment with, 156–157
objective with, 157
review on, 156–158
two-firefighter portable ladder placement, 268
two-hoseline attack, 118–119
two-team concept. *See also* engine company operations; truck companies
to look, 242–243
policies for, 218–219
standard operating guidelines for, 235
success of, 236
for truck company assignments, 221–223
for vent, enter, isolate, and search, 288

U

urgent reports, 34–35
U.S. Fire Administration (USFA), xiv
U.S. *Fire Administration Firefighter Fatalities in the United States in 2010*, 299
U.S. *Firefighter Disorientation Study*, 45

V

VEIS tactical skill review
enter in, 289–290
isolate and search in, 290–293
leapfrog rooms in, 293
positioning in, 287–288
safety in, 286–287
vent in, 288–289
vent, enter, isolate, and search (VEIS), 21, 31, 69, 228
doors in, 290–292
ladders for, 287–288
positioning in, 287–288
probability in, 285–286
reason to, 295
review of, 294–295
roof firefighter for, 232
safety in, 286–287
smoke in, 289
staff for, 294
team for, 231
tools for, 289
two-team concept for, 288
use of, 286
vent in, 288–289
what to, 294
when to, 295
where to, 294–295
windows for, 233
vent firefighter, 231–232
vent for extinguishment (VFE), 275
vent for search (VFS), 275
ventilation, 115, 207. *See also* roof ventilation
coordination of, 142
from doors, 197
fans for, 208–209, 274
for fire, 208–209
forcible entry as, 92–93
horizontal, 233, 273, 275, 277–279, 284
hydraulic, 274–275
for life, 208
NIST study on, 143–144
positive pressure ventilation, 208–209
review on, 284
science on, 145
staffing and, 276
training on, 145
vertical, 233, 273, 277–279
window bars and, 205
from windows, 231
victim locations
adults, 300–302
bathrooms as, 303
bedrooms as, 301–302
bunk beds as, 302
children, 300–302
closets as, 301

doors as, 301
windows as, 301
victim removal
from basement fires, 297–298
horizontal, 308
ladders for, 292–293, 308
misinformation about, 297–299
responsibility for, 297–299
rope for, 308
tubular webbing for, 307
through windows, 292–293, 307
victims
jumping by, 252–253
packaging of, 306–307
vinyl siding, 72–73
volunteer firefighters, 5, 251
advantages of, 78–79
deaths of, 7–8

W

walkout basements, 112
watches, Timex, 22
water, 92
control of, 214
meter reader for, 214
supply plan for, 171
water delivery
laying out for, 167–168
as mission, 167–168
weather, 125–126
webbing, 165–166
webbing, tubular, 307
well hole, 99–100
lack of, 97–98
measurement of, 98
Wilson, Kyle, 126
wind, 125–126
windows, 83
A/C units in, 125
bars on, 125, 203–205
casement, 82
clearing of, 200–201, 278
door placement and, 291
fire in, 92, 127
forcible entry of, 200–202
guards on, 200–202
Halligan bar for, 201–202
hydraulic ventilation in, 275
jumping from, 252–253
ladder entry through, 289–290
stairs and, 99
360-degree check of, 124–125
for vent, enter, isolate, and search, 233
ventilation from, 231
as victim locations, 301
victim removal through, 292–293, 307

Z

zero visibility, 41
drill about, 36–37
nozzle obstruction and, 186
training for, 185–187